MONOGRAPH OF THE MONOCOTYLEDON-INHABITING SPECIES OF *LOPHODERMIUM*

CAB *International*

CAB *International* is an intergovernmental organization providing services world-wide to agriculture, forestry, human health and the management of natural resources.

CABI *Publishing* is one of the world's foremost publishers of databases, books and journals in agriculture, forestry, veterinary science and related disciplines, including human health and diseases.

CABI *Information for Development* helps design, build and sustain information and knowledge management systems in developing countries.

CABI *Bioscience*

On 1 January 1998 the four scientific institutes of CAB INTERNATIONAL, including the International Mycological Institute, were integrated into CABI *Bioscience*. The research, training and service activities have been restructured to create a stronger science base and are arranged into ten programmes in three sectors. The division has two centres in the UK and five overseas, incorporating specialists in biosystematics, biotechnology, ecology, parasitology, crop protection and biological control. CABI *Bioscience* builds on and will enhance the high international reputation established by the CABI scientific institutes over the course of much of this century. Its unique multidisciplinary scientific capacity positions it ideally for tackling some of the world's most challenging problems.

Mycological Publications from CABI *Bioscience*

Mycological Papers is an irregular series dedicated to publishing longer papers in fungal systematics and diversity. Submissions by external authors will be considered for publication, and anyone interested should contact the address below. This is just one aspect of CABI *Bioscience*'s continuing involvement in fungal biosystematics. Other serials issued are the *Bibliography of Systematic Mycology*, *Descriptions of Fungi and Bacteria* and *Index of Fungi*, whilst books include *Ainsworth & Bisby's Dictionary of the Fungi*.

Mycological Papers is produced by:

CABI *Bioscience* UK Centre (Egham)
Bakeham Lane, Egham
Surrey TW20 9TY, UK

Tel.: +44 (0)1491 829800
Fax: +44 (0)1491 829100
E-mail: bioscience@cabi.org

For the full range of our products and services see CAB *International* at http://www.cabi.org.

Issued 22 January 2001 *Mycological Papers* No. 176

MONOGRAPH OF THE MONOCOTYLEDON-INHABITING SPECIES OF *LOPHODERMIUM*

by

Peter R. Johnston
Herbarium PDD
Landcare Research
Private Bag 92170
Auckland
New Zealand

CABI *Publishing*

CABI *Publishing* is a division of CAB *International*

CABI Publishing
CAB International
Wallingford
Oxon, OX10 8DE
UK

Tel.: +44 (0)1491 832111
Fax: +44 (0)1491 829292
Email: cabi@cabi.org
Web site: http://www.cabi.org

CABI Publishing
10 East 40th Street
Suite 3203
New York, NY 10016
USA

Tel.: +1 (212) 481 7018
Fax: +1 (212) 686 7993

E-mail: cab-nao@cabi.org

A catalogue record for this book is available from the British Library.

ISSN 0027–5522
ISBN 0–85199–558–6

Edited by Ken Hudson and Dr Paul M. Kirk

Printed and bound in the UK at Information Press, Eynsham

CONTENTS

SUMMARY

The species of *Lophodermium* originally described or reported from monocotyledonous hosts are monographed. Seventy-seven specific or subspecific epithets in *Lophodermium* are reported from the literature, together with those of six other species at present placed in the genera *Clithris*, *Colpoma*, *Hysterium* or *Rhytisma* but closely related to one or more of the species described in *Lophodermium*.

Of the 83 epithets treated, 24 are retained in *Lophodermium s. lat.*, two appear to represent good species of *Lophodermium* but are of uncertain disposition because of missing type material, 15 have been placed in synonymy with accepted species of *Lophodermium*, and 11 have been transferred to other genera, mostly to *Terriera*. Two species of *Clithris* have been transferred to *Terriera*, and *Colpoma eucalypti* to *Lophodermium*; the earlier transfer of *Lophodermium scirpinum* to *Hypohelion* is accepted. *Rhytisma juncicola* has not been transferred because of doubts over its correct position, although it is considered to be misplaced in *Rhytisma*. Type material was not available for 12 of the taxa, and 11 were never validly published. An additional six new species are recognized in *Lophodermium* and three in *Terriera*.

The genus *Terriera* (also referred to as *Lophodermium* Group B here) has been accepted as distinct from *Lophodermium*. Most of the species of *Lophodermium* described on monocotyledons from tropical regions have been transferred to this genus.

Several groups are recognized among the species retained in the heterogeneous *Lophodermium s. lat.* The largest of these contains the type of *Lophodermium* (*L. arundinaceum*) and is treated here as *Lophodermium s. str.* (or *Lophodermium* Group A). The other three groups contain few species and morphological and anatomical evidence suggests that they are closely related to species at present placed in other genera within the *Rhytismataceae*. The species in Group C are very similar to *Coccomyces tumidus* (the type of *Coccomyces*), those in Group D to *C. coronatus* and some species of *Duplicaria*, and those in Group E to a number of tropical and Southern Hemisphere species of *Coccomyces* and *Lophodermium* from dicotyledonous hosts.

A cladistic analysis, based solely on morphological characters, provided some support for these postulated groups, indicating that current generic concepts within the *Rhytismataceae* have little phylogenetic meaning. The analyses carried out provided preliminary hypotheses for relationships among the leaf-inhabiting species of *Lophodermium* but, because of poorly resolved bootstrap analyses, the groups within *Lophodermium s. lat.* (apart from Group B, the genus *Terriera*) are not recognized formally here.

ACKNOWLEDGEMENTS

This study is based on a PhD thesis submitted to Victoria University, Wellington. Ann Bell, my supervisor, provided ongoing assistance and encouragement, while Landcare Research provided funding and time to complete this work. Jo Berry, Tingkui Qin and Dianne Gleason, colleagues at Landcare Research, provided assistance and discussion regarding methods and interpretation of cladistic analyses. Paul Cannon (CABI *Bioscience* UK Centre, Egham) and Ove Eriksson (University of Umeå) provided useful taxonomic and nomenclatural advice.

Liliane Petrini made available a beautiful collection of recent material of grass-inhabiting species of *Lophodermium* from Europe, together with a copy of her unpublished MSc thesis from ETH, Zürich. The librarians at the Royal Botanic Gardens, Kew, Natural History Museum (London), Farlow Herbarium (Harvard University), New York Botanical Gardens, and the (then) International Mycological Institute (Kew, now CABI *Bioscience* UK Centre, Egham) provided access to their holdings and willing assistance in finding obscure references. Egon Horak and Christian Scheuer provided copies of publications not otherwise available to me, and Irma Gamundi facilitated a loan of type material from Spegazzini's collections in **LPS**. Pavel Lizon, Ivona Kautmanova, J. Gönczöl and Lucia d'Avila Freire provided advice on the possible locality of Hazslinsky's type material; Burghard Hein on the locality of Hennings and Nyman material; Chiara Nepi on locality of Graniti material; Patricia Rogers on locality of Lyon material; Jean Mouchacca on the locality of Feltgen material; Zdeněk Pouzar on the locality of Baudyš material; and Leonard Hutchinson on the material of *Rhytismataceae* available from Canadian herbaria. Tom May clarified the possible identity of '*Spinifex hirsutus*'.

I thank the curators and staff of the following herbaria for allowing me to visit their institutions and to freely examine material in their care: **BPI**, **DAR**, **FH**, **HO**, **IMI**, **K**, **MEL**, **NY**, **P**, **TNS**, **UPS**, **W**, **ZT**. In addition loans were provided by the curators of **BISH**, **BR**, **BRA**, **CUP**, **GZU**, **HBG**, **ILL**, **LPS**, **LUX**, **PAD**, **PRM**, **RO**, **S** and **TRTC**.

Brett Robertson and Larissa Morris, Victoria University, printed many of the photographs; Paul Sutherland and Ian Hallett, HortResearch, Mt Albert Research Centre, kindly provided darkroom facilities and advice.

INTRODUCTION

Lophodermium Chevall. includes stem-inhabiting and leaf-inhabiting ascomycetes having immersed, discomycete-like fruiting bodies which open by a single longitudinal slit, asci with little or no apical differentiation and filiform ascospores.

This paper reports a type study of those species and infraspecific taxa of *Lophodermium* described or reported from monocotyledons. The history of the taxonomy and classification of these fungi and methods used in their study are given, together with an outline of the morphology and ontogeny of the *Rhytismataceae* as a whole. The majority of the monocotyledon-inhabiting species can be placed in one of five broad morphological groups. The morphology and ontogeny of these groups are described in detail, together with notes on those morphological characters most useful at both the specific and supraspecific level. The biology and geographical distribution of the monocotyledon-inhabiting species is discussed. A cladistic analysis compares the relationships of the groups recognized among the monocotyledon-inhabiting species with *Rhytismataceae* from other host substrata. Keys are presented to all species accepted.

With over 250 specific or infraspecific epithets published, *Lophodermium* comprises the largest genus in the *Rhytismataceae*. Little progress has been made towards achieving a classification which reflects phylogenetic relationships among these fungi at any taxonomic level. Classifications in the group remain based on opinion, various authors proposing alternative schemes usually based on a few 'key' characters that they considered the most significant at the time. No cladistic analyses have been attempted for these fungi, and no molecular studies have been carried out to test independently the often conflicting relationships suggested by the various morphological characters used in their classification to date.

> '*Lophodermium* Chev. ... though admittedly containing a widely divergent group of organisms, is the most difficult to reorganize in any satisfactory manner ... Although certain broad divisions might be suggested it seems premature to attempt to make any worthwhile reorganization at this time.' (DARKER, 1967).

> 'Although it is considered unsatisfactory to use one or two 'key' characters such as spore shape ... and ascomatal shape ... to define genera, such a system is still used by most authors for the *Rhytismataceae* because a workable alternative is not yet available. Thus, in the present paper, *Lophodermium* is retained in its traditional sense, although the genus defined in this way is considered artificial.' (JOHNSTON, 1989*b*).

The monocotyledon-inhabiting species remain one of the most poorly-known groups among these taxa.

> 'The species of *Lophodermium* on grasses are difficult to separate or define ... Terrier even thought that morphological characters alone were insufficient to define the graminicolous species of the genus: 'la systématique des formes recontrées sur les Graminées est encore fort obscure. Pour l'établir avec certitude, il est absolument nécessaire de procéder à la culture des formes envisagées, et d'établir, au moyen d'expériences d'infection quels sont leur hôtes spécifiques. Si l'on n'examine que le côté morphologique de la question on se trouve en présence de grandes difficultés que l'on ne peut tourner que dans une certaine mesure par des mensurations des spores et par la structure des fructifications, abstraction faite des

modifications apportées par la morphologie de l'hôte lui-même.' (DENNIS, 1980, quoting TERRIER, 1942).

TERRIER's synopsis is a little pessimistic. Although species limits among the grass-inhabiting taxa are difficult to define with certainty, clear groups can be recognized among these species on the basis of ascomatal structure. The differences between some of the species recognized in this study are somewhat cryptic, but the difficulty in distinguishing them relates in some degree to lack of well-preserved individual collections available for examination.

HISTORY OF THE GENUS *LOPHODERMIUM*

Lophodermium was established by CHEVALLIER (1826), who included nine species in the genus: *L. arundinaceum* (Schrad.) Chevall., *L. gramineum* (Fr.) Chevall., *L. nervisequium* (DC.) Chevall., *L. pinastri* (Schrad.) Chevall., *L. petiolare* Chevall., *L. scirpinum* (DC.) Chevall., *L. rubi* (Pers.) Chevall., *L. herbarum* Chevall. and *L. xylomoides* (DC.) Chevall.

Several of these species had been previously treated by CHEVALLIER (1822) in the genus *Hypoderma* DC., distinguished from the superficially similar *Hysterium* Pers. by having immersed ascomata, with the hymenium becoming exposed following a longitudinal split in the covering host tissue. CHEVALLIER (1826) did not mention *Hypoderma*, but apparently used the name *Lophodermium* in the same sense that he (1822) and other authors (LAMARCK & CANDOLLE, 1805, 1815) had earlier used *Hypoderma*. A type species was not designated for *Lophodermium*, and the species accepted in the genus by CHEVALLIER (1826) include what are now considered to be the type species of both *Lophodermium* (*L. arundinaceum*) and *Hypoderma* (*H. rubi* (Pers.) DC. ex Chevall.).

DE NOTARIS (1847) was the first author to separate *Hypoderma* and *Lophodermium* on the basis of ascospore shape, a distinction still used today. Species with filiform spores were placed in *Lophodermium*, those with elliptical to cylindrical spores in *Hypoderma* (POWELL, 1974). DE NOTARIS included 13 species in *Lophodermium* in two sections, *Phacidioidea* and *Hysterioidea*. The three species in section *Phacidioidea* are now regarded as species of *Coccomyces* De Not., while the other 10 species, five of which occur on monocotyledons, remain in *Lophodermium*. Although DUBY (1862) placed *Lophodermium* and *Hypoderma* in different groups, distinguished by mode of ascus dehiscence, he appeared to follow DE NOTARIS's arrangement: all the species of *Lophodermium* listed by DE NOTARIS which DUBY treated were retained in the genus.

Most subsequent authors followed the DE NOTARIS circumscription of *Lophodermium*. An exception was KUNTZE (1898), who rightly considered that *Lophodermium* was synonymous with *Hypoderma* as originally described and the *Lophodermium* name hence superfluous. He transferred all species then accepted in *Lophodermium* to *Hypoderma* and erected a new genus, *Hypodermopsis*, to accommodate the then-accepted species of *Hypoderma*. Although KUNTZE's approach was nomenclaturally correct, his treatment has been universally rejected because of the confusion it would cause. This problem was not resolved until 1983, when CANNON & MINTER proposed conservation of *Lophodermium* Chevall., 1826, over *Hypoderma* DC., 1805, and of *Hypoderma* De Not., 1847, over *Hypoderma* DC., so maintaining the almost universal usage of the names *Lophodermium* and *Hypoderma*. These proposals were subsequently accepted (KORF, 1988).

HÖHNEL (1917*a*) selected a lectotype for *Lophodermium*, *L. arundinaceum*, which has been accepted by subsequent authors, TEHON (1935) designating a neotype: Mougeot and Nestler, *Stirpes cryptogamae vogeso-rhenanae* no. 655. CANNON & MINTER (1983) cited specific examples of this series to fix unambiguously the application of this name.

A new classification for immersed discomycete-like fungi proposed by HÖHNEL (1917*b*) defined genera primarily on the basis of the position at which fruiting bodies develop within the host tissue. This system restricted *Lophodermium* to species with subepidermal ascomata. HÖHNEL transferred other species to his new genera *Lophodermina* and *Lophodermellina*, characterized by subcuticular ascomata and intra-epidermal ascomata, respectively. The only other author to adopt this scheme fully for the *Rhytismataceae* was TEHON (1935), who included nine species in *Lophodermellina* and 24 in *Lophodermina*. He slightly modified the concept of these genera, including the presence or absence of so-called aliform mycelium in the ascomatal walls as significant features at the generic level. This required another segregate genus, *Dermascia* Tehon (*nom. inval.*, *ICBN* Art. 36.1), to accommodate species with intra-epidermal ascomata which lacked aliform tissue. HILITZER (1929) rejected HÖHNEL's arrangement of *Lophodermium* and related genera, considering that the depth at which ascomata develop in host tissue is dependent on the anatomical structure of the host. HÖHNEL's scheme was also strongly criticized by DARKER (1932) and TERRIER (1942), and the genera *Lophodermina*, *Lophodermellina* and *Dermascia* were rejected by all subsequent authors.

DARKER (1932, 1967) made major contributions to the taxonomy of the conifer-inhabiting *Rhytismataceae*. In his 1932 paper he rejected HÖHNEL's (1917*b*) classification as '... artificial in the extreme ... unusable in practise ...', but noted that 'there do exist some discrepancies and subdivisions within these genera that eventually may warrant certain realignments.' With regard to *Lophodermium* he noted that it could be separated into two sections on the basis of presence or absence of a well-defined band of tissue delineating the future line of opening of unopened ascomata.[1] DARKER (1932) placed heavy emphasis on ascospore shape as a generic character in the *Rhytismataceae*. He was the first author to note the potential of using features associated with ascospore germination as taxonomic characters. These have been largely ignored since, but were discussed in detail recently by OSORIO & STEPHAN (1989).

While retaining ascospore shape as an important generic character, DARKER (1967) developed a classification based on a synthesis of features associated with ascomata, asci, ascospores, conidiomata and biology. This work concentrated on conifer-inhabiting species and proposed several new genera to accommodate many species previously placed in the genera *Hypodermella* Tubeuf, *Hypoderma*, *Lophodermium* and *Bifusella* Höhn. DARKER treated few species of *Lophodermium*, but accepted the genus *Lophomerum*, established by OUELLETTE & MAGASI (1966), for those species with septate ascospores.

ERIKSSON (1970) erected the segregate monotypic genus *Terriera* B. Erikss., with type species *T. cladophila* (Lév.) B. Erikss. (syn. *Lophodermium cladophilum* Lev.), distinguished by its paraphyses swollen at the apex to form an agglutinated epithecium, and the lack of lip cells. JOHNSTON (1988*a*, under *L. multimatricum*; 1989*a*, under *L. minus*) listed several other features associated with ascomatal development and structure which can be used to distinguish the two genera. Although ERIKSSON remains the only author to take up this name, it appears to represent a morphologically quite distinct group of species within *Lophodermium*, and its adoption would be a start to the long overdue rearrangement of this heterogeneous genus.[2]

CANNON & MINTER (1986) largely accepted DARKER's (1967) classification, apart from rejecting the separation of *Lophomerum* from *Lophodermium*. They noted that *Lophodermium* was

---

[1] NANNFELDT (1932) noted the lack of an opening mechanism in the upper wall of the ascomata of another member of the *Rhytismataceae*, *Hypoderma scirpinum* DC. He compared this to the normally well-developed opening mechanism of other species of *Hypoderma* and speculated that this feature might be used at the generic level by future investigators. JOHNSTON (1990*c*) used this, along with several other features associated with ascomatal development, to separate *H. scirpinum* into the segregate genus *Hypohelion* P.R. Johnst.

[2] Several species described by TEHON (1918, 1939) in the genus *Clithris* P. Karst. also have an ascomatal structure typical of *Terriera*. TEHON discussed the relationship between *Clithris* and other members of the *Hysteriales* and, in 1939, transferred two species of *Lophodermium* to this genus. Subsequently, two of the species treated by TEHON as *Clithris* have been placed in synonymy with *Lophodermium platyplacum* (Berk. & M.A. Curtis) Sacc. (JOHNSTON, 1989*a*). ERIKSSON & HAWKSWORTH (1993) listed *Clithris* as a synonym of *Colpoma* Wallr.

well defined within the 'present artificial system of classification of the *Rhytismataceae*', but also noted apparent similarities between some species of *Lophodermium* and the genus *Coccomyces*. *Coccomyces* as currently accepted was regarded as an unnatural group in a major study of this genus by SHERWOOD (1980). She noted that *Coccomyces tumidus* (Fr.) De Not. (the type species of the genus) and *C. coronatus* (Schumach.) De Not. were 'at least as different from each other as *Hypoderma rubi*, the type of *Hypoderma*, and *Lophodermium arundinaceum*, the type of *Lophodermium*, two genera which have been considered distinct since 1826'. She suggested that DE NOTARIS's (1847) scheme, which grouped *C. coronatus*, *C. delta* (Kunze) Sacc. and *C. dentatus* (J.C. Schmidt & Kunze) Sacc. separately from *C. tumidus* in *Lophodermium* subgen. *Phacidioidea*, may be preferable to the present arrangement.

JOHNSTON (1988*a*, *b*, 1989*b*, 1990*a*, *d*), in discussing primarily non-coniferous species, emphasized the artificial nature of *Lophodermium* as accepted and suggested characters associated with ascomatal development which might be useful for defining more natural groups within the genus. Use of these characters would closely align various species of *Lophodermium* with *Meloderma* Darker, *Hypoderma* and *Lophodermium* subgen. *Phacidioidea sensu* DE NOTARIS (1847) (= *Coccomyces p.p.*) and require the establishment of several new genera.

Therefore, despite several authors over the past 60 years questioning the homogeneity of *Lophodermium* (DARKER, 1932, 1967; TEHON, 1935; JOHNSTON, 1989*b*) and an attempt by TEHON (1935) to rearrange the genus, DE NOTARIS's concept is largely intact today. The genus continues to be defined within the *Rhytismataceae* primarily on the basis of ascospore shape, a patently unsatisfactory system, as noted by DICOSMO *et al.* (1984).

THE MONOCOTYLEDON-INHABITING SPECIES OF *LOPHODERMIUM*

The first species now placed in *Lophodermium* to be described from a monocotyledon was *Hysterium arundinaceum* Schrad. in 1799. FRIES (1823) listed seven species of *Lophodermium* (as *Hysterium*) on monocotyledons; since then there has been a steady and more or less continual increase in the number of monocotyledon-inhabiting species described.

DUBY (1862) grouped the *Poaceae*-inhabiting species as infraspecific taxa under *L. arundinaceum*. Several other authors followed DUBY in listing most monocotyledon-inhabiting species of *Lophodermium* as infraspecific taxa under *L. arundinaceum*, retaining only *L. typhinum* (Fr.) Lambotte (ELLIS & EVERHART, 1892; REHM, 1912), *L. herbarum* (FUCKEL, 1873) and *L. caricinum* (Roberge ex Desm.) Duby (DUBY, 1862; FUCKEL, 1870) as distinct species. More recent authors have returned to treating as distinct species the infraspecific taxa which these authors included under *L. arundinaceum*.

Three authors this century have treated the monocotyledon-inhabiting species in some detail: HILITZER (1929), TEHON (1935) and TERRIER (1942). HILITZER provided descriptions for 11 *Poaceae*-inhabiting species, including eight he described as new (one of which he regarded as a new combination of '*Hysterium culmigenum* a. *airarum* Fries' although it should be treated as a new species; see Notes under this epithet in the species descriptions). HILITZER considered each of these species to be biologically distinct, host-specialized forms which could also be distinguished by subtle morphological differences. From the key HILITZER published on p. 49, the table on p. 97, and the notes accompanying his new descriptions, it is clear that he placed much weight on host substratum to differentiate species. However, the series of *Poaceae*-inhabiting species described by HILITZER have been taken up in few publications.

In his monographic treatment of the genus *Lophodermium*, TEHON (1935) treated most of the monocotyledon-inhabiting species which had been described. Several species were placed in the segregate genera *Dermascia*, *Lophodermellina* and *Lophodermina*, but most were treated as *Lophodermium*. A few species and infraspecific taxa were placed in synonymy, but most names were accepted as good species. Three species were described, invalidly (*ICBN* Art. 36.1, no Latin description was provided), as new. Examination of TEHON's descriptions reveals few features of use to distinguish many of the species he accepted. Perusal of his keys shows that the features of ascomatal size, ascus length and width, ascospore width and paraphysis shape were most often used to differentiate species. All these features need to be used with care; for example, ascomatal size appears to vary with host substratum; ascus length and ascospore width vary greatly with maturity, and ascospores appear wider after being released from the asci. There are many inconsistencies in TEHON's publication; species here considered typical of *Terriera* were included in three of the four genera he accepted.

TERRIER (1942) criticized the approaches of both HILITZER and TEHON. He considered HILITZER to have placed too much emphasis on host substratum, and while there might be some host specialization among species of *Lophodermium* he did not regard it to be as strict as HILITZER had maintained. He rejected TEHON's segregate genera, and provided full descriptions of *Lophodermium arundinaceum*, *L. alpinum* (Rehm) Weese and *L. festucae* (Roum.) Terrier, together with notes on *L. culmigenum* (Fr.) De Not. and *L. apiculatum* (Wormsk. ex Fr.) De Not. He also emphasized the significance of some characters not often used in the taxonomy of this group, including mode of development of the ascomatal opening and structure of the lower wall of the ascomata in grass-inhabiting species.

The importance of the structure of the lower wall of the ascoma was also emphasized by PETRINI (1980) in an unpublished MSc thesis on the monocotyledon-inhabiting species of *Lophodermium*. In addition, she discussed for the first time the heterogeneous nature of the gelatinous sheath of these species, the spores having a firm, small gelatinous cap at both ends as well as being surrounded by a looser, wider gelatinous sheath. She noted that both the gelatinous sheath and caps dissolved slowly in KOH and were difficult to observe without phase-contrast optics. Structure of the gelatinous sheaths of *Rhytismataceae* was later discussed by JOHNSTON (1994).

Until the early 1880s almost all species of *Lophodermium* had been described from Europe, but from 1882 the number of new species being described from other parts of the world matched those described from Europe over the same period. The extra-European species were mainly from tropical and southern South America and tropical Asia and, in recent years, a number from New Zealand. Few species have been described from North America, but many of the European species names have been applied to North American collections.

Today the names most widely applied to monocotyledon-inhabiting species of *Lophodermium* in both European and North American literature (FARR *et al.*, 1989; CANNON *et al.*, 1985) continue to be six of the seven first listed by FRIES (1823) – *L. apiculatum*, *L. arundinaceum*, *L. culmigenum*, *L. gramineum*, *L. herbarum* and *L. typhinum* – together with the later-described *L. caricinum* (DESMAZIÈRES, 1847) and *L. alpinum* (REHM, 1881). An examination of the literature shows that several of these names have been applied in different senses by different authors, and TERRIER (1942) discussed this point with reference to *L. alpinum* and *L. apiculatum*. Most other species have been described either as host-specific forms or from exotic locations and have usually been applied to one or a few specimens by the describing author alone.

In the present study, 77 epithets in species of *Lophodermium* from monocotyledons are listed from the literature. Fifty of these names were based on material collected from north-temperate regions, 45 of them from western and northern Europe. Of the other species, about five have been described from each of tropical America, tropical Asia, temperate South America, Australasia and Hawaii. Although many of the European species have since been placed in synonymy and some have since been reported from several different regions, the numbers do represent the relative effort which has been devoted to studying these fungi in those areas. Likewise, most of the material examined during this study was collected from Europe or North America.

MATERIALS AND METHODS

Morphological description

Most material was examined from rehydrated herbarium specimens. Ascus and ascospore size and shape and paraphysis shape were examined from excised hymenium teased apart in 3% KOH and crushed gently under a cover-slip. Where possible (in reasonably recent collections or in well-preserved material), released ascospores were examined in water mounts to determine the arrangement of the spores following release and the structure of the gelatinous sheaths surrounding the spores. Microscopic observations were made using bright field, differential interference contrast and phase contrast illumination.

Ascomatal structure was described from vertical sections 8–10 µm thick, cut from rehydrated herbarium material using a freezing microtome and mounted in lactic acid without staining. Only sections from the widest, central part of the ascomata were selected for examination. Observations on changes during ascomatal development were made from single collections in which ascomata were present at a range of stages of maturity. Ascomatal primordia (e.g. fruiting bodies sectioned for **Figs 11C**, **13A**, etc.) were barely visible macroscopically, their presence often noted only after a piece of host leaf tissue with several immature ascomata was rehydrated in 3% KOH; primordia in the same piece of leaf were then visible using light transmitted through the saturated tissue. Structure of the ascomatal walls was described further from squash mounts in 3% KOH, and from horizontal sections 8–10 µm thick; to assess lower wall structure properly, all three methods were used. Thickness of the lower wall of the ascomata was best determined from vertical sections, but a squash mount could be used to give a quick indication. In species with several layers of dark-walled cells, cell shape could be difficult to assess in squash mounts and horizontal sections were often required.

Ascus and ascospore measurements provided are of those considered to be mature asci and spores. In most species of *Lophodermium s. lat.* the asci are more or less cylindrical, the developing spores initially extending to the base, but rapid elongation of the asci immediately before spore release results in empty basal, stalk-like region. (Note that some species of *Rhytismataceae* characteristically have clavate-stipitate asci, with spores confined to the upper part of the ascus at all stages of development.) Ascus length and width measurements are of asci in which differentiated ascospores are visible and (in almost all instances) in which a basal stalk is evident. Ascus width was measured at the widest point, mostly just below the apex. Whenever possible, ascospores were measured after release from the asci. Spores appeared narrower when within asci, and length of unreleased spores could rarely be judged accurately because of the difficulty of tracking individual spores down the ascus. Hymenial elements examined from old herbarium specimens often appeared a little smaller than those from more recent collections.

Shape of paraphyses was always recorded from those near the centre of the hymenium in opened, mature ascomata.

Biological features

Host and geographical distributions are based almost totally upon information from the herbarium labels. Host range information from this source may be expected to be particularly inaccurate and should be taken only as a guide to the host range of individual species. These fungi usually fruit on fallen dead leaves, and hosts such as the grasses and sedges are often difficult for a non-specialist in these groups to identify accurately. Herbarium specimens usually contain only small pieces of dead leaf material from the host, requiring the identification provided by the original collector to be taken at face value.

Species definition

The units recognized as species in this study are defined purely on the basis of patterns of morphological variation, with what appear to be consistent discontinuities being recognized as species boundaries. Whether the groups recognized in this way are 'real' is uncertain. Although most collections can be placed in one or other of the species with confidence, there are a few collections apparently 'intermediate' between the accepted species. Whether these represent other undescribed or unrecognized species, hybrids between two other putative species or simply demonstrate the inapplicability of the species concept, remains to be seen.

As discussed by MISHLER & DONOGHUE (1982), the complex patterns of sexuality in fungi, with poorly-understood incompatibility systems and parasexual cycles, make understanding of the biological meaning of morphologically distinct groups of individuals currently impossible. The biological reality of 'species' among different groups of organisms will vary depending on the morphology, ecology and breeding systems of the particular group. There is no information on genetic diversity, breeding systems, community structure, gene flow, potential for hybridization, etc., among the *Rhytismataceae*. There is little understanding of whether the species are recognizable because of genetic or environmental/adaptive factors or are an artifact of the sampling procedure used[3], whether they are 'real' units in an evolutionary sense, and whether they will retain their distinctiveness over time.

Despite these uncertainties, distinct morphological groups were recognizable among the individual collections examined during this study, and it is assumed that these are related in some way to common ancestry.

Cladistic analyses

Cladistic analyses were carried out using PAUP version 3·1 (SWOFFORD, 1993) with the intention of testing the relationships between the groups of monocotyledon-inhabiting species recognized in *Lophodermium s. lat.* (see p. 23), and between these groups and other leaf-inhabiting taxa of *Rhytismataceae*. Other taxa represented include species of *Lophodermium* characteristic of other groups of plants, as well as species from other genera of *Rhytismataceae*. Those included are discussed more fully on p. 53.

The list of characters used is given on pp. 50–53 (see **Table 11**). Some of the characters useful for distinguishing species were excluded, as they were considered uninformative at the supraspecific level. Falling into this category were features such as ascus and ascospore size, and structure of the darkened lower wall of the ascoma. Anamorph characters were also excluded, as the presence or absence of an asexual state is also uninformative above the species level. Although the method of proliferation of the conidiogenous cell may provide phylogenetically significant

---

[3] This may be the case for some of the tropical species of *Terriera*. Although accepted in this study, several of these species are known from only one or a few collections, and the choice of 'species level' morphological discontinuities may be found to be inappropriate in future.

information, this feature was excluded too because it was applicable only to those few species having an anamorph.

To root the trees, the genus *Terriera* was designated as the outgroup. As discussed by REYNOLDS (1986), decisions on valid outgroups for cladistic studies involving fungi are difficult. It was impossible to select an outgroup outside the *Rhytismataceae*, because morphological differences between the groups means that few homologous characters would be available on which to compare them.

Species descriptions

All species and infraspecific taxa are treated alphabetically by final epithet. For each taxon, information is provided on host substratum, geographical distribution and type specimen. All specimens examined are cited.

MORPHOLOGY AND ONTOGENY

Since NANNFELDT (1932), most authors have interpreted the teleomorph fruiting body (ascoma) of *Rhytismataceae* as an apothecium (with a very reduced excipulum) developing within a stroma (TERRIER, 1942; POWELL, 1974; MORGAN-JONES & HULTON, 1979; SHERWOOD, 1980; UECKER & STALEY, 1973). Although not commonly recognized, the stroma is sometimes (possibly always) covered by a separate clypeus-like layer. The ascoma always develops immersed within host tissue (becoming erumpent in some genera – see, for example, LIVSEY & MINTER, 1994), the hymenium being exposed after the covering host and fungal tissue (the 'covering layer') splits by one or more elongate slits.

Observations in this section are based on examination of the monocotyledon-inhabiting taxa, as well as representatives of other species and genera listed in the Appendix.

Ascomatal development

The ascomata of all *Rhytismataceae* possess the same basic structural plan (**Fig. 1**). Variation occurs in the time and extent of development of the various layers of tissue within the ascoma, and in some instances a particular tissue layer may be lacking. This variation is useful taxonomically at both the species level and higher ranks.

The ascomatal primordium always develops immersed in host tissue. A cavity forms within the primordium, within which the apothecium develops. Stromatic wall layers form across the top and the bottom of the primordium, the apothecium remaining surrounded by stromatic and primordial tissue. As they mature the ascomata expand radially and increase in height as the paraphyses and asci of the hymenium elongate.

The hymenium and excipular tissue of the apothecium develop from a subhymenial layer of hyaline, thin-walled, angular cells which forms across the base of the primordial cavity. The hymenium initially comprises paraphyses, with asci developing later. Asci have a single wall layer but, as discussed by MINTER & CANNON (1984), the appearance of asci of *Rhytismataceae* varies in different species following spore release, indicating differences in ascus wall structure within this group. Asci do not develop to the edge of the apothecium, the margin of which is sterile and comprises elements often barely distinguishable from paraphyses. These paraphyses-like elements are typically more closely septate than the paraphyses. This layer is retained in mature, opened apothecia and has been interpreted here as homologous with the excipulum of *Leotiales*, albeit much reduced.

The stromatic wall layers start to differentiate at about the same time as the hymenium, distinct layers typically forming across both the top and bottom of the primordium. The upper wall of the stroma comprises more or less globose, thin-walled cells, the walls of which become encrusted with dark material to a greater or lesser extent. The upper wall is lined with downward-projecting, cylindrical, hyaline, thin-walled cells in most species. The term 'periphysoids', used by ERIKSSON (1981) to describe filaments of similar structure and position in fruiting bodies of bitunicate ascomycetes, is employed here to refer to these descending cylindrical cells, although whether these structures are homologous in the two groups is debatable. The lower wall of the stroma is usually narrower than the upper wall and is variable with respect to cell shape and the number of layers of dark-walled cells present. In some species separate stromatic layers are not formed, the darkened upper and lower walls of the ascoma comprising darkened cells of the primordium itself.

A separate, clypeus-like layer comprising narrow, interwoven hyphae with thickened and darkened walls forms within, or adjacent to, the host tissue above the developing ascoma. This layer appears to be functionally equivalent to the structure termed a clypeus in many fungi. The narrow, dark-walled hyphae of the clypeus can be clearly distinguished from the more or less globose cells of the upper wall of the stroma. Note that when describing *Rhytismataceae*, several authors (e.g. DARKER, 1932; ZILLER, 1968; MINTER, 1981; MORGAN-JONES & HULTON, 1977) have used the term 'clypeus' in the sense that 'covering layer' is used in this study.

Ascomata open by one or more elongate slits, the arrangement of the opening slits usually reflecting the shape of the ascoma: a single elongate slit in elliptical or oblong ascomata, several radiate slits in round or angular ascomata. What appear to be fully differentiated ascospores are usually present at the time of opening, with the ascomata themselves ceasing to increase in size from then. Following opening, a palisade of thin-walled, cylindrical cells often forms across the newly exposed part of the broken upper wall of the stroma and are here referred to as lip cells, following MINTER (1981). The periphysoids or the excipulum may continue to develop following opening, forming extensive layers in some species.

The terms defined and illustrated by LETROUIT-GALINOU *et al.* (1994) for the structure of ascomata of ascomycetes in general could equally be applied to the *Rhytismataceae*, the same ontogenetic stages they described being easily recognized. Thus, 'primordium' is used in the same sense, while 'stroma' is equivalent to their 'ébauche' differentiated into several layers. Their terms 'roof' and 'floor' are equivalent to the 'upper wall' and 'lower wall' of stroma here, and both the roof and 'sushymenial bell' layers are included in the upper wall of the stroma. 'Periphysoids' is used here for their 'descending filaments'. It is also probable that the term 'reduced excipulum', the layer of paraphysis-like elements which develop between the hymenium and the stromatal wall in the *Rhytismataceae*, is equivalent to the 'parathecium' of LETROUIT-GALINOU *et al.*, although they did not describe a tissue layer equivalent to the lip cells of *Rhytismataceae*.

Morphology of ascomata and conidiomata

Macroscopic appearance

Ascomatal shape is clearly defined in most species because the ascomata generally develop close to the surface of the host. Although highly variable throughout the family, within most species it is uniform. Ascomata may be elliptical to oblong in outline with a single longitudinal opening slit, angular (3–5-sided) or more or less round, in the latter two cases with several opening slits radiating from a central point (radiate opening). A few species have less regularly shaped ascomata, the opening slit either single and longitudinal or radiate depending on ascomatal shape. The genus *Rhytisma* Fr. is distinct in typically having large, multiloculate ascomata.

Ascomatal shape has traditionally been an important generic character in the family (DARKER, 1967). However, some groups of species can be recognized in which most characters are shared, but in which ascomatal shape is variable. An example is the species of *Lophodermium* Group E described below and morphologically very similar species in *Coccomyces*: the former traditionally with elongate ascomata and a single opening slit, the latter with angular ascomata and a radiate opening. Most groups, however, are uniform in their overall ascomatal shape, two good examples being the species of *Lophodermium s. str.* (*Lophodermium* Group A; see p. 24) on grasses and sedges, and *Propolis* (Fr.) Corda (*sensu* SHERWOOD, 1977).

Ascomatal size and colour can be useful species-level characters, but they must be used with care. Consideration should be given to the possibility of macroscopic appearance being affected by factors such as the age of the specimen collected, the physical nature of the host tissue and

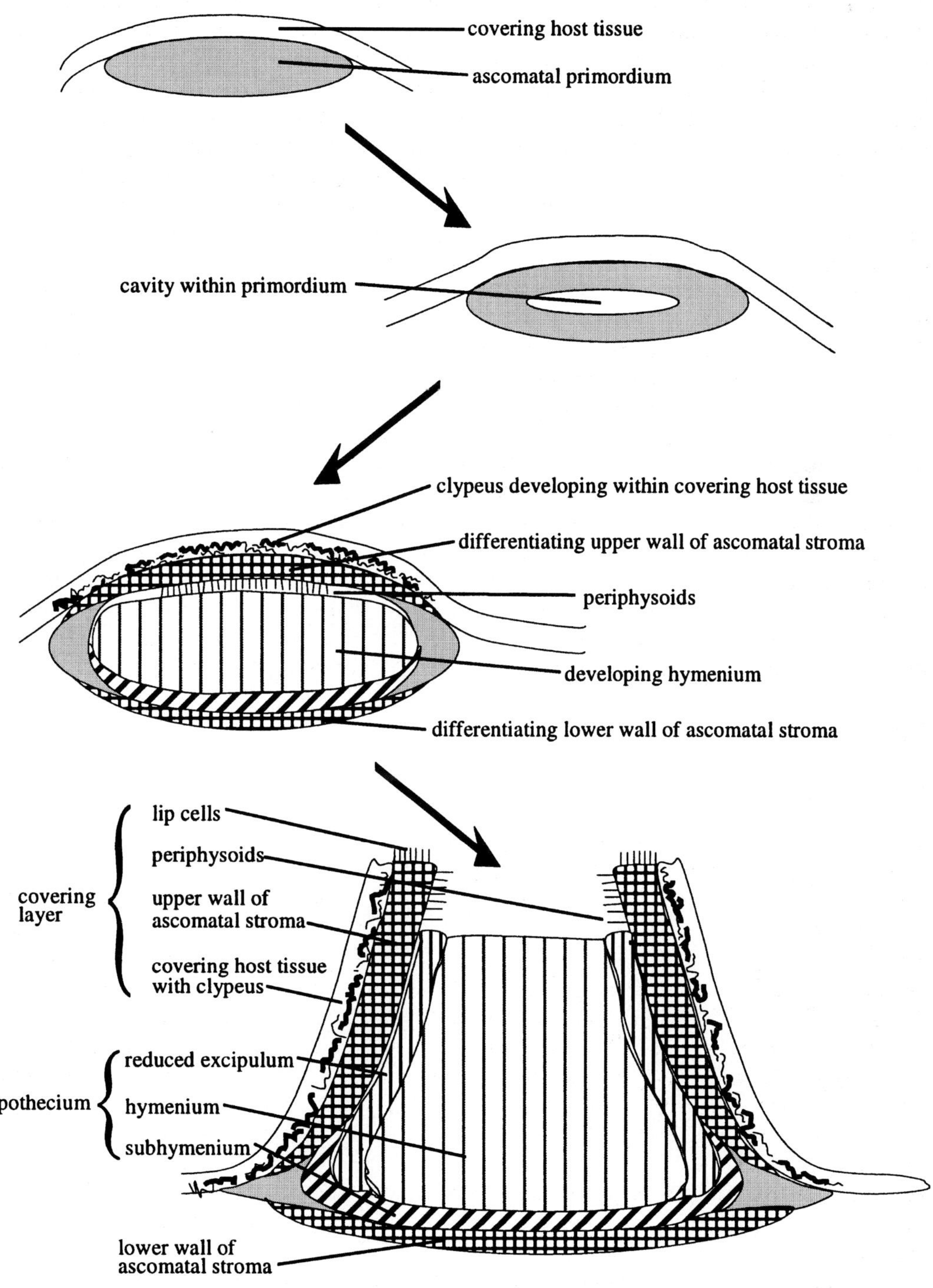

Fig. 1. Generalized scheme of ascomatal development for *Rhytismataceae*. Groups can be defined within the family on the basis of the extent and time of development of the various tissue layers.

other environmental parameters..

Primordium

In members of the *Rhytismataceae* the primordium assumes one of two kinds of cellular structure: either *textura intricata* (following definition in KORF, 1958), with cells arranged more or less parallel to the host surface, or *textura prismatica*, of short-cylindrical to brick-shaped cells arranged in rows more or less at right angles to the host surface. *Lophodermium* as currently circumscribed includes species with both kinds of structure (**Fig. 2**). This variation occurs so early in the ontogeny of ascomata that it is likely to be phylogenetically highly significant and in this study is used as one of the arguments for recognizing the genus *Terriera* as distinct from *Lophodermium s. lat.*

The level at which the primordium develops within the host tissue varies between species: between the cuticle and epidermis; within the epidermal layer, the invaded cells of the epidermis being split in half; or immediately beneath the epidermal cells. The universally rejected taxonomic schemes of HÖHNEL (1917*b*) and TEHON (1935) were based heavily on this character and it has been treated with suspicion by subsequent mycologists. Despite this, it rarely varies within a species and, if used in combination with other characters, appears to be useful to help define some groups above the species level in the *Rhytismataceae*.

Stroma

An anatomically distinct stroma has been seen in all species of *Rhytismataceae* examined, except for those in the genus *Terriera*. In *Terriera* the upper and lower walls appear to comprise darkened but otherwise undifferentiated primordial cells (see **Fig. 2B**, dark upper and lower walls comprising cells of primordium, dark clypeus within epidermal cells above upper wall).

Species differ with respect to whether the upper or lower wall becomes darkened first (*cf.* **Figs 7A**, **13A**), this feature helping to define groups above the species level within *Lophodermium s. lat.* Several conifer-inhabiting species form only a darkened upper wall; the lower wall, if present at all, is poorly developed and comprises one or two layers of hyaline, thin-walled cells.

The upper wall of the ascoma can have a complex internal differentiation, the pattern of which is characteristic for a species and, in some instances, is useful for defining broader groups within *Lophodermium s. lat.* In most species, all cells of the upper wall are more or less uniformly darkened, although characteristically the cells toward the inner part of the wall are somewhat paler than those in the outer part. However, in some species, restricted parts of the wall are much darker than others, a good example of this being provided by *Lophodermium s. str.* (*Lophodermium* Group A; see p. 28). In many species the upper wall of the unopened ascoma has a zone of paler, thin-walled cells along the future line of opening, representing a preformed line of weakness. Depending on the species this can be toward the inside or outside of the wall, or can extend all the way through the wall. This line of weakness is often visible macroscopically in species where it forms in the outer part of the wall. However, in those species with a well-differentiated clypeus-like layer, the clypeus usually does not extend across the future line of opening, and this can be confused macroscopically with the preformed line of weakness in the wall itself.

Periphysoids are present in most, possibly all, species. They often remain poorly developed (e.g. **Fig. 7**) and may be visible only in unopened ascomata, being lost after the ascomata open. However, in some species they form well-developed, distinct layers within the upper wall (e.g. **Fig. 8**). Depending on the species, this layer may become differentiated either before or after the

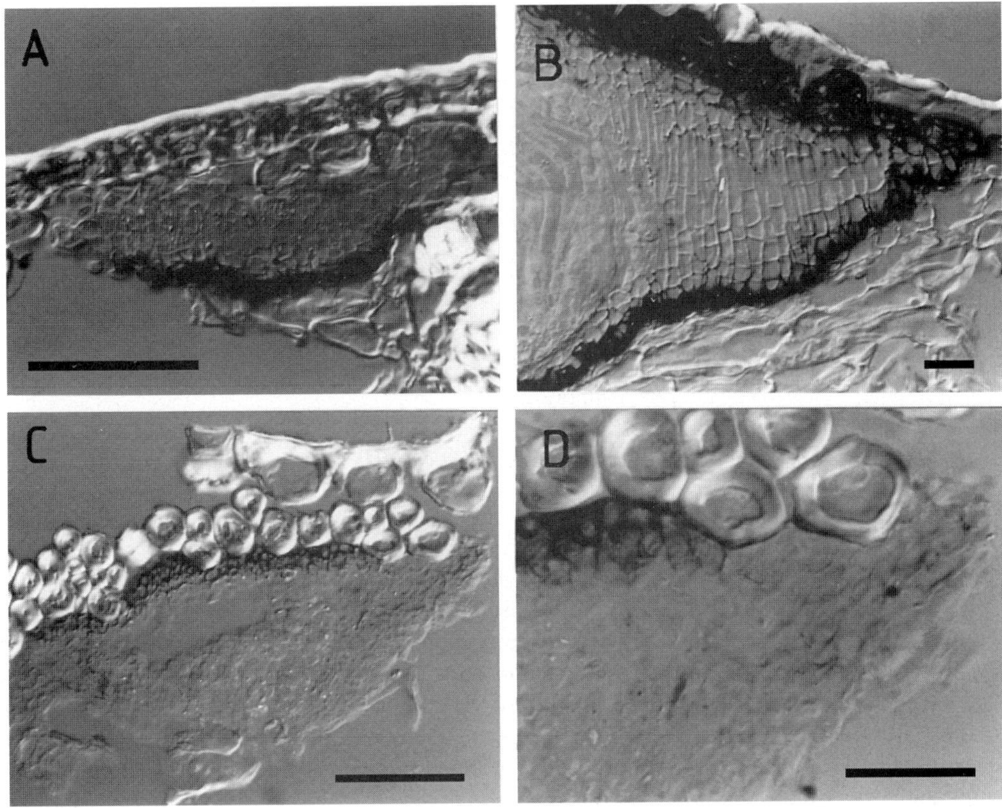

Fig. 2. Structure of ascomatal primordia. **A–B.** Species with primordia comprising *textura prismatica*, characteristic of the genus *Terriera*. **A**, *Lophodermium minus* (**PDD** 46127), ascomata before development of stromatal cavity and paraphyses (bar = 50 µm); **B**, *Terriera breve* (**PDD** 47493), unopened ascomata showing primordial structure at expanding margin (bar = 20 µm). **C–D.** Species with primordia comprising *textura intricata* (*Lophodermium alpinum*, **PDD** 59605). **C**, ascomata with primordial cavity starting to develop, with darkened upper wall developing, but before development of darkened lower wall (bar = 50 µm); **D**, detail of expanding margin of ascoma in **C**, showing primordial cellular structure (bar = 20 µm).

ascomata open (*cf.* **Figs 8**, **12**). The periphysoids may comprise a compact palisade of cylindrical cells or a less regular, looser layer of more or less monilioid elements. In unopened ascomata the layer of periphysoids often does not extend across the future line of opening. Variation in the timing and extent of periphysoid development is useful to help define groups above the species level.

Ascomata usually open by one or more elongate slits along preformed lines of weakness. In some species lip cells form along the edge of the opening slit. Lip cells are normally unbranched, but in *Lophodermium s. str.* they are characteristically dichotomously branched (**Fig. 3**). Those species lacking any preformed line of weakness in unopened ascomata generally do not develop

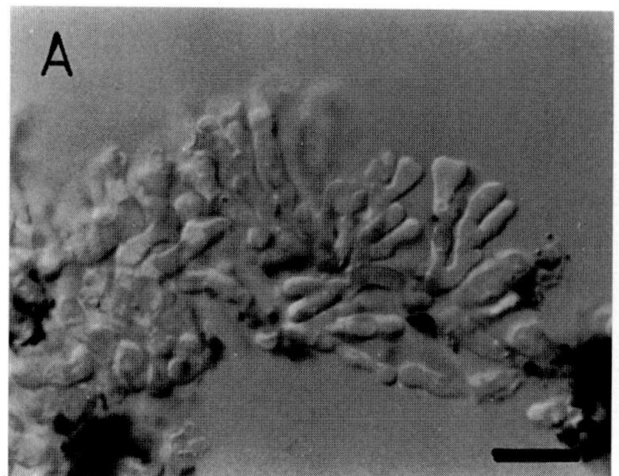

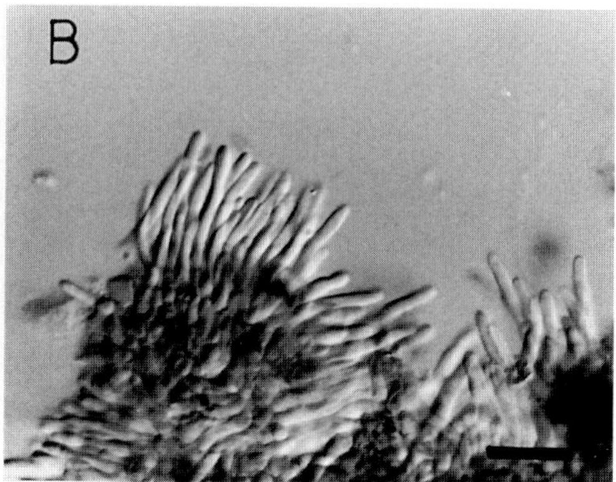

Fig. 3. Variation in branching pattern of lip cells (from squash mounts). **A**, *Lophodermium actinothyrium* (**PRM** 820250); **B**, *Lophodermium eucalypti* (*Sherwood et al.*, **FH**) (bars = 20 μm).

lip cells. The presence or absence of lip cells is useful to help define groups above the species level, with other differences, such as the colour of the lips, sometimes useful at the species level. The lower part of the stroma usually becomes darkened to form a differentiated lower wall. In some species this layer comprises cells of the ascomatal primordium itself becoming darkened, but in most a distinct wall develops below the tissue of the ascomatal primordium. The wall develops outwards from the centre of the ascoma and comprises one to several rows of angular to globose cells, or several layers of narrow-cylindrical branched cells forming a layer of *textura intricata*. In some instances (possibly partly dependent on host substratum) the *textura intricata* is patchy in development across the ascoma. Variation in lower wall structure is taxonomically useful at the species level (**Fig. 9**).

Clypeus

The extent to which the clypeus develops depends on the depth at which stroma develops within the host, being most easily observed in subepidermal or intrahypodermal species (**Fig. 4A**, **B**). Although difficult to observe, several species with intra-epidermal ascomata appear to have a reduced clypeus-like layer immediately below the remains of the broken-down epidermal cells (**Figs 4C**, **D**; **117**). This layer is immediately adjacent to the angular/globose cells of the stroma. In subcuticular species a single layer of cells with a *textura epidermoidea*-like arrangement is often present (**Fig. 4E**, **F**) and is here interpreted as an extremely reduced clypeus. Alternatively, it could be argued that the differentiated shape of these cells is simply due to the physical effect of the adjacent, unyielding cuticle. However, all species of *Rhytismataceae* with more deeply immersed ascomata have a clypeus above the ascoma itself. These species are otherwise morphologically diverse, suggesting that they represent widely divergent lineages within the family. It is reasonable to expect that this layer will also be present in those lineages which have subcuticular ascomata.

Apothecium

Although the position at which the asci develop and the arrangement of the ascogenous cells have

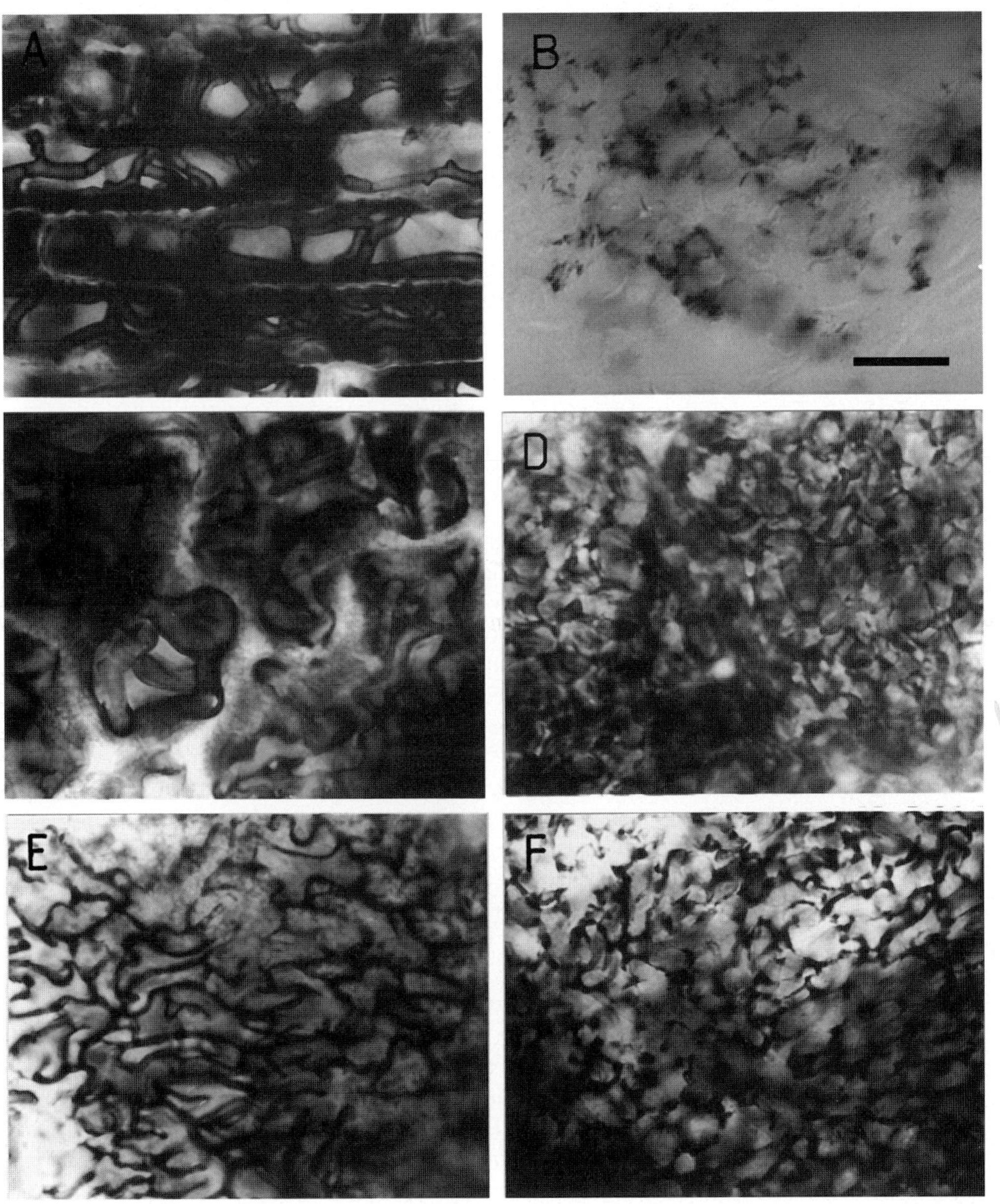

Fig. 4. Comparison between structure of clypeus and upper wall of stroma, from horizontal sections. **A–B.** Subepidermal ascomata, *Lophodermium* Group A. **A**, clypeus (*L. gramineum*, Rehm *Ascomyceten* no. 775, **K**); **B**, structure within wall (*L. typhinum*, **TRTC** 24514). **C–D.** Intra-epidermal ascomata (*Coccomyces limitatus*, **PDD** 53644). **C**, clypeus; **D**, structure within wall. **E–F.** Subcuticular ascomata (*Hypoderma rubi*, **PDD** 48949). **E**, cells adjacent to cuticle; **F**, cells within wall (bar = 20 µm).

not been investigated in this study, other authors (GORDON, 1966; UECKER & STALEY, 1973; CAMPBELL & SYROP, 1975; MORGAN-JONES & HULTON, 1977, 1979) have shown apparent variation between some conifer-inhabiting species. While these authors disagreed somewhat over the interpretation of their observations, future detailed studies may show that these features provide taxonomically useful characters.

The excipulum forms a distinct layer between the hymenium and the adjacent stromatal wall, and is here termed a 'reduced excipulum', following SHERWOOD (1980). The degree to which this layer develops varies between species. It is usually poorly developed, comprising paraphysis-like elements differentiated by having no asci intermixed and in being more closely septate than the paraphyses themselves (see **Fig. 10**). However, in some species such as *Coccomyces radiatus* Sherwood (SHERWOOD, 1980; JOHNSTON, 1986) and *C. parasiticus* P.R. Johnst. (JOHNSTON, 1993) a broad excipular layer is present in opened ascomata. Those species in which the excipulum becomes well developed generally have the upper part of the stroma poorly developed, the excipulum here possibly carrying out the protective function performed by the stroma in other species.

Paraphyses extend into the ascomatal cavity before the development of asci. They are closely septate near the base, and anastomoses form near the base of adjacent paraphyses. In opened, mature ascomata the apices of the paraphyses often become differentiated: swollen, circinate or branched, depending on the species. The differentiated part of the paraphyses often extends above the level of mature asci, forming an epithecium. In all species the paraphyses are surrounded by a gelatinous layer, and in some species are embedded together in a common thick gelatinous matrix. Paraphysis shape is taxonomically useful at the species level and in some instances at higher levels. Some care must be exercised, however, when assessing this character. Paraphyses become differentiated only after the ascomata open, and even then those near the margins often remain undifferentiated, especially in species with paraphyses circinate at the apex (**Fig. 10**). In a few conifer-inhabiting species paraphyses are either never present or are lost early in ascomatal development (MINTER, 1985). This feature correlates with synchronous ascus development and is likely to be phylogenetically significant.

Ascus shape varies: typically cylindrical, clavate, clavate-stipitate or saccate, with the apex rounded, subtruncate or truncate, features which are all diagnostic at the species level. The ascus wall is usually undifferentiated at the apex, but is sometimes slightly thickened and with what appears to be a small central pore. Asci may be 4- or 8-spored and the spores may extend to the base of the ascus or be confined to the upper portion, with an empty basal stalk region. Ascus size may be diagnostic at species level, although this must be used with care as both length and width increase with maturity. In many species the basal stalk of the ascus elongates rapidly immediately before spore release, pushing the ascus apex above the surrounding paraphyses.

Ascospore shape is extremely variable in the *Rhytismataceae* and has traditionally been used as a distinguishing feature at the generic level. If other characters are considered at the same time, it is clear that ascospore shape cannot be used uncritically above the species level (JOHNSTON, 1988*b*, 1990*d*). In spite of this, in some groups of the *Rhytismataceae* ascospore shape is uniform. An example is the species of *Lophodermium s. str.* from grasses and sedges, all with filiform, non-septate ascospores. Another example may be the monotypic genus *Parvacoccum* R.S. Hunt & A. Funk (type species *P. pini*) and other undoubtedly closely related species occurring on conifer bark in North America, such as *Coccomyces heterophyllae* A. Funk, *C. pseudotsugae* A. Funk and *C. mertensianae* Sherwood (M.A. SHERWOOD, pers. comm.).

In some species with filiform ascospores, released spores are characteristically bent or curved in various ways (JOHNSTON, 1994). The spores in some species are septate; in many of the

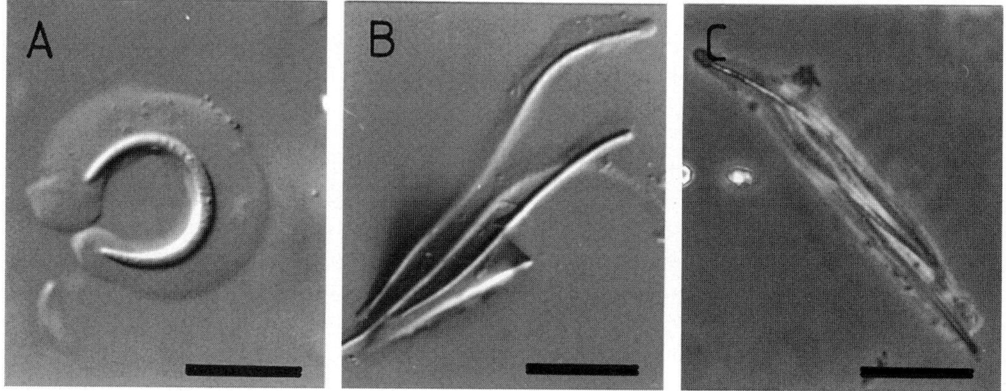

Fig. 5. Ascospore sheaths of species with two structurally distinct parts, firm gelatinous caps at the ends and another loose sheath surrounding the entire spore. **A**, *Lophodermium alpinum* (**PDD** 59526); **B**, *L. fusiforme* (**PDD** 65725); **C**, *Colpoma* sp. from *Populus* (**PDD** 64128 ex *F. Candoussau*) (bars = 20 μm).

filiform-spored species a proportion of the spores develop a single septum at maturity. Ascospore size is a useful character at the species level although, in some species, length especially may be quite variable.

Ascospores surrounded by a gelatinous sheath are generally considered characteristic of the *Rhytismataceae* (HAWKSWORTH *et al.*, 1995). However, some authors have noted the apparent lack of a sheath in some species (e.g. SPOONER, 1991). BARAL (1992) discussed the advantages of examining fungi in the living state, including observations on gelatinous sheaths of *Leotiales*. PETRINI (1980) noted that gelatinous sheaths in *Poaceae*-inhabiting species of *Lophodermium* often dissolve in KOH. Following the observations of PETRINI and BARAL a number of species of *Rhytismataceae* were re-examined in water mounts (JOHNSTON, 1994). As a result, the shape and distribution of the gelatinous sheaths observed around the spores were often at variance with previously published descriptions and illustrations.

As noted by PETRINI, the gelatinous sheaths of most *Poaceae*-inhabiting species of *Lophodermium* comprise two structurally distinct parts. At both the base and apex of the spore is a small, usually either globose or cylindrical, firm, gelatinous cap, and surrounding both this and the remainder of the spore is a looser gelatinous sheath (**Fig. 5A, B**). The shape and size of the gelatinous caps often differ between species. Although all species of *Lophodermium s. str.* have this kind of sheath structure, other filiform-spored species have similar sheaths, e.g. *Lophodermium pinastri*, *L. eucalypti* (Rodway) P.R. Johnst. (JOHNSTON, 1994, as *L. mahuianum* P.R. Johnst.) and at least one species of *Colpoma* Wallr. (**Fig. 5C**).

Other species of *Lophodermium* have small gelatinous appendages along the sides of the spores, for example *L. hauturuanum*, in which there is a small, 'barb-like' appendage near the distal end, at the point where the spores of this species characteristically bend at *c.* 120° (JOHNSTON, 1994).

None of the species examined with an ascomatal structure matching *Terriera cladophila* (the type species) had gelatinous sheaths visible when mounted in water (JOHNSTON, 1994). Two species of *Hypoderma* have been described as lacking ascospore sheaths: *H. shimanense* Suto (SUTO, 1983) and *H. labiorum-aurantiorum* H.C. Evans & Minter (MINTER, 1985). Both have the well-developed lip cells, elliptical to short-cylindrical ascospores and clavate-stipitate asci typical of this genus (JOHNSTON, 1990*d*). However, it is interesting that both species are found on members of the *Pinaceae*, an unusual substratum for *Hypoderma sensu* POWELL (1974) or JOHNSTON (1990*d*). Reports of the type specimen of *Hypoderma rubi* lacking an ascospore sheath (MINTER, 1984; CANNON & MINTER, 1986) are probably due to the sheath being lost following prolonged herbarium storage.

Appressoria or appressorial-like structures have been described for some species (DARKER, 1932; SAHO & ZINNO, 1972). OSORIO & STEPHAN (1989) illustrated these for collections identified as *Lophodermium arundinaceum* as well as a number of *Pinaceae*-inhabiting species. They found that each species had a characteristic form of appressorium, and that similarity in appressorial shape between different species may suggest a close relationship between them. They found that not all species of *Rhytismataceae* developed appressoria and that this might be related to the mode of host penetration. *Cyclaneusma minus* (Butin) DiCosmo, Peredo & Minter, previously shown to infect through stomata, did not form darkened appressoria. Note, however, that CHOI & SIMPSON (1995) described hyaline-walled structures in this species which they interpreted as appressoria.

Conidiomata

Many leaf-inhabiting species of *Rhytismataceae* are associated with conidial states, usually placed in the genus *Leptostroma* Fr. These anamorphs are generally considered to be spermatial rather than to have a distributive function (CANNON & MINTER, 1986).

Most species form separate conidiomata, the macroscopic appearance of which can vary between species. Typically they are small (< 0·3 mm diam.) and round in outline. They may be very pale brown or concolorous with the surrounding host tissue, usually with a narrow, darker edge, or may be dark, either brown or black. If the latter, they are usually subcuticular and pustulate. The paler conidiomata are usually immersed within the host plant and are flattened, being more or less lenticular in vertical section. Commonly these conidiomata lack any upper wall layer. A few species have more or less globose, immersed conidiomata.

Anamorphs in the genus *Rhytisma* are distinct, developing within the same fruiting body as the teleomorph and, in some species at least, within the same locules, the hymenium developing after the conidiogenous layer has matured and been lost.

In most species the conidiogenous layer develops on the lower wall only. The conidiogenous cells are typically solitary, forming a compact palisade across the base of the conidioma. The conidiogenous cells proliferate either sympodially or percurrently, some species having both kinds of proliferation in a single conidioma. Conidia are typically non-septate, small and cylindrical with rounded ends. Occasionally conidial shape and length can be useful characters at the species level, e.g. in *Lophodermium litseaeoides* Spooner (SPOONER, 1991).

MORPHOLOGICAL GROUPS

Lophodermium s. lat. is characterized within the *Rhytismataceae* by the possession of elliptical ascomata opening by a single, longitudinal slit, cylindrical rather than saccate asci, and filiform ascospores (DARKER, 1967). Most species of *Lophodermium* found on monocotyledons can be placed in one of five broadly-defined morphological groups (**Table 1**). A few other species remain for which no clear relationship can be postulated on the basis of morphology. Each of the groups is fully described and illustrated below, together with a list of the species accepted in the groups and, where relevant, notes regarding species limits applied to each of them.

The groups recognized are hypothesized to reflect phylogenetic relationships among these fungi. Group B is accepted as the genus *Terriera*, with the required combinations proposed later. However, the cladistic analyses provided little support for the other groups recognized and, until additional characters, either morphological or molecular, become available to test them, they are presented as informal groups within a heterogeneous *Lophodermium s. lat.* The features characteristic of each group are summarized in **Tables 2–4**.

Table 1. Species in each of the groups recognized among the monocotyledon-inhabiting species of *Lophodermium s. lat.*

Group A (*Lophodermium s. str.*)	**Group B** (*Terriera*)	**Group C** (cf. *Coccomyces p.p.*)	**Group D** (cf. *Coccomyces p.p.*)
L. actinothyrium	'*L. andropogonis*'	*L.* cf. *juncinum*	*L.* cf. *luzulae*
L. alliaceum	*T. arundinacea*	*L. tumidulum*	*Rhytisma juncicola*
L. alpinum	*T. asteliae*		
L. arundinaceum	*T. breve*		
L. caricinum	*T. clithris*		
L. culmigenum	*T. dracaenae*		
L. danthoniae	*T. fourcroyae*		
L. eximium	*T. fuegiana*		
L. fusiforme	*T. javanica*		
L. gramineum *L. grandialpinum*	*T. latiascus* *T. longissima*	**Group E** (cf. *Coccomyces p.p.*)	**Miscellaneous spp.**
L. herbarum	'*L. miscanthi*'	*L. alienum*	*L. agathidis*
L. iridicolum	*T. nematoidea*	*L. hauturuanum*	*L. eucalypti*
L. nitidum	*T. pandani*	*L. unciniae*	*L. inclusum*
L. nonramosum	*T. javanica* var. *pandani*		*L. raapianum*
L. petriniae			*L. vrieseae*
L. robergei	*T. sacchari*		
L. rubrum	*T. samuelsii*		
L. seriatum	*T. stevensii*		
L. sesleriae			
L. sieglingiae			
L. typhinum			

Lophodermium Group A (*Lophodermium s. str.*)

This group contains *L. arundinaceum*, the type species of *Lophodermium*, and is also here referred to as *Lophodermium s. str.* Most species in this group are confined to *Poaceae* and most collections from this host family, at least from temperate regions, will be members of this group. A few species are known from other host families such as *Cyperaceae*, *Typhaceae* and *Liliaceae*. Most species have been reported from northern temperate regions, with a few also from temperate regions of the Southern Hemisphere. The difference in species numbers between northern and southern regions probably reflects the relative amount of taxonomic work carried out on this group in the two regions. However, the group appears to be genuinely rare in tropical regions (JOHNSTON, 1997).

Fruiting bodies almost always develop on dead leaves. OSORIO & STEPHAN (1989) reported the formation of dark appressoria following ascospore germination in *L. arundinaceum*, suggesting that this species may invade the living leaves of its host, although fruiting bodies do not form until the leaf dies. Other grass-inhabiting species may have the same biology. *Lophodermium seriatum* (Lib.) De Not. is distinctive among the grass-inhabiting species as its fruiting bodies develop within necrotic lesions on otherwise green leaves.

Species within this group are distinguished on the basis of macroscopic appearance of the ascoma, presence or absence of lip cells and periphysoids, structure of the lower wall of the ascoma, ascus and ascospore size, ascus and ascospore shape, paraphysis shape, appearance of conidiomata and host substratum (see p. 47). Some of these features can be used to place the species in a number of broad subgroups (see p. 33).

Species differentiation can be problematic because most of the characters used are either difficult to measure accurately or their variation within a species is still relatively poorly understood. Macroscopic appearance can be a useful guide to species identification (see **Fig. 6**), but variation is evident in some species. Ascus size (especially length) varies greatly with maturity and may do so depending on the length of time the specimen has been stored. Ascospore size can be measured accurately only after spore release, and spore width increases with maturity. Paraphysis shape must be measured from near the centre of mature ascomata. Lower wall structure can be difficult to measure, requiring thin sections, and although it appears stable on a broad scale, minor variation within a species is not well understood. The presence or absence of anamorphs can vary between collections and between groups of fruiting bodies within individual collections; they may be missing by chance in small collections.

No single character can be considered 'diagnostic' of any of the species in this group. All characters must be considered together and a decision made as to whether to interpret an apparently unique combination of characters as indicative of a separate taxon, or whether they are best seen as an atypical variant of a previously recognized species (see Notes under *L. actinothyrium*, *L. culmigenum* and *L. nitidum*). Whether or not the species concepts applied to the group are appropriate remains uncertain. Twenty-two species are accepted, all known from Europe (**Table 5**), in part reflecting the large number of species which have been previously described from the region. A conservative position has been taken regarding the species-level taxonomy and, except where the type specimen clearly matches an earlier described species, the variation represented by the type (and other collections examined by the describing author) has been accepted here, at least tentatively, at the species level. The validity of most of the species accepted from Europe has been supported by more recent collections which fit this species concept well and, with few exceptions, collections examined from outside Europe fit well into one of the species accepted here.

Table 2. Summary of features characteristic of *Lophodermium* Groups A and B, together with postulated relationships to other *Rhytismataceae*.

Feature	Group A (*Lophodermium s. str.*)	Group B (*Terriera*)
Ascomata: Macroscopic appearance	Elliptical, with single longitudinal slit; lips pale and macroscopically distinct, indistinct (because dark or poorly developed) or absent	Elliptical to linear, with single longitudinal opening slit, well-developed pale zone along future line of opening, dark shelf-like zone lining opening slit
Clypeus	Well developed within epidermal cells	Well developed within epidermal cells
Primordium: Position within host	Subepidermal or intrahypodermal	Subepidermal or intrahypodermal
Structure	*Textura intricata*	*Textura prismatica*
Stroma: Time of development of darkened lower wall	After upper wall becomes darkened	Before upper wall becomes darkened
Pigmentation of cells in upper wall	Mostly pale, characteristic small group of very dark cells near opening	Upper wall comprises slightly darkened cells of primordium
Preformed zone of weakness	Clypeus lacking over future line of opening; dark cells within upper wall forming two groups, one either side of future line of opening	Clypeus lacking over future line of opening
Periphysoids	In some species poorly developed, lost after ascoma opens; in other species well developed, monilioid, retained in open ascomata	Few near opening slit, elongating after ascomata open; extending across top of hymenium, becoming very dark with cellular structure obscured
Lip cells	Present or absent	Absent
Structure of lower wall	In most species several layers of angular to cylindrical cells, one or more layers darkened; some species with dark-walled hyphae forming *textura intricata*	One or two lowermost rows of ascomatal primordium become dark-walled
Apothecium: Reduced excipulum	Relatively poorly developed	Poorly developed
Shape of paraphysis apex	Undifferentiated or circinate (rarely slightly swollen), not forming epithecium	Usually swollen or branched, often tangled and in gel, forming epithecium
Shape of asci	Cylindrical, subclavate or subfusoid, rarely subsaccate	Cylindrical to subcylindrical
Shape of ascospores	Filiform, straight when released, non-septate	Filiform, usually 1-septate, often curved or sigmoid on release
Ascospore sheath	Complex; firm gel caps at apex and base, surrounded by additional loose gel sheath	None
Hosts: Monocotyledonous	Most on *Poaceae*, some on *Cyperaceae*, *Liliaceae* and *Typhaceae*	Numerous families
Other	None	Many dicotyledons, few conifers
Relationship with other genera in *Rhytismataceae*	Unknown	Unknown

Table 3. Summary of features characteristic of *Lophodermium* Groups C and D, together with postulated relationships to other *Rhytismataceae*.

Feature	Group C	Group D
Ascomata: Macroscopic appearance	Broadly elliptical to ovate in outline, erumpent, *Hysteriales*-like in appearance, with wall either raised or sunken along opening slit; hymenium orange-brown	Angular to rounded in outline, with irregular, usually radiate opening slits; no macroscopically distinct lip cells; hymenium orange-brown
Clypeus	Reduced clypeus possibly present	Reduced clypeus possibly present
Primordium: Position within host	Subepidermal or subcuticular	Subcuticular or intra-epidermal
Structure	*Textura intricata*	*Textura intricata*
Stroma: Time of development of darkened lower wall	Possibly before upper wall becomes darkened	Possibly before upper wall becomes darkened
Pigmentation of cells in upper wall	More or less uniform, slightly paler toward inside of wall	More or less uniform
Preformed zone of weakness	Absent	Absent
Periphysoids	Innermost cells of wall forming loose tissue; periphysoid-like layer	More or less persistent layer of short-cylindrical cells embedded in gel
Lip cells	Absent	Absent
Structure of lower wall	Several rows of angular to cylindrical cells	Several rows of dark-walled, cylindrical to hyphal cells
Apothecium: Reduced excipulum	Relatively poorly developed	Relatively poorly developed
Shape of paraphysis apex	Branched, often circinate	Swollen, either tangled or embedded in gel, forming epithecium above asci
Shape of asci	Clavate to subclavate	Clavate to subclavate
Shape of ascospores	Filiform, straight when released	Long-cylindrical to filiform, 0–3-septate, more or less straight when released
Ascospore sheath	Gelatinous caps	Gelatinous caps
Hosts: Monocotyledonous	*Cyperaceae*, *Juncaceae* and *Poaceae*	*Cyperaceae*, *Juncaceae* and *Poaceae*
Other	Some dicotyledons, especially deciduous trees	Some dicotyledons, especially deciduous trees
Relationship with other genera in *Rhytismataceae*	Ascomatal structure and hymenium morphologically similar to *Coccomyces tumidus*	Ascomatal structure and hymenium morphologically similar to *Coccomyces coronatus*

Table 4. Summary of features characteristic of *Lophodermium* Group E, together with postulated relationships to other *Rhytismataceae*.

Feature	Group E
Ascomata: Macroscopic appearance	Oblong to sublinear; main longitudinal slit, with short side-slits near ends; narrow dark line around edge of opening slit; hymenium often exposed in dry ascomata
Clypeus	Well developed within epidermis, reduced in intra-epidermal species
Primordium: Position within host	Subepidermal or intra-epidermal
Structure	*Textura intricata*
Stroma: Time of development of darkened lower wall	Before upper wall becomes darkened
Pigmentation of cells in upper wall	More or less uniformly pigmented, pale to dark brown
Preformed zone of weakness	Clypeus lacking over future line of opening, and a few pale cells in outer part of wall near centre of ascoma
Periphysoids	A few adjacent to ascomatal opening elongating after ascoma opens, often with apex embedded in dark brown material
Lip cells	Present but poorly developed
Structure of lower wall	Several rows of dark-walled cells, either cylindrical or hyphal in shape
Apothecium: Reduced excipulum	Present, often forming broad layer, especially in parts of apothecium adjacent to side slits in upper wall
Shape of paraphysis apex	Swollen or tangled at apex, often embedded in gel, forming epithecium
Shape of asci	Subclavate or cylindrical
Shape of ascospores	Filiform, 1-septate, often characteristically bent or curved on release
Ascospore sheath	Small gel caps or appendages at ends of spores, or at points where spores bend
Hosts: Monocotyledonous	*Cyperaceae* and *Poaceae*
Other	Numerous dicotyledons
Relationship with other genera in *Rhytismataceae*	Morphologically similar to a number of tropical and Southern Hemisphere species of *Coccomyces* and *Lophodermium* from dicotyledons

Evidence from species known from large numbers of collections, such as *L. alpinum* and *L. gramineum*, suggests that ascomatal wall structure is a more reliable feature than ascus and ascospore size for defining relationships around the species level in the grass-inhabiting species of *Lophodermium*. If similar levels of ascus and ascospore size variation were accepted for all species in *Lophodermium* Group A then, for example, species such as *L. alliaceum* Feltgen, *L. robergei* (Desm.) Sacc. and *L. typhinum* would be placed in a more broadly defined *L. actinothyrium* and, similarly, the various *L. nitidum*-like collections from North America would fit easily within this species. Whether or not the species accepted in this study represent 'real' biological units awaits some understanding of the population structure and breeding systems of these fungi, together with independent testing of the morphological species using molecular data.

Although few extra-European species have been described, some of those from Europe are also known from other regions (**Table 5**). No new extra-European species are accepted here, but there are collections from Greece, New Zealand, Chile and North America which do not fit well into any of these accepted species (see p. 36 and Notes under *L. actinothyrium*, *L. culmigenum* and *L. nitidum*). New species are not proposed for these collections, primarily because of the uncertainty of species limits resulting from the paucity of extra-European collections. However, as more data become available, it is likely that the level of diversity present in Europe will also be found to occur elsewhere and, if the species limits applied to the European collections continue to be accepted as valid, then many more species await description from other parts of the world.

Morphological features

In summary, Group A is characterized by elliptical ascomata with a single opening slit, which may or may not be lined with lip cells. Ascomata are subepidermal with a well-developed clypeus and in vertical section the upper wall of the ascoma characteristically has a restricted patch of very dark cells adjacent to the lips (or to the opening slit in those species without lips), in contrast to the otherwise quite pale wall. Ascospores remain non-septate and have gelatinous caps at both ends as well as a separate, looser gelatinous sheath.

Macroscopic appearance. In surface view, ascomata elliptical to oblong-elliptical in outline, with ends either rounded, ± acute or apiculate, opening by single longitudinal slit (**Fig. 6**). Wall grey to dark grey, with edge of ascoma sharply defined or diffuse. Opening slit often lined with lip cells but, even when present, these may be poorly developed or dark and macroscopically indistinct; in all but one species, lip cells are dichotomously branched, unopened ascomata sometimes with paler zone along future line of opening. Ascomata 0·4–3 × 0·2–0·5 mm but, within individual collections or species, do not usually vary more than *c.* 0·5 mm in length.

Primordium. Ascomatal primordium plectenchymatous; insertion subepidermal or within upper layers of hypodermis.

Stroma. Both upper and lower walls present, upper wall darkening before lower (**Fig. 7A**). Upper wall mostly pale brown, comprising few rows of angular to globose cells, with walls irregularly encrusted with dark brown material. In unopened ascomata, two groups of cells with thickly encrusted walls form on either side of future line of opening (**Fig. 7B**, **C**), each group of cells roughly triangular, with base of triangle towards inside of wall, these two groups often meeting near inside of wall, leaving paler area in outer part of wall along future line of opening. In opened ascomata, groups of dark cells are adjacent to lip cells, or to opening slit in those species without lips (**Fig. 7D**). Periphysoids always present in unopened ascomata, although extent to which they develop varies, in most species poorly developed and lost after ascomata open (**Fig. 7C**, **E**). In several species there is broad layer of periphysoids in unopened ascomata, which do not extend across future line of opening, but which may be persistent after ascomata open.

Most species with broad layer of periphysoids also lack lip cells (**Fig. 8**), but few have both well-developed periphysoids and lip cells (e.g. *L. caricinum*). Lower wall of ascomata comprises several layers of cells, shape of which and number of layers of darkened cells varies between species. Darkened cells of lower wall are arranged as single layer of angular/globose cells, several layers of cylindrical to short-cylindrical cells or several layers of long-cylindrical to hyphal cells tangled together to form layer of *textura intricata* (**Fig. 9A–F**). In species with thickened lower wall, extent of development of additional layers of darkened cells varies between ascomata and even across base of individual ascomata.

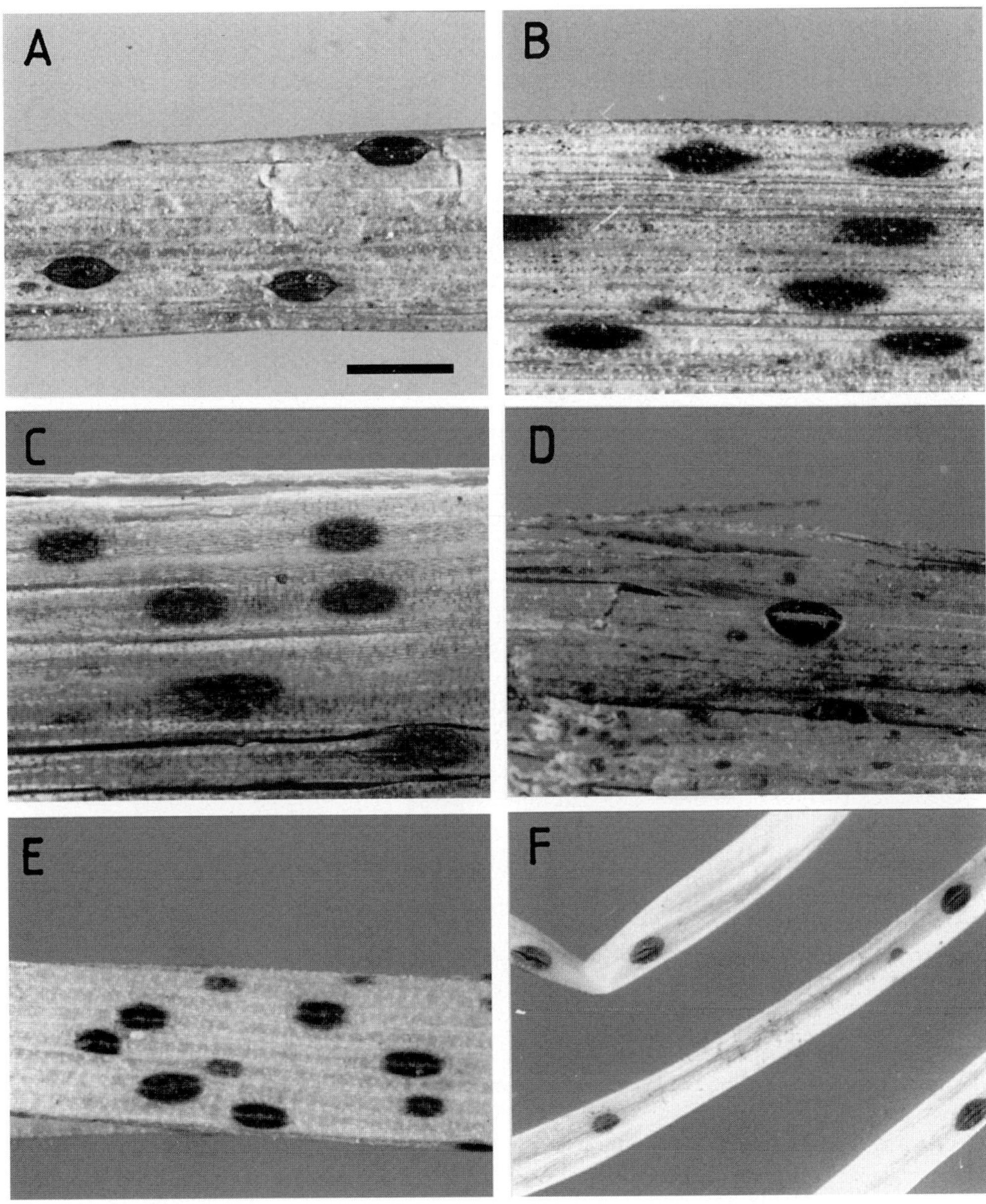

Fig. 6. Taxonomically useful variation in macroscopic appearance between species in *Lophodermium* Group A. **A**, *L. actinothyrium* (**PRM** 734499), oblong-elliptical ascomata, ends rounded with small apiculus; **B**, *L. sesleriae* (**PRM** 756746), elliptical ascomata, margins poorly defined, ends acute; **C**, *L. culmigenum* (Rabenhorst, *Fungi Europaei* no. 4366, **HBG**), large, broadly elliptical ascomata, ends rounded, margins poorly defined; **D**, *L. nitidum* (**PDD** 56286), broadly elliptical ascomata, ends rounded, margins sharply-defined, lips well developed; **E**, *L. grandialpinum* (**PDD** 65057), well-developed pale zone along future line of opening characteristic of many species in the *alpinum*-group of *Lophodermium* Group A; **F**, *L. alpinum* (Rehm, *Ascomyceten* no. 319, **HBG**), small ascoma characteristic of this species (bar = 1 mm).

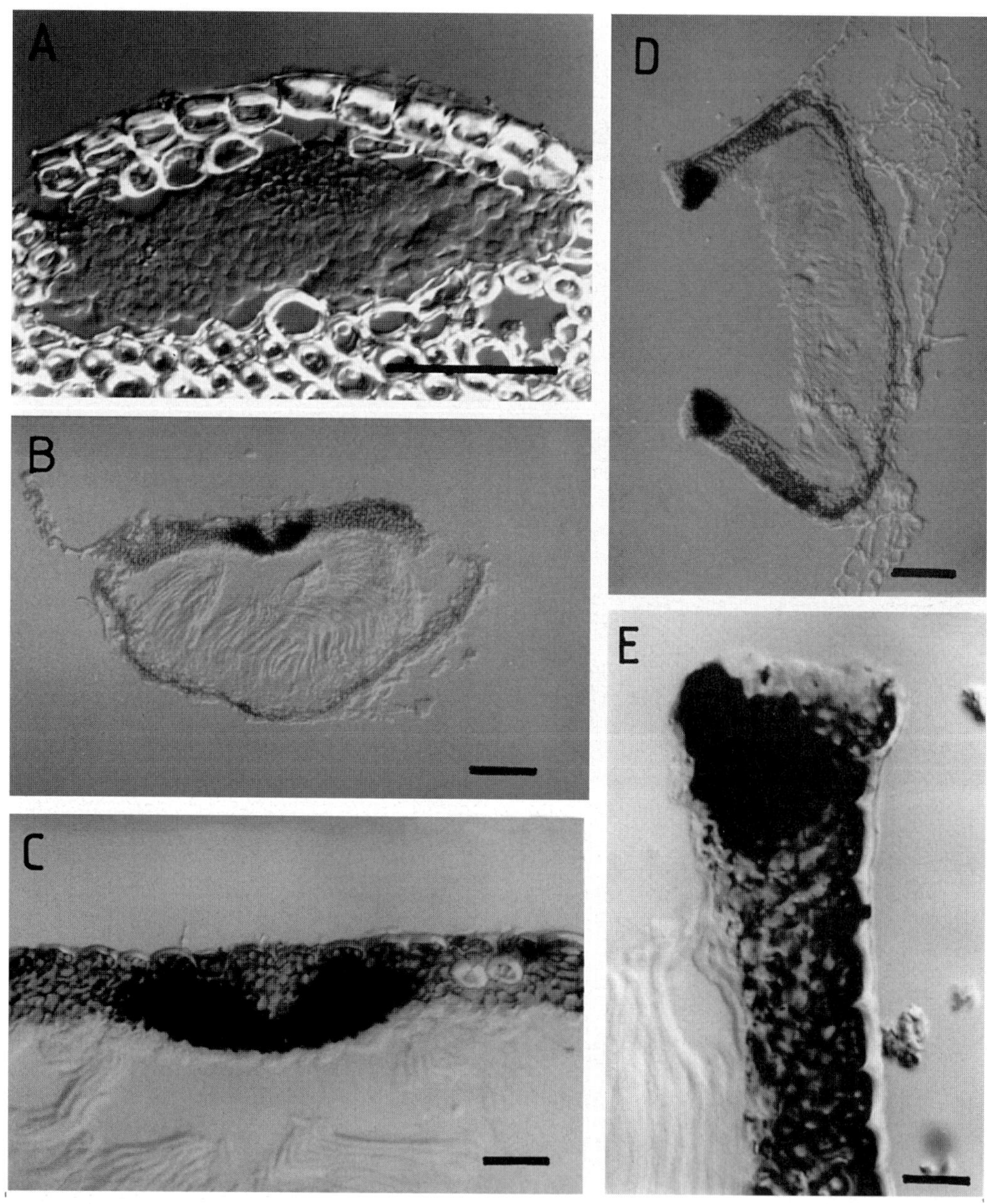

Fig. 7. Ascomatal development in the *actinothyrium*-group of *Lophodermium* Group A (**A**, *L. culmigenum*, **PRM** 705172; **B–E**, *L. robergei*, Desmazières, *Pl. crypt. N. France* no. 169, **K**). **A**, ascomatal primordium, upper stromatal wall starting to form; **B**, unopened ascoma, with dark groups of cells on either side of future line of opening (bars = 50 μm); **C**, detail of upper wall of unopened ascoma, lined with poorly-developed layer of periphysoids (bar = 20 μm); **D**, opened ascoma (bar = 50 μm); **E**, detail of upper wall of opened ascoma, with lip cells (bar = 20 μm).

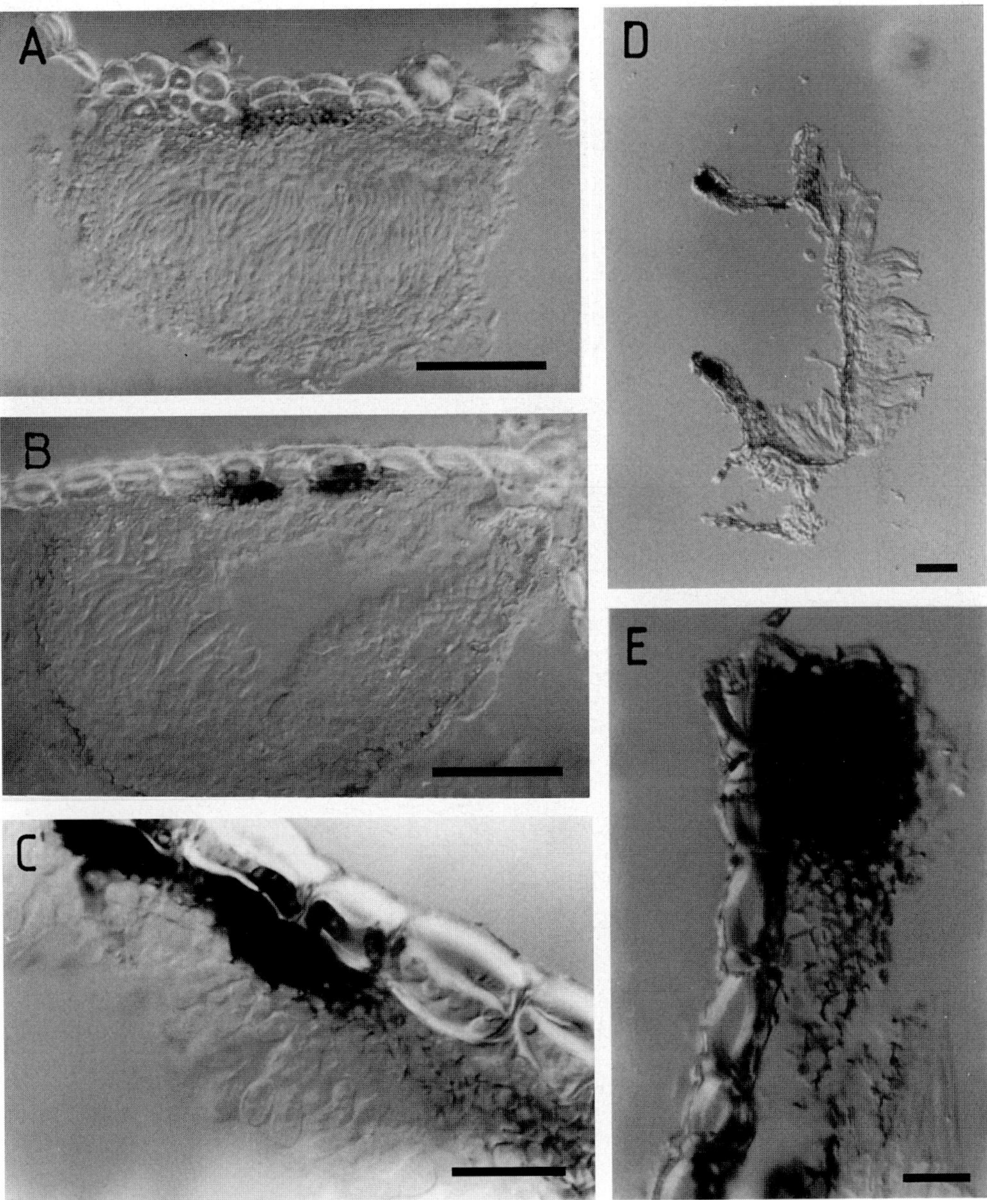

Fig. 8. Ascomatal development in the *alpinum*-group of *Lophodermium* Group A. **A**, ascoma before asci start to elongate, with upper wall darkening before lower wall (*L. sesleriae*, *Libert* 73, **K**); **B**, unopened ascoma, with dark patches in upper wall, on either side of future line of opening and with well-developed, monilioid periphysoids (*L. seriatum*, *Fungi Exs. Suecici Upsaliensis* no. 3490, **K**) (bars = 50 µm); **C**, detail of periphysoids (*L. seriatum*, *Libert* 371, **K**) (bar = 20 µm); **D**, opened ascoma (*L. seriatum*, *Libert* 371, **K**) (bar = 50 µm); **E**, detail of upper wall (*L. seriatum*, *Libert* 371, **K**) (bar = 20 µm).

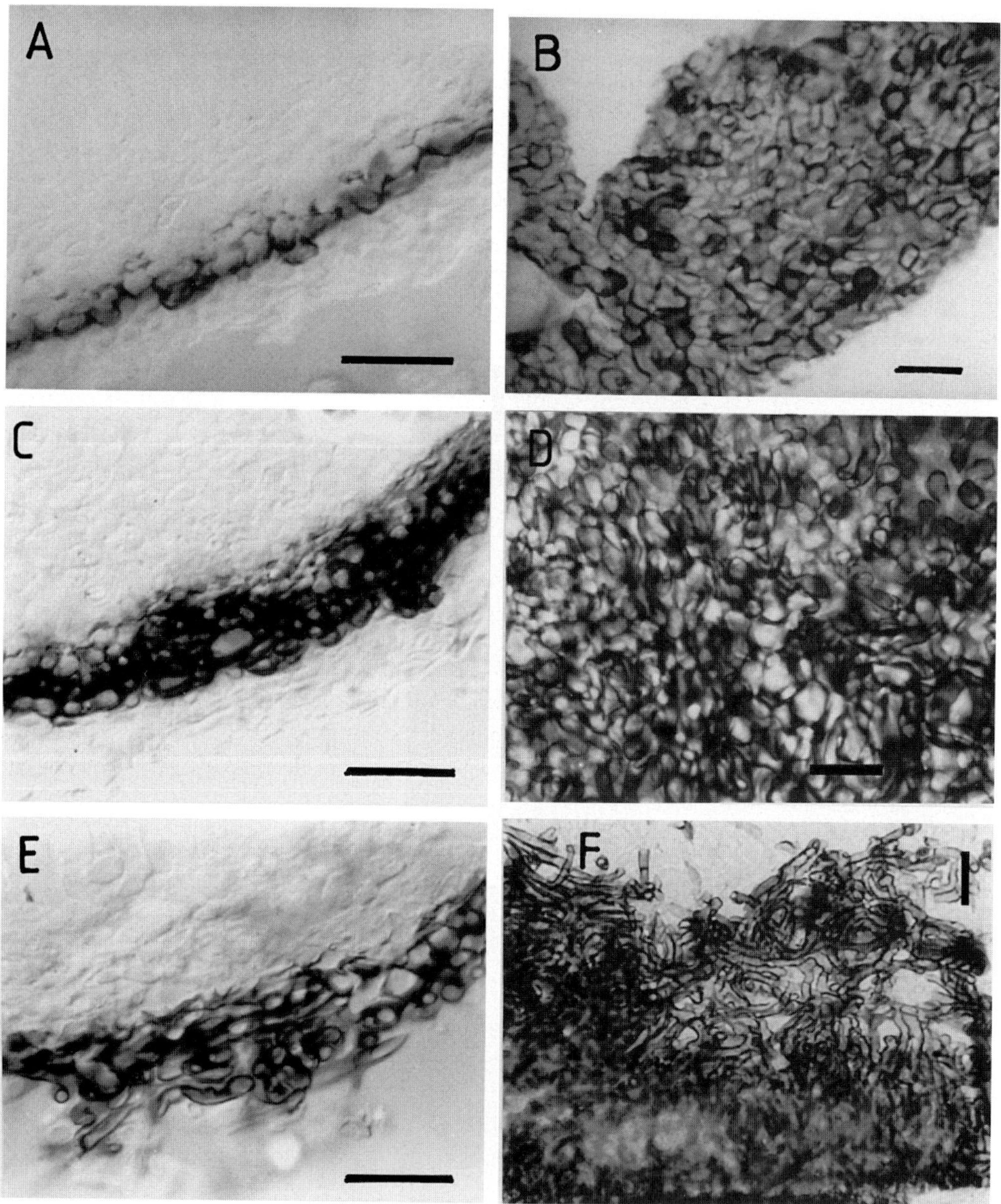

Fig. 9. *Lophodermium* Group A, arrangement of darkened cells within lower wall of ascomatal stroma. **A–B.** *L. typhinum*, lower wall with single layer of angular-globose cells. **A**, in vertical section (**DAOM** 86505); **B**, in squash mount (**TRTC** 24514). **C–D**. *L. nitidum*, lower wall with several layers of short-cylindrical to angular cells. **C**, in vertical section (**PDD** 59627); **D**, in squash mount (**PDD** 65286). **E–F**. *L. culmigenum*, lower wall of several layers of hyphae forming *textura intricata*. **E**, in vertical section (Fries, *Scleromyceti suecici* no. 97, **K**); **F**, in squash mount (*Magnus*, 8 Jul. 1871, **HBG**) (bars = 20 μm).

Clypeus. Normally well developed, with epidermal cells above developing ascoma becoming packed with narrow, dark-walled hyphae.

Apothecium. Reduced excipulum poorly developed but always present, comprising several rows of closely septate paraphysis-like elements not intermixed with asci (**Fig. 10A**, **B**). Paraphyses normally differentiated near apex, commonly circinate, sometimes slightly swollen (**Fig. 10D**). Asci cylindrical, subclavate or subfusoid, tapering to rounded or subtruncate apex with ± undifferentiated wall, 8-spored, with spores extending to base in immature asci but, in most species, asci extending in length immediately before spore release, forming empty basal stalk with spores remaining in upper part of ascus. Ascospores filiform, non-septate, almost always straight when released, with small gelatinous caps at base and apex and entire spore surrounded by additional, loose gelatinous sheath (**Fig. 5A**, **B**).

Conidiomata. Present in some species, usually subepidermal, lenticular, with upper wall ± lacking, lower wall lined by palisade of ± cylindrical conidiogenous cells with sympodial proliferation. Conidia short-cylindrical, hyaline, non-septate.

Relationships within *Lophodermium* Group A

Distinct subgroups can be recognized within *Lophodermium* Group A (**Table 5**), primarily on the basis of the presence or absence of lips, development of periphysoids and structure of the lower wall of the ascomata.

1. *actinothyrium*-group

- lips present
- periphysoids poorly developed in unopened ascomata, lost in open ascomata
- lower wall of ascoma thin; darkened cells angular to short-cylindrical, initially arranged in single layer across base of ascoma, with additional layers of similarly shaped cells subsequently forming in some species
- rarely with hypodermal or fibre cells embedded in upper wall of ascoma
- ascomata in vertical section usually ± triangular in outline

2. *arundinaceum*-group

- lips present
- periphysoids poorly developed in unopened ascomata, lost in open ascomata
- lower wall of ascoma thick; darkened cells hyphal, arranged as layer of *textura intricata* several cells thick across base of ascoma
- commonly with hypodermal or fibre cells embedded in upper wall of ascoma
- ascomata in vertical section usually ± triangular in outline

3. *alpinum*-group

- lips absent
- periphysoids forming broad, well-developed layer in unopened ascomata, sometimes persistent in opened ascomata
- lower wall of ascoma thin; darkened cells angular to short-cylindrical, arranged in single layer across base of ascoma
- rarely with hypodermal or fibre cells embedded in upper wall of ascoma
- ascomata in vertical section usually ± orbicular in outline

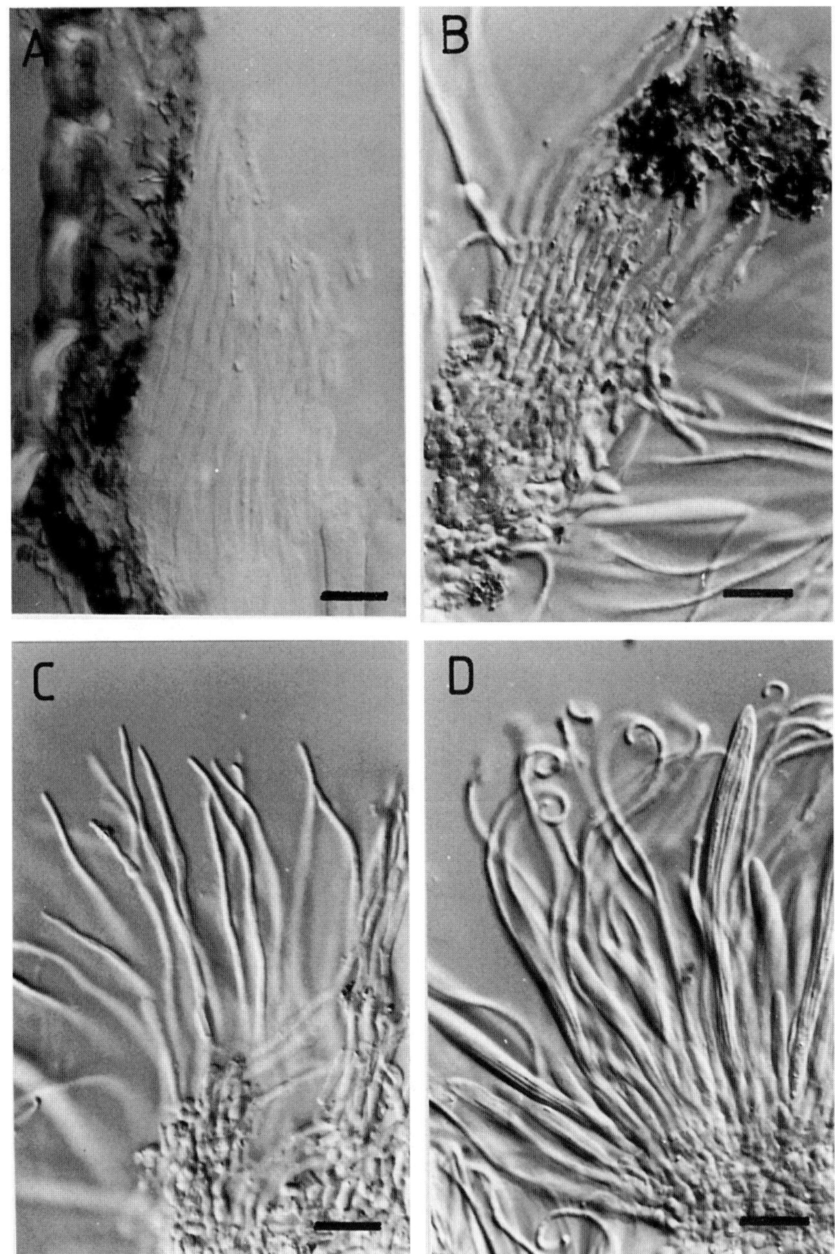

Fig. 10. *Lophodermium* Group A, paraphyses, marginal paraphyses and reduced excipulum (**A**, *L. sieglingiae*, **PRM** 693134; **B–D**, *L. nitidum*, **PDD** 65286). **A**, reduced excipulum between hymenium and upper wall of stroma, comprising closely septate, paraphysis-like elements, in vertical section; **B**, reduced excipulum, in squash mount; **C**, paraphyses near edge of hymenium, undifferentiated at apex; **D**, paraphyses near centre of hymenium, with circinate apices characteristic of this species (bars = 20 μm).

Table 5. Host and geographical distribution of species placed in *Lophodermium* Group A.

Species	Host	Geographical distribution
L. actinothyrium[1]	*Agrostis, Anthoxanthum, Arrhenatherum, Bromus, Calamagrostis, Dactylis, Deschampsia, Festuca, Koeleria, Molinia, Pentoschistis, Sesleria* (*Poaceae*)	Europe, North America
L. alliaceum	*Allium* (*Alliaceae*)	Luxembourg
L. alpinum[3]	*Carex* (*Cyperaceae*); *Dactylis, Elymus, Festuca, Nardus, Phleum, Poa, Sesleria, Trisetum* (*Poaceae*)	Europe, North America, Turkey, India, North Africa, Argentina
L. arundinaceum[2]	*Phragmites* (*Poaceae*)	Europe, North America
L. caricinum	*Carex* (*Cyperaceae*)	Europe
L. culmigenum[2]	*Agropyron, Agrostis, Aira, Ammophila, Dactylis, Deschampsia, Elymus, Festuca, Holcus, Phleum, Poa, Secale, Triticum* (*Poaceae*); *Typha* (*Typhaceae*); *Juncus* (*Juncaceae*)	Europe, North America, Chile, Australia, New Zealand
L. danthoniae[1]	*Danthonia* (*Poaceae*)	USA
L. eximium[2]	*Ampelodesmos* (*Poaceae*)	Southern Europe, North Africa
L. fusiforme[1]	*Carex* (*Cyperaceae*); *Agrostis, Anthoxanthum, Brachypodium, Calamagrostis, Dactylis, Festuca, Helictotrichon, Koeleria, Poa, Sesleria* (*Poaceae*)	Switzerland
L. gramineum[3]	*Carex* (*Cyperaceae*); *Agrostis, Anthoxanthum, Calamagrostis, Elymus, Festuca, Nardus, Poa, Sesleria* (*Poaceae*)	Europe, Russia, Madeira, North America, India
L. grandialpinum[3]	*Elytrigia, Festuca, Sesleria* (*Poaceae*)	Europe
L. herbarum[1]	*Convallaria* (*Liliaceae*)	Europe
L. iridicolum[1]	*Iris* (*Iridaceae*)	Albania
L. nitidum[1]	*Carex* (*Cyperaceae*); *Calamagrostis, Deschampsia, Festuca, Molinia, Nardus, Sesleria* (*Poaceae*)	Europe
L. nonramosum[1]	*Carex* (*Cyperaceae*); *Festuca, Nardus* (*Poaceae*)	Switzerland
L. petriniae	*Agrostis, Anthoxanthum, Calamagrostis* (*Poaceae*)	Switzerland
L. robergei[1]	*Agropyron, Brachypodium, Bromus, Festuca, Oryzopsis* (*Poaceae*)	Europe, Australia
L. rubrum[2]	*Cortaderia* (*Poaceae*)	New Zealand
L. seriatum[3]	*Festuca* (*Poaceae*)	Europe
L. sesleriae[3]	*Calamagrostis, Deschampsia, Sesleria* (*Poaceae*)	Europe
L. sieglingiae[1]	*Danthonia, Koeleria, Rytidosperma* (*Poaceae*)	Europe, New Zealand
L. typhinum[1]	*Typha* (*Typhaceae*)	Europe, North America

[1] = *actinothyrium*-group; [2] = *arundinaceum*-group; [3] = *alpinum*-group (see p. 33).

Within each of these subgroups, species are distinguished on the basis of differences in ascus and ascospore size and shape, shape of the paraphysis apex, occurrence of conidiomata, host substratum and macroscopic appearance of the ascomata.

Some species cannot be placed with confidence in any one of these subgroups. For example, *L. caricinum* has a broad layer of periphysoids in unopened ascomata, but also has well-developed lip cells. *Lophodermium petriniae* is very similar to species of the *alpinum*-group but, at least in recently opened ascomata, poorly-developed lip cells are present.

In squash mounts, the lower ascomatal wall of those species in the *actinothyrium*-group with several layers of darkened cells can appear similar to species in the *arundinaceum*-group. However, horizontal sections clearly show differences in cell shape: globose to angular in the *actinothyrium*-group, hyphal in the *arundinaceum*-group. The difference is also clearly seen in very young ascomata, the first-differentiated darkened cells showing the characteristic angular or hyphal shape, respectively.

Lophodermium nitidum is similar to species of the *arundinaceum*-group in having a thick, dark lower wall in mature ascomata, but differing in the darkened cells of the lower wall being angular to short-cylindrical in shape, rather than hyphal as typical for this group. However, development of the lower wall of *L. nitidum* suggests a closer relationship to the *actinothyrium*-group than to the *arundinaceum*-group. As in the species of the *actinothyrium*-group, when first differentiated the lower wall of the ascoma comprises a single row of dark-walled, angular cells across the base; the additional layers of cylindrical cells (usually narrower than the first-formed cells) develop as the ascomata mature. In contrast, the lower wall of the species in the *arundinaceum*-group comprises narrow hyphal cells from the time it first becomes differentiated.

Extra-European taxa

A number of the European species are known also from other regions (**Table 5**). However, as noted on p. 24, much of the variation represented in collections from outside Europe does not fit easily into the taxa listed in the table. Examples of some of the groups representing this variation are discussed under *L. actinothyrium*, *L. culmigenum* and *L. nitidum* in the species descriptions.

Lophodermium Group B (*Terriera*)

Most of the species within Group B have traditionally been placed in *Lophodermium*, but are referred in this study to *Terriera*. Although the type species of the genus, *T. cladophila*, is widespread on *Vaccinium* in northern temperate regions, this genus is predominantly tropical in distribution (JOHNSTON, 1997). Species are common on monocotyledons and dicotyledons, with a few also on conifers. Monocotyledon-inhabiting species are found on hosts representing many families, including *Arecaceae*, *Cyperaceae*, *Juncaceae*, *Liliaceae*, *Pandanaceae* and *Poaceae*. Fruiting bodies develop on dead leaves and twigs. Species are distinguished primarily on the basis of variation in ascus and ascospore size, shape of the ascus apex, shape of the paraphysis apex and features of the conidiomata. Most species are known from the tropics, often from few collections, and their concepts remain poorly understood. A conservative position has been taken regarding these species, all currently valid names being retained. Additional collections may reveal some of them to be better placed in synonymy. Some may be found to occur on dicotyledonous hosts as well, under yet further species names. Species treated are listed in **Table 6**.

Table 6. Host and geographical distribution of species treated in *Lophodermium* Group B (*Terriera*)*.

Species	Host	Geographical distribution
'*L. andropogonis*'	*Andropogon* (*Poaceae*)	Costa Rica
T. arundinacea	*Bambusa* (*Bambusaceae*)	Java
T. asteliae	*Astelia* (*Asteliaceae*)	New Zealand
T. breve	*Carex*, *Gahnia*, *Uncinia* (*Cyperaceae*)	New Zealand
T. clithris	monocotyledon	Brazil
T. dracaenae	*Dracaena* (*Agavaceae*)	California
T. fourcroyae	*Furcraea* (*Agavaceae*)	Sri Lanka
T. fuegiana	*Rostkovia* (*Juncaceae*)	Tierra del Fuego
T. javanica	*Elettaria* (*Zingiberaceae*)	Java
T. latiascus	*Heliconia* (*Heliconiaceae*); *Arecaceae*	Brazil
T. longissima	*Poaceae*	Guyana
'*L. miscanthi*'	*Miscanthus* (*Poaceae*)	Philippines
T. nematoidea	*Gahnia* (*Cyperaceae*)	New Zealand
T. pandani	*Pandanus* (*Pandanaceae*)	Puerto Rico
T. javanica var. *pandani*	*Pandanus* (*Pandanaceae*)	Java
T. sacchari	*Saccharum* (*Poaceae*)	Hawaii
T. samuelsii	monocotyledon	Brazil
T. stevensii	*Vincentia* (*Cyperaceae*)	Hawaii

* A single collection from Gabon (**Gabon:** Libreville, Forêt de la Mondah, on *Ancistrophyllum secundifolium*, 22 Jul. 1979, *G. Gilles s. num.*; **FH**) is the only known representative of this genus from Africa. This specimen could not be placed with certainty in any of the accepted species.

Morphological features

In summary, this group is characterized by oblong to sublinear ascomata, sometimes coalescing and then appearing branched. The single longitudinal opening slit is lined with a black, shelf-like zone along both sides. Paraphyses form an epithecium, asci are usually narrow-cylindrical and spores are usually 1-septate, tapering slightly at both ends, often becoming gently sigmoid on release and lacking a gelatinous sheath.

Macroscopic appearance. Ascomata in surface view oblong to sublinear in outline, often irregular in orientation on leaf surface and sometimes appearing branched when two or more coalesce during development. In most collections, immature ascomata commonly appear macroscopically as two parallel blackened areas separated by broad zone concolorous with surrounding host surface (**Fig. 11A**). Pale zone marks future line of opening and is poorly developed in some collections, probably reflecting differences in host leaf structure rather than taxonomic differences between fungi. Ascomata open by single, longitudinal slit. Wall dark grey to black, often with ascomata raising host surface so that wall becomes steeply angled to surrounding leaf surface. Ascomata develop distinctive black, flattened zone along both sides of opening slit (**Fig. 11B**).

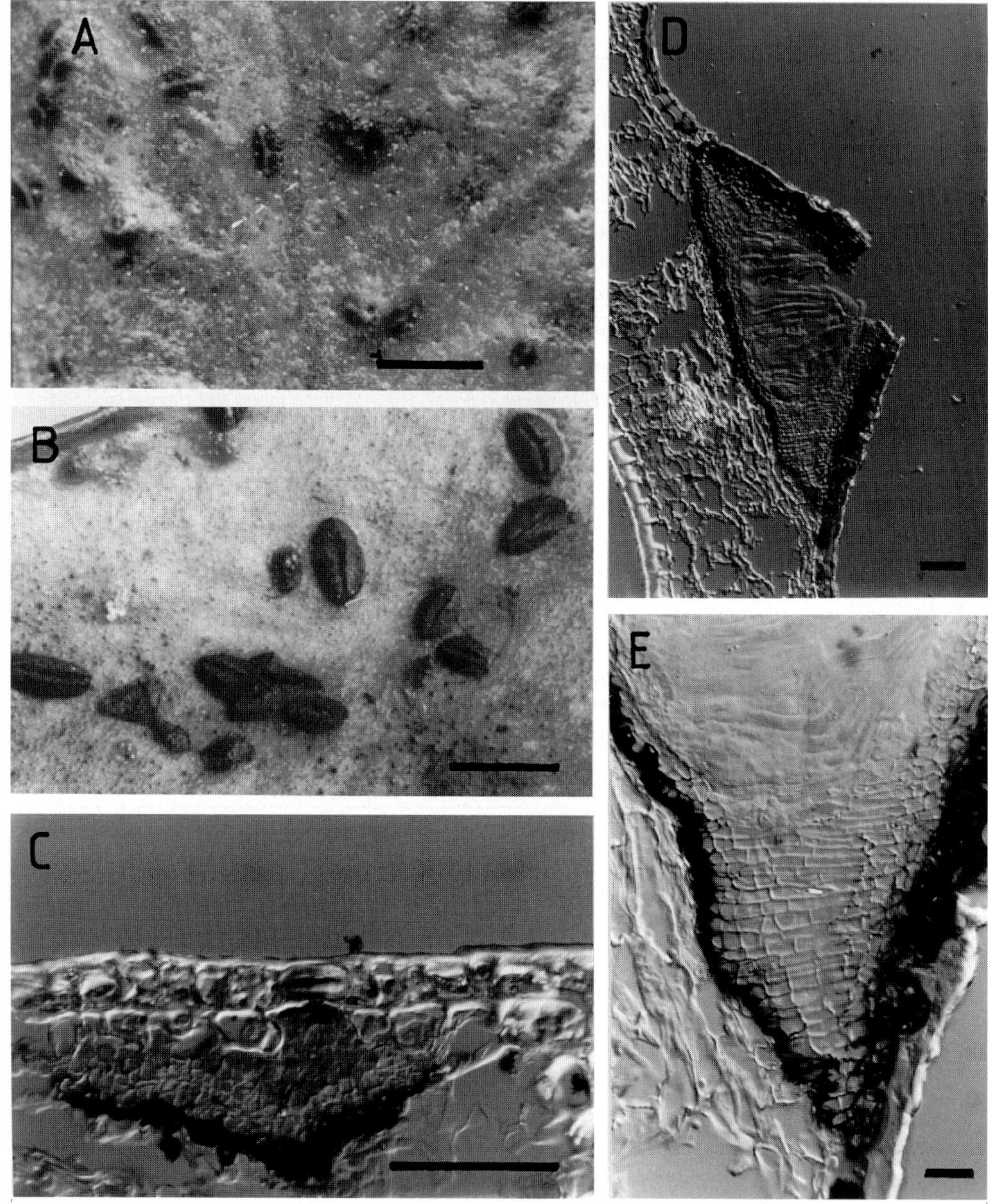

Fig. 11. *Terriera*, macroscopic appearance of ascomata and structure of ascomatal primordium. **A**, *L. minus* (**PDD** 40553), unopened ascomata with broad paler zone along future line of opening; **B**, *L. minus* (**PDD** 43972) opened ascomata with black, flattened zone along sides of opening slit (bars = 1 mm); **C**, *L. minus* (**PDD** 46127) palisadic structure of ascomatal initial; **D**, *T. breve* (**PDD** 47493) ascoma before opening (upper wall broken during sectioning), in vertical section, (bars = 50 μm); **E**, *T. breve* (**PDD** 47493) detail of margin of ascoma, showing retention of palisadic tissue at edges of expanding ascoma and detail of structure of lower wall of stroma, darkened cells of ascomatal primordium rather than additional wall layer (bar = 20 μm).

Primordium. Ascomatal primordium comprises *textura prismatica* (**Fig. 11**); insertion subepidermal (rarely intra-epidermal), often also with scattered hypodermal cells or fibre bundles above upper wall of ascomata.

Stroma. Separate upper and lower stromatic walls appear to be lacking, darkened walls of ascoma comprising cells from part of ascomatal primordium remaining above hymenium and retaining their columnar arrangement, walls of primordial cells themselves becoming slightly thickened and darkened (**Fig. 11E**). Although covering layer of ascomata often appears very dark, this is due to darkening of clypeus and periphysoid layer which develops after ascomata open. Following development of primordial cavity, remains of broken columns of vertically oriented cells can be seen above developing paraphyses (**Fig. 12A**). As it matures, expanding hymenium causes sides of ascomata (comprising covering layer of unopened ascomata) to become ± vertical in orientation. Narrow-cylindrical, periphysoid-like cells develop from part of upper wall adjacent to ascomatal opening, these elements growing across top of expanding hymenium (**Figs 12B–D**). Walls of periphysoids become very thick and darkened, cellular structure of layer soon becoming almost completely obscured (**Fig. 12D**). Periphysoids grow from both sides of ascomata, two layers meeting (but not joining) along centre of ascomata. Slit-like opening of ascomata is along line at which two layers meet (**Fig. 12E**), forming black, flattened zone, visible macroscopically along sides of opening slit.

Clypeus. Well developed, with most of epidermal cells above developing ascoma becoming packed with narrow, dark-walled hyphae. Epidermal cells down centre of ascoma, along future line of opening, usually remain empty.

Apothecium. Reduced excipulum present but poorly developed, in vertical section remaining closely adhering to sides of stroma, even if hymenium breaks up and falls away during sectioning. Paraphyses characteristically form distinct epithecium above asci, often being differentiated near apex, either branched or swollen and usually tangled together and embedded in gel. Asci cylindrical to subclavate; top of ascus variable in shape, truncate, broadly rounded or tapering to ± acute or rounded apex, this feature being useful at species level. Asci 8-spored, initially with spores extending ± to base, usually with short basal stalk developing before spore release. Ascospores filiform, usually becoming 1-septate, in most species characteristically curved when released, with no gelatinous sheath or appendages.

Conidiomata. Present in some species, morphologically matching those of *Lophodermium* Group A.

Lophodermium Group C

This macroscopically distinctive group, known only from Europe, has been collected most commonly from *Cyperaceae* and *Juncaceae* but occurs also on *Poaceae*.

The putative species in this group are distinguished by differences in ascus and ascospore size and ascomatal insertion. Species limits are again difficult to define, at least partly because of the small number of collections available. Most of the collections referred to this group were discussed by Müller (1977) under the name *L. luzulae* (see Notes under this species). The species treated are listed in **Table 7**.

Similarity in ascomatal structure and in the general appearance of the hymenial elements suggests that this group is closely related to the dicotyledon-inhabiting *Coccomyces tumidus* (see Notes under *L. tumidulum*) and possibly two other monocotyledon-inhabiting species: *Lophium eriophori* Henn. and *Coccomyces insignis* P. Karst. (see Notes under *Lophium eriophori*).

Table 7. Host and geographical distribution of species in *Lophodermium* Group C.

Species	Host	Geographical distribution
L. cf. *juncinum*	*Juncaceae*	Europe
L. tumidulum	*Cyperaceae, Juncaceae, Poaceae*	Europe

Morphological features

In summary, this group is characterized by ascomata which become erumpent at maturity and which are ± *Hysteriales*-like in appearance. Ascomata are broadly elliptical to ± round in outline, with one or more irregular opening slits, the number and arrangement depending on ascomatal shape. The upper wall of the ascoma is characteristically very loose in structure, comprising ± globose, hyaline cells with slightly and irregularly encrusted walls. The hymenium contains an orange-brown pigment, asci and ascospores are generally large and paraphyses typically are branched near the apex.

Macroscopic appearance. Ascomata in surface view broadly elliptical to ± round in outline, becoming erumpent and distinctly arched above surrounding leaf tissue. Opening irregular in shape, either single longitudinal slit or several radiate slits. Ascomata becoming *Hysteriales*-like in appearance, wall usually either sunken or raised along opening slit, black, with no paler zones associated with future lines of opening and no lip cells along edge of opening slit. Immature, unopened ascomata of this group can appear macroscopically similar to those of Group D, but two groups can be distinguished by differences in structure of upper wall of ascoma.

Primordium. Early stages of ascomatal development have not been observed. However, structure of stromatal tissue surrounding developing hymenium in unopened ascomata shows that ascomatal primordium is likely to comprise *textura intricata*. Insertion subepidermal or subcuticular.

Stroma. Both upper and lower darkened walls present, lower wall probably developing before upper. Upper wall comprises mostly globose cells with slightly and irregularly encrusted walls, ± uniform in pigmentation, although cells towards inside of wall a little paler than those towards outside. Inside of upper wall comprises loose tissue of globose cells forming persistent, periphysoid-like layer. Lower wall comprises several layers of dark-walled, cylindrical to hyphal cells.

Clypeus. A reduced clypeus appears to be present. In subcuticular species, cells within ascomatal wall adjacent to cuticle are long-cylindrical and irregular in shape.

Apothecium. Hymenium with pale orange-brown pigment. Reduced excipulum poorly developed. Paraphyses unswollen, irregularly branched and circinate at apex. Asci large, subclavate to clavate, tapering to rounded or rostrate apex, wall undifferentiated at apex, 8-spored, with spores confined to upper part of ascus, with long basal stalk. Ascospores filiform, straight when released, with gelatinous cap at apex and at base.

Conidiomata. Not seen.

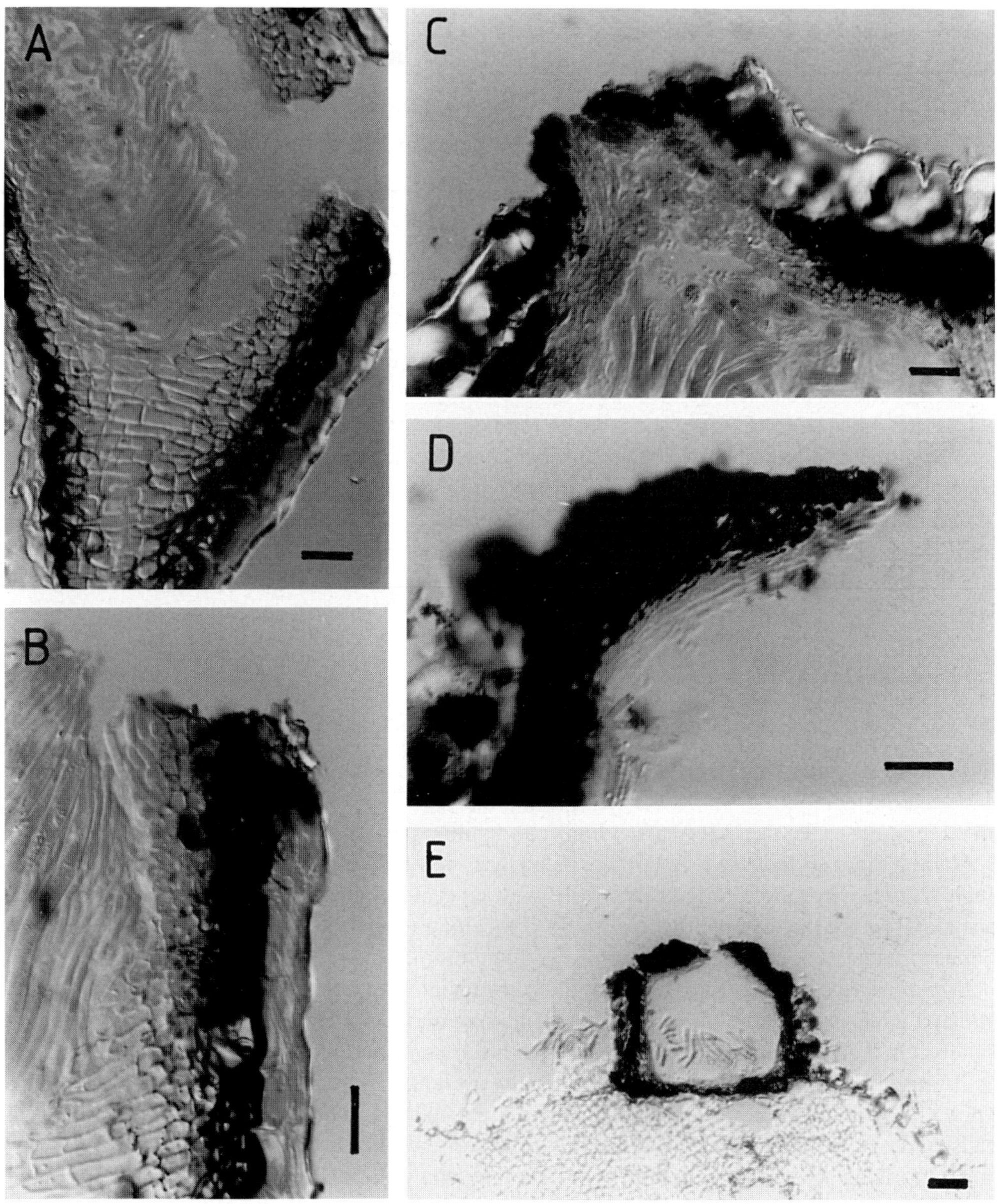

Fig. 12. *Terriera breve* (**PDD** 47493), development of periphysoids, forming black shelf-like zone across top of hymenium. **A**, upper wall of unopened ascoma in vertical section (upper wall broken during sectioning), showing broken columns of palisadic tissue and lack of periphysoids; **B**, detail of upper wall of one side of recently opened ascoma in vertical section, showing periphysoids starting to elongate; **C**, slightly later stage of development, detail of opening slit, periphysoids further elongated, walls of cells starting to become encrusted with dark brown material; **D**, opened ascoma, detail of top of wall in vertical section, showing shelf-like nature of elongated periphysoids, with structure almost completely obscured by encrusting material (bars = 20 μm); **E**, open ascoma in vertical section, with shelf-like areas extending across hymenium from upper wall to meet, but not join, in a line down the centre of the ascoma (bar = 50 μm).

Table 8. Host and geographical distribution of species in *Lophodermium* Group D.

Species	Host	Geographical distribution
L. cf. *luzulae*	*Cyperaceae, Juncaceae, Poaceae*	Europe
*Rhytisma juncicola**	*Juncaceae*	Europe

* *R. juncicola* and *C. coronatus* may be conspecific; see Notes under the former

Lophodermium Group D

This group is known from only a few collections from central European alpine regions, where it has been found on dead leaves of *Cyperaceae*, *Juncaceae* and *Poaceae*.

Species have been distinguished by differences in dimensions of the hymenial elements. Because of the small number of collections available, species limits in the group remain uncertain (see Notes under *L.* cf. *luzulae*). Species treated in this group are listed in **Table 8**.

Similarity in ascomatal appearance and structure suggests that the monocotyledon-inhabiting fungi in this group are closely related to the dicotyledon-inhabiting *Coccomyces coronatus* and some species of *Duplicaria* (see Notes and illustrations under *L.* cf. *luzulae* and *Duplicaria antarctica*).

Morphological features

In summary, this group is characterized by irregularly shaped, black-walled ascomata opening by one or more irregular slits, often with the upper wall of the ascoma eroding with age, so exposing the orange-brown hymenium. In vertical section the outer half of the upper wall comprises dark-walled, angular cells, the inner half cylindrical, hyaline periphysoids embedded in gel.

Macroscopic appearance. Ascomata in surface view ± round to irregular in outline; opening irregular, either by single longitudinal slit or by several radiate slits. Wall black, with no paler zone along future line of opening and no lip cells along edge of opening slit. Wall often eroding after ascomata open, exposing orange-brown hymenium.

Primordium. Early stages of ascomatal development have not been observed, but structure of stromatal tissue surrounding developing hymenium in unopened ascomata suggests that ascomatal primordium would comprise *textura intricata*. Insertion subcuticular to intra-epidermal.

Stroma. Both upper and lower darkened walls present, lower wall probably developing before upper. Upper wall with dark-walled angular to globose cells in outer half; inner half comprising persistent layer of hyaline, thin-walled, cylindrical periphysoids, normally embedded in gel. Lower wall comprising several layers of dark-walled, tangled cylindrical to hyphal cells.

Clypeus. A reduced clypeus appears to be present, with long-cylindrical to irregularly shaped cells adjacent to cuticle.

Apothecium. Reduced excipulum poorly developed. Paraphyses swollen at apex, unbranched, either embedded in gel or tangled, forming epithecium. Asci clavate to subclavate, wall often slightly thinner across small, truncate apex, 8-spored, with spores confined to upper part of ascus when mature. Ascospores filiform, 0–3-septate, with small apical and basal gelatinous caps.

Conidiomata. Not seen.

Lophodermium Group E

There are only three monocotyledon-inhabiting species in this group, but other species with a similar kind of ascomatal structure occur on dicotyledons (e.g. *L. medium* on *Nothofagus*, *L. brunneolum* on *Dracophyllum*).

Fruiting bodies typically form on dead leaves, rarely on necrotic tissue within lesions on living leaves. Most members of the group are confined to tropical and Southern Hemisphere regions (JOHNSTON, 1997).

Species are distinguished by differences in ascomatal shape, ascus and ascospore size, arrangement of the released spores and their gelatinous appendages, paraphysis shape and features of the conidiomata.

Some species of *Coccomyces* may be related, including members of the *C. leptosporus* complex of SHERWOOD (1980) and *Coccomyces* Groups 1 and 2 of JOHNSTON (1986). Apart from a difference in ascomatal shape, the species of *Coccomyces* differ in generally having a better developed excipular layer, the excipulum forming a wall-like layer around those parts of the hymenium adjacent to the several radiate splits in the upper wall (see, for example, the illustration of *C. parasiticus* in JOHNSTON, 1993) and ascospores ± straight with gelatinous caps at the apices, as compared with the strongly bent spores of the species of *Lophodermium* in this group (see illustrations in JOHNSTON, 1994). Species treated are listed in **Table 9**.

Species in this group share a number of features with *Terriera* (Group B), although they do not have the characteristic primordial structure of this genus. Features shared include: some species with crystals embedded in the hymenial gel; characteristically with narrow, cylindrical asci; paraphyses usually swollen, often embedded in gel and forming a well-defined epithecium, and periphysoids near ascomatal opening elongating after ascoma opens (although to a much lesser extent in Group E than in *Terriera*).

Morphological features

In summary, this group is characterized by ± cylindrical ascomata with a longitudinal opening slit together with short side-slits near the ends and the hymenium often remaining exposed in dry ascomata. The periphyses elongate after the ascomata open and the tips of these, along with the tips of the excipulum and poorly-developed lips, become encrusted with dark brown material. Paraphyses are swollen and tangled or branched at the apex, forming an epithecium.

Macroscopic appearance. Ascomata in surface view oblong to cylindrical (to sublinear) in outline, with rounded ends. Wall pale to dark grey, usually with paler zone along future line of longitudinal slit, commonly with short side-slits near ends of ascomata. Dark line sometimes seen

Table 9. Host and geographical distribution of species in *Lophodermium* Group E *.

Species	Host	Geographical distribution
L. alienum	*Andropogon* (*Poaceae*)	USA
L. hauturuanum	*Gahnia* (*Cyperaceae*)	New Zealand
L. unciniae	*Uncinia*, *Carex* (*Cyperaceae*); *Juncus* (*Juncaceae*)	New Zealand

* Another species likely to belong in this group was present on part of the type collection of '*L. miscanthi*' and is briefly described and discussed under that species description.

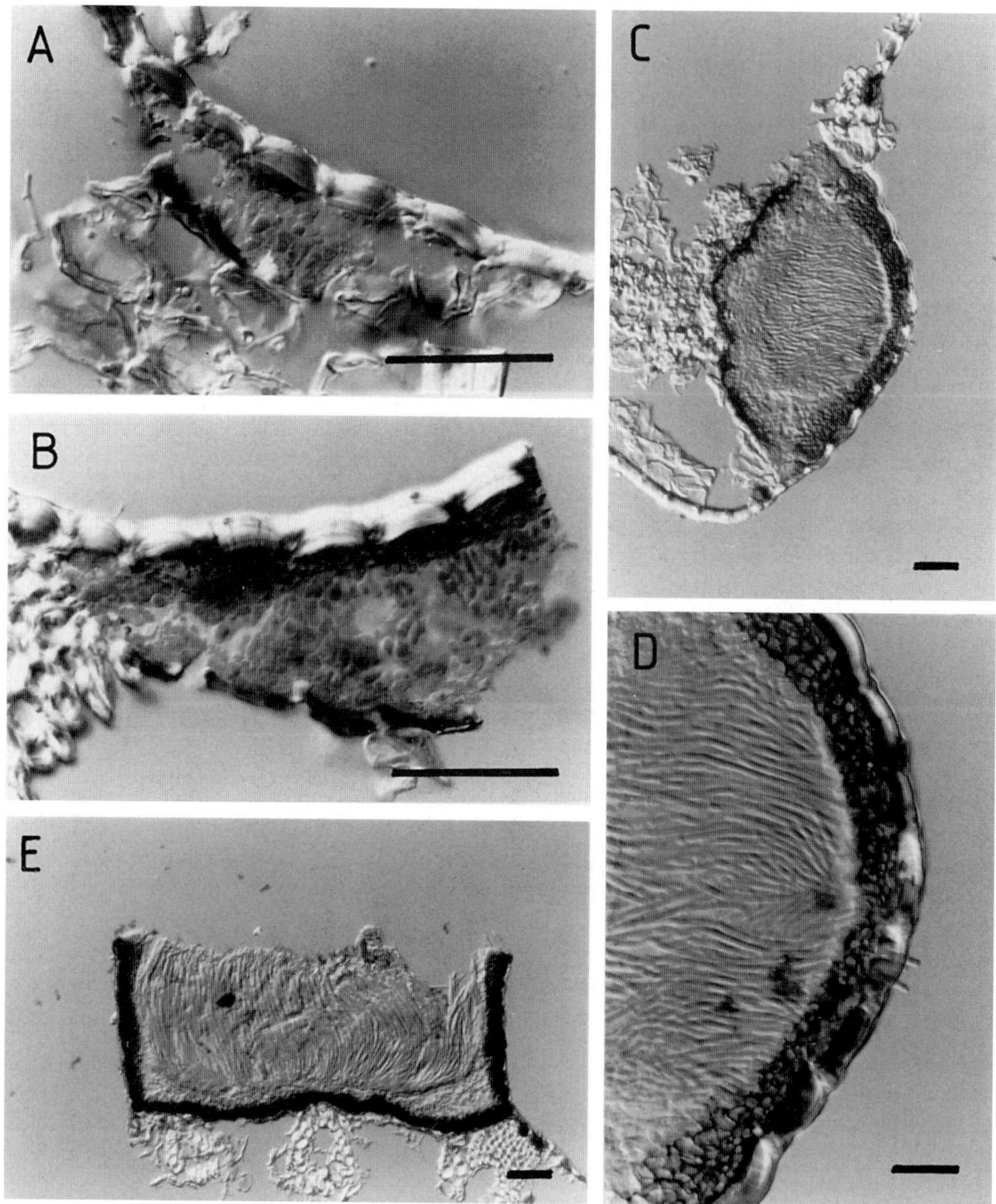

Fig. 13. *Lophodermium* Group E, ascomatal development (*L. unciniae*, **PDD** 54534). **A**, ascomatal initial, with lower wall darkened; **B**, ascomata with paraphyses starting to develop and upper wall starting to darken; **C**, unopened ascoma (bars = 50 μm); **D**, detail of upper wall of unopened ascoma, with poorly-developed periphysoids and slightly paler cells along future line of opening (bar = 20 μm); **E**, opened ascoma (bar = 50 μm).

around edge of opening slit, comprising poorly-developed lip cells which have become encrusted with dark material at their apices. Hymenium in open ascomata often remains exposed, even when dry.

Primordium. Ascomatal primordium comprises *textura intricata*. Insertion intra-epidermal or subepidermal, rarely within upper layer of hypodermis.

Stroma. Both upper and lower darkened walls present, lower wall differentiating before upper (**Fig. 13**). Upper wall of ascomata poorly developed, comprising several rows of angular cells, their walls becoming slightly and irregularly encrusted with dark brown material; cells in outer part of wall, along future line of opening, remain ± hyaline. Periphysoids poorly developed in unopened ascomata; most are lost after opening, but a few usually remain near ascomatal opening, these elongating after ascomata open. Unbranched lip cells present, but poorly developed (**Fig. 14**). Apices of lips and tops of persistent periphysoids usually become encrusted with dark brown material. Lower wall comprises several rows of angular cells with thick, dark walls.

Clypeus. Well developed in subepidermal species, reduced but clearly present in intra-epidermal species (see **Figs 63**, **117**). Clypeus does not extend across future line of opening.

Apothecium. Reduced excipulum always present, usually forming broad, distinct layer, especially in those parts of apothecium adjacent to opening slits in covering layer. Paraphyses differentiated at apex, commonly swollen or irregularly knobbled and tangled (**Fig. 14**), often embedded in common gelatinous matrix, gel sometimes containing numerous small crystals, forming well-defined epithecium. Asci usually cylindrical, variable in shape at apex, 8-spored. Ascospores filiform, usually 1-septate, often bent in characteristic way when released, usually with small gelatinous appendages, either at apices or at angles when bent.

Conidiomata. Present in some species; structure as described for Group A.

Miscellaneous species

Of the monocotyledon-inhabiting species, *L. agathidis* Minter & Hettige, *L. eucalypti*, *L. inclusum* P.R. Johnst., *L. raapianum* Penz. & Sacc. and *L. vrieseae* Rehm cannot be placed with other species of *Rhytismataceae* on the basis of characters used to distinguish the above five groups. The morphology of each of these is discussed in detail in the species descriptions.

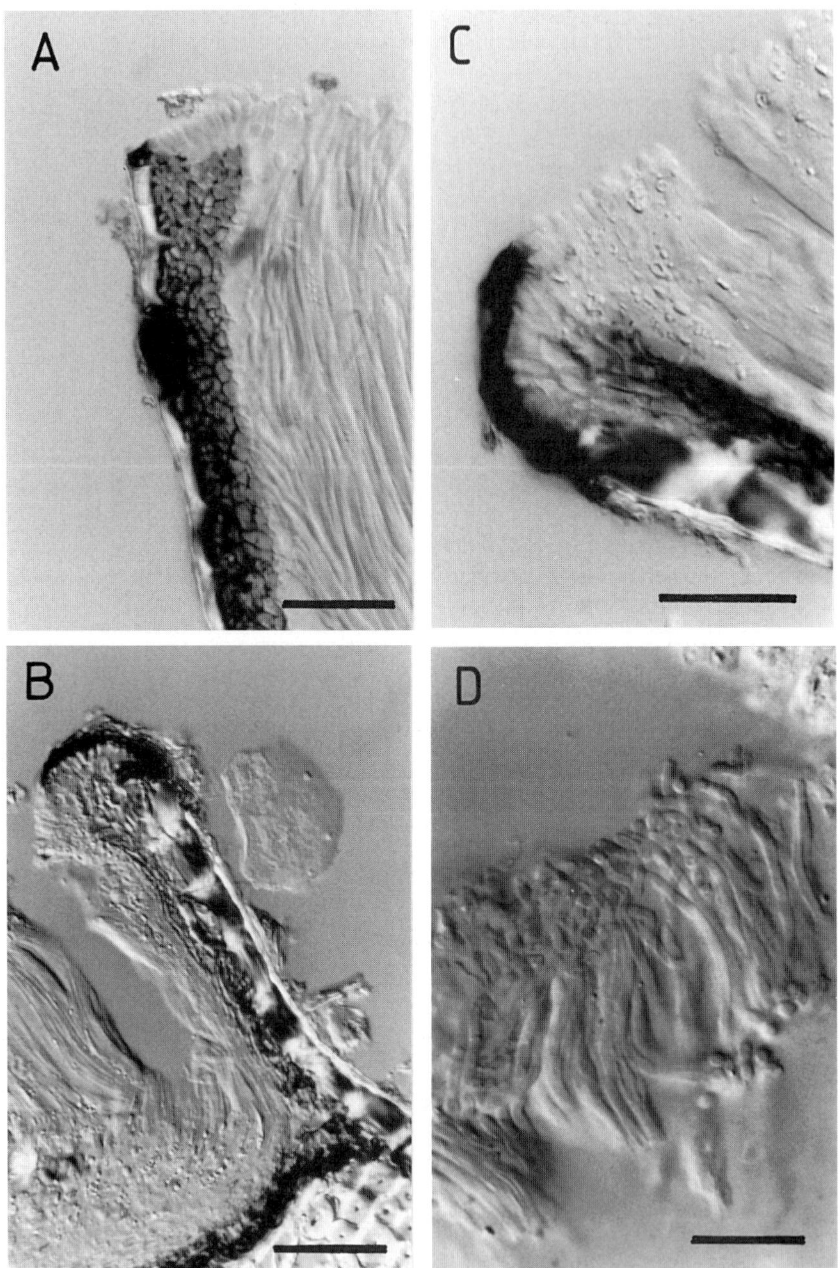

Fig 14. *Lophodermium* Group E. **A–C.** Differentiated cells along edge of opening slit. **A**, lip cells and few elongate periphysoids (*L. unciniae*, **PDD** 54534); **B**, lip cells, periphysoids and/or reduced excipulum, becoming encrusted with dark material (*L. unciniae*, **PDD** 54534); **C**, lip cells, periphysoids and/or reduced excipulum, becoming encrusted with dark material (*L. hauturuanum*, **PDD** 49353). **D**. Paraphyses tangled to form epithecium (*L. alienum*, **CUP** 8625) (bars = 20 μm).

BIOLOGY AND DISTRIBUTION

The *Rhytismataceae* contains both wood-inhabiting and leaf-inhabiting species, with very few species known from both woody and herbaceous substrata. Most species form fruiting bodies on dead tissue, but a number of species have been shown to live as endophytes within the symptomless tissue of their hosts before the development of ascomata (CARROLL & CARROLL, 1978; BARKLUND, 1987). The production of dark-walled appressoria by other species (OSORIO & STEPHAN, 1989) suggests that an endophytic phase may occur in the life cycle of many of the apparently 'saprophytic' species. The ascomata of *Rhytisma* develop on living leaves, producing tar-spot symptoms. A few other species form ascomata within necrotic lesions on otherwise green leaves and may be parasitic (e.g. *Coccomyces parasiticus*, *C. vilis* Syd., P. Syd. & E.J. Butler and *Lophodermium seriatum*).

Many of the leaf-inhabiting species form ascomata in groups within bleached areas on the fallen host tissue. The bleached area can be assumed to be that part of the leaf invaded by the fungus and in some species is surrounded by a narrow, dark stromatic line, termed a 'zone line' by MINTER (1981). On some leaves several distinct bleached areas can be seen, all associated with the same fungal species. Each area may represent a genetically distinct fungal colony (CANNON & MINTER, 1986). In those species which form zone lines, separate ones are formed around each area. If two areas are adjacent, a double zone line may be seen between them.

Biology as a Taxonomic Character

Many species of *Rhytismataceae* have been described on the basis of host specialization. While it is true that species may be restricted to one or a few hosts, this feature must be used with care, as there are a number of species which occur on a wide range of plants. Accurate host identification can also be a problem, since these fungi often fruit on fallen dead tissue. This is especially true for hosts such as the grasses and sedges, which are difficult for a non-specialist to identify accurately. Herbarium specimens usually contain only small pieces of dead leaf material from the host and hence the identification provided by the original collector cannot be verified. Of those species with a wide host range, most are restricted to monocotyledons, dicotyledons or conifers.

On a broader scale, host preference can be used to help define supraspecific groups within the *Rhytismataceae* and *Lophodermium* in particular. Examples include groups of apparently closely related species on *Pinaceae* and on some families in the *Ericales* (JOHNSTON, 1988*b*). Most species on *Poaceae* form another closely related group, discussed as *Lophodermium* Group A (*Lophodermium s. str.*) in this paper. Members of such host-related groups share a range of morphological similarities apparently independent of the anatomy of the host itself. However, not all species of *Rhytismataceae* found on each of these hosts belong in the same group.

Geographical distribution can be a clue to taxonomic relationship. As discussed by JOHNSTON (1990*b*, 1992, 1997) few species of *Rhytismataceae* are widely distributed naturally in both hemispheres. Some morphologically distinct, putatively monophyletic groups of species in the family are distributed primarily either in the Northern Hemisphere or in tropical regions and the Southern Hemisphere (JOHNSTON, 1997).

Among the monocotyledon-inhabiting species of *Lophodermium* there appears to be some host specialization (**Table 10**). Several species are known from only a single host genus. As noted

Table 10. Hosts of monocotyledon-inhabiting species of *Lophodermium* which provide useful indications as to species identity. Note, although, that other species may occur on the hosts listed. Possible identifications indicated by host preference should be checked morphologically.

Host genus	Most common species
Aira	*L. culmigenum*
Ammophila	*L. culmigenum*
Ampelodesmos	*L. eximium*
Brachypodium	*L. robergei*
Calamagrostis	*L. actinothyrium, L. gramineum, L. nitidum*
Carex	*L. caricinum, L.* cf. *luzulae*
Convallaria	*L. herbarum*
Danthonia	*L. sieglingiae*
Deschampsia	*L. actinothyrium, L. nitidum*
Elymus	*L. culmigenum*
Festuca	*L. alpinum, L. gramineum*
Gahnia	*L. hauturuanum, L. inclusum*
Juncus	*L.* cf. *juncinum, L. tumidulum*
Luzula	*L.* cf. *luzulae, L. tumidulum*
Molinia	*L. actinothyrium, L. nitidum*
Nardus	*L. alpinum, L. gramineum*
Phragmites	*L. arundinaceum*
Poa	*L. alpinum, L. gramineum*
Scirpus	*L. tumidulum*
Secale	*L. culmigenum*
Triticum	*L. culmigenum*
Typha	*L. typhinum*
Uncinia	*L. unciniae*

under *L. culmigenum*, some of the European grass-inhabiting species may be specialized at the level of tribe within the *Poaceae*.

CLADISTIC ANALYSES

Cladistic analyses were carried out to test the putative groups recognized among the monocotyledon-inhabiting species of *Lophodermium s. lat.* on the basis of general morphological similarity. The relationship of these groups to species in other genera of the *Rhytismataceae*, or to species from other hosts and postulated to be related to some of the monocotyledon-inhabiting species (see p. 23), was also tested. Finally, an analysis was carried out which included several other groups of species within the *Rhytismataceae* not thought to be related to any of the monocotyledon-inhabiting species, but which had been suggested as phylogenetically distinct groups in earlier publications (see p. 53). The analyses were restricted to leaf-inhabiting members of the *Rhytismataceae*. Including a wider range of organisms in a morphologically-based study such as this was not feasible because of the difficulty in recognizing homology between the various features of taxa with widely divergent morphology.

Although routinely applied in analysis of DNA sequence data, cladistic methods have been used only rarely in morphological studies of fungi, some examples being REYNOLDS (1986), MORTON (1990), MUELLER (1992) and LANGER (1994). These methods are difficult to apply to studies on fungi because of the small numbers of characters available in these morphologically simple organisms, the difficulty in making decisions on character homology and character state polarity and in the recognition of suitable outgroups. Although *a priori* character polarity decisions are a central step in many published cladistic studies, a number of authors argue that this is not required (see below) and, despite the other problems being real, it can be argued that cladistic analysis is the only acceptable way to derive testable taxonomic groups.

The elegant and intellectually satisfying basis of cladistic methods demands that they be applied more often to the fungi. However, as noted by CHRISTOFFERSEN (1995), these methods are based on the existence of nested hierarchical patterns of relationship, so that it may not be appropriate to apply cladistic methods to entities that are not expected to be related hierarchically. No attempt has been made to analyse 'species-level' relationships among the monocotyledon-inhabiting species of *Lophodermium*. As discussed earlier (see p. 10) the entities recognized as species in this study are defined purely on the basis of patterns of morphological variation. There is no information on genetic diversity, breeding systems, gene flow, the potential for hybridization, etc., among the *Rhytismataceae*, nor is there evidence that the entities recognized as species are real in an evolutionary sense, that they will retain their distinctiveness over time, or that the relationship between them is hierarchical. In addition, the number of characters available for distinguishing taxa at the species level is very small and most are continuously variable features (such as spore and ascus size) which were not used in the present analysis. The use of continuous characters has been avoided, since this approach requires them to be divided into essentially arbitrary character states with no evidence of homology (PIMENTEL & RIGGINS, 1987; STEVENS, 1991).

A number of authors have argued against the use of cladistic methodology, especially for analysing morphological data in plant taxa (e.g. HEDBERG, 1995). The objections relate to problems in defining valid outgroups and character polarities, as well as the phylogenetic inferences often drawn from the cladograms, the problems associated with including hybridization or reticulate patterns of evolution on a cladogram, the insistence on accepting only monophyletic groups as taxonomically sound and the blind reliance on a computer program, using algorithms often poorly understood by the user, to give the answer. These objections are both taxonomic and methodological.

The taxonomic objections are a matter of debate. For example, some authors insist that cladograms be interpreted as representations of the actual phylogenetic history of the organisms under study, yet there is no inherent reason why the branching pattern of a cladogram must be interpreted in this way. Hybridization certainly occurs in plant and fungal taxa traditionally regarded as species. Such a relationship cannot be illustrated easily on a cladogram but, for example, it may be postulated as an explanation for a cladogram in which a number of terminal taxa remain unresolved. Genetic or molecular evidence will probably be required to test the hypothesis, but without the cladistic analysis no hypothesis would have been available for testing. Although arguments may be advanced for accepting only strictly monophyletic groups in a classification (e.g. HUMPHRIES & FUNK, 1984; TEHLER, 1994), usually on the basis that classifications should reflect as closely as possible actual phylogenetic groups, this again is a matter for debate. For example, to avoid the multiplicity of categories inevitable if only monophyletic groups are recognized, some authors argue that convex (paraphyletic) groups are acceptable when classifications are constructed from the information contained in a cladogram (MEACHAM & DUNCAN, 1987).

The methodological objections are less valid. Cladistic methodology certainly relies on a number of assumptions,[4] but these are known and defined and the results of the analysis must be assessed in light of the assumptions applied. Results from different studies can be validly compared only if they are based on the same methodological principles. The alternative, so-called 'traditional evolutionary systematics' (ELDREDGE & CRACRAFT, 1980) relies on subjective decisions based primarily on personal preferences and biases. Because the assumptions on which such studies are based cannot be challenged, their results cannot be tested. The 'answer' provided by a cladistic study is not necessarily 'correct' and does not have to be interpreted on the basis of any particular evolutionary model. Its value lies in providing an hypothesis of relationship available for testing.

Methods

The cladistic analyses were carried out using PAUP version 3·1. MacClade was used at times to manipulate the branching patterns of the shortest trees, in order to compare the effect of different patterns of relationship on tree length.

Characters included

Characters are listed in **Table 11**. Those such as ascus and ascospore size, ascomatal size, colour of lip cells and conidiomatal features were excluded from the analysis. Although useful to distinguish species, methods of defining discrete states for continuous characters are essentially arbitrary and the shared possession of the others was considered unlikely to indicate common ancestry above the species level.

All characters were treated as unordered, a change from one state to any other having the same 'cost', i.e. no assumptions were made about allowable character state changes. Traditionally, a cladistic analysis has started with the *a priori* allocation of derived (apomorphic) versus ancestral

[4] The most critical of these are that a hierarchical pattern of relationship exists among the taxa being studied, that different states recognized for a single character are homologous, that different characters are truly independent and that parsimony is a valid method for deciding between alternative patterns of relationship.

Table 11. Characters and character states used in cladistic analyses.

Character	Character states
1. Shape of ascomata	1. Uniloculate; ovate, elliptical, to oblong-elliptical; single longitudinal opening slit 2. Uniloculate; oblong; long central opening slit and irregular side slits 3. Uniloculate; angular; several regularly radiate slits 4. Uniloculate; irregular in shape; opening slits irregular, often radiate 5. Multiloculate, several locules developing within large ascomata
2. Ascomatal insertion	1. Subepidermal 2. Intra-epidermal 3. Subcuticular 4. Initially subcuticular, subepidermal towards edge of ascoma
3. Structure of ascomatal primordium	1. *textura prismatica*, cylindrical cells oriented vertically to host surface 2. *textura intricata*, narrow hyphae oriented parallel to host surface
4. Darkened upper wall	1. Comprises darkened cells of ascomatal primordium 2. Comprises separate layer of *textura angularis/globosis* developing above ascomatal primordium
5. Preformed line of dehiscence	1. Lacking 2. Present, towards inside of upper wall 3. Present, initially to inside of upper wall, becoming confined to centre 4. Present, towards outside of upper wall 5. Extending through entire upper wall
6. Pigmentation of cells in upper wall	1. ± uniform, either pale or dark 2. Mostly pale, but with a restricted group of dark-walled cells near ascomatal opening
7. Periphysoids	1. Poorly developed in unopened ascomata, short-cylindrical, lost in opened ascomata 2. Poorly developed in unopened ascomata, short-cylindrical, a few near ascomatal opening elongating after ascomata open, tips becoming encrusted with dark material 3. Absent in unopened ascomata, developing near ascomatal opening after ascomata open, becoming very thick-walled, forming black, shelf-like layer across top of hymenium 4. Well developed in unopened ascomata, comprising hyaline, cylindrical elements arranged in vertical columns, often persistent in open ascomata 5. Well developed in unopened ascomata, cylindrical, embedded in gel, persistent in open ascomata 6. In both opened and unopened ascomata inner part of upper wall comprising loose tissue of hyaline, thin-walled, globose cells, forming a periphysoid-like layer
8. Lip cells	1. Absent 2. Present, unbranched 3. Present, branched

9. Time of development of darkened lower wall	1. Before upper wall darkens 2. After upper wall darkens
10. Darkened lower wall	1. Comprising darkened cells of the ascomatal primordium. 2. Comprising an additional layer of angular, globose or hyphal cells developing beneath the primordium 3. Darkened lower wall absent
11. Paraphyses	1. Persistent 2. Lost as ascomata mature
12. Shape of paraphysis apex	1. Undifferentiated 2. Circinate 3. Swollen (unbranched) 4. Branched (often also swollen) 5. Propoloid
13. Epithecium	1. Present 2. Absent
14. Reduced excipulum	1. Forming a narrow, indistinct layer, apex not embedded in dark material 2. Forming a broad layer, apex becoming embedded in dark brown material
15. Ascus development	1. Sequential 2. Synchronous
16. Ascus shape	1. Saccate; developing asci with spores extending to base; no basal stalk at maturity 2. Cylindrical or subclavate; developing asci with spores extending to base; basal stalk developing immediately before spore release 3. Clavate or clavate-stipitate; developing asci with spores only in upper part; clavate-stipitate before spore maturity
17. Ascospore shape	1. Filiform, ± straight when released 2. Filiform, characteristically bend or curved when released 3. Elliptical 4. Bifusiform
18. Septation of filiform ascospores	1. 1- or more septate when released from asci 2. Non-septate
19. Ascospore gelatinous sheath	1. Lacking 2. Single sheath surrounding whole spore 3. Gelatinous caps at one or both ends of spore 4. Gelatinous appendage along side of spore (at angles of bent spores) 5. Two structurally distinct sheaths, firm gelatinous caps at each end of spore, with whole spore surrounded by separate, loose sheath
20. Hymenial gel	1. With embedded crystals 2. Without embedded crystals

(plesiomorphic) states for the characters recognized as homologous. A number of methods have been proposed for recognizing apomorphic versus plesiomorphic character states, including outgroup comparison, ingroup comparison and the palaeontological and ontogenetic methods (Kitching, 1992; Maddison *et al.*, 1984; Wiley, 1981). However, there are problems inherent in applying any of these methods and, given the uncertainty over whether the character polarity being applied is valid or not, a number of authors recommend inferring the character state polarities and the topology of the tree simultaneously (BRYANT, 1991; CARPENTER, 1993; CLARK & CURRAN, 1986; MEACHAM, 1984; NIXON & SLOWINSKI, 1993; SWOFFORD & BEGLE, 1993). With this method no assumptions are made about allowable character state changes: those represented in the cladogram are simply the ones allowing the most parsimonious arrangement of branching patterns for the data. Thus, the tree produced is unrooted and may be rooted subsequently, usually on the basis of outgroup comparison. The authors cited above argue that the results of undirected analyses such as these are methodologically as robust as methods requiring prior character polarity assumptions. Because the number of *a priori* assumptions required is reduced, this method should minimise user bias.

Taxa included

Several analyses were carried out. Members of the genus *Terriera* (*Lophodermium* Group B) were designated as the outgroup.

1. *Relationship between the monocotyledon-inhabiting species only.* Two species representing each of the five groups (see p. 23) within the monocotyledon-inhabiting species of *Lophodermium s. lat.* were included, together with all of the ungrouped species.

2. *Relationship between the monocotyledon-inhabiting Lophodermium s. lat. groups and other species within the Rhytismataceae.* Relationships between the *Lophodermium s. lat.* groups and species in other genera of *Rhytismataceae* have been hypothesized in this study. These include:

- *Lophodermium* Group C with *Coccomyces tumidus* (see p. 39)
- *Lophodermium* Group D with *Coccomyces coronatus* (see p. 42)
- *Lophodermium* Group E with the species of *Coccomyces* in Groups 1 and 2 of JOHNSTON (1986) and dicotyledon-inhabiting species of *Lophodermium* in *Lophodermium* Groups 4 and 5 of JOHNSTON (1989*b*) (see p. 43)
- *Lophodermium eucalypti* with *Hypoderma* and the dicotyledon-inhabiting species of *Lophodermium* in *Lophodermium* Group 3 of JOHNSTON (1989*b*).

The relationships between these taxa were tested together with *Lophodermium* Group A, *L. agathidis* and *L. raapianum*, the initial analysis suggesting that the two last-named species may be related to *L. eucalypti.*

3. *Relationship between the monocotyledon-inhabiting Lophodermium s. lat. groups and other species of leaf-inhabiting Rhytismataceae on diverse hosts.* These analyses included all the monocotyledon-inhabiting species together with representatives of other major genera of the leaf-inhabiting *Rhytismataceae* and of several informal, putatively monophyletic, groups within the *Rhytismataceae* recognized by various authors in earlier publications. The groups included in the analysis are labelled as follows:

- Group A: species of *Lophodermium* Group A, *Lophodermium s. str.* (p. 24).
- Group B: species of *Lophodermium* Group B, the genus *Terriera* (p. 36).
- Group C: species of *Lophodermium* Group C (p. 39) together with *Coccomyces tumidus.*
- Group D: species of *Lophodermium* Group D (p. 42) together with *C. coronatus.*
- Group E: species of *Lophodermium* Group E (p. 43) together with the species of *Coccomyces* of Groups 1 and 2 of JOHNSTON (1986), which includes the *C. leptosporus* group of SHERWOOD (1980), represented by *C. limitatus*, and dicotyledon-inhabiting species of *Lophodermium* of *Lophodermium* Groups 4 and 5 of JOHNSTON (1989*b*), represented by *L. medium.*

- Group M: the group of *Ericales*-inhabiting species discussed by JOHNSTON (1988*b*), represented by *Meloderma dracophylli* and *L. sphaerioides*.
- Group N: the group of *Pinus*-inhabiting species discussed by JOHNSTON (1988*b*), represented by *L. pinastri* and *M. desmazieresii*.
- Group O: *Hypoderma* and the morphologically similar species of *Lophodermium* discussed by JOHNSTON (1989*b*) as *Lophodermium* Group 3, represented by *H. rubi* and *L. maculare*.
- Group P: *Rhytisma*, represented by *R. acerinum*.
- Group Q: *Propolis* (Fr.) Corda *sensu* SHERWOOD (1977), represented by *P. quadrifida* and *P. emarginata*.
- Group R: the conifer-inhabiting species with synchronous ascus development and lacking paraphyses when mature (see MINTER, 1985), represented by *Bifusella linearis* and *Hypodermella laricis*.

Options used for analysis

Analyses 1 and 2. These were restricted to less than 20 taxa, so allowing the use of the Branch and Bound tree-building algorithm, ensuring discovery of the shortest tree or trees. The default options were used in these analyses.

Analysis 3. This included 32 taxa, requiring the use of heuristic search options. An initial analysis was carried out using Stepwise Addition, Addition Sequence Random, with 100 replicates, tree bisection-reconnection (TBR) branch swapping, with a maximum of 100 trees saved at each replicate. The minimum length trees derived from this procedure were then used as the starting trees for a TBR Branch Swapping analysis with all shortest trees saved.

Whenever the analysis resulted in more than one shortest tree, strict and majority rule consensus trees were constructed.

The option to scale character weights in relation to the number of states was invoked in all analyses. The number of states defined for each character affects the effective weight of that character and hence has an influence on the analysis when different characters have different numbers of states. Thus, an unscaled 3-state character has twice the effective weight of a two-state character (SWOFFORD & BEGLE, 1993).

Bootstrap analyses were carried out using either the Branch and Bound or the Heuristic option, depending on the number of taxa involved.

Results and Discussion

The results of the analyses, confined to morphological characters only, were largely inconclusive. This was due in part to the small number of characters available, most branches on the cladograms being supported by one or a small number of characters. This was reflected in the poorly-resolved bootstrap analyses. Arbitrary exclusion of some of the taxa from the analyses also affected the results.

All the cladograms presented had *Terriera javanica* alone as the designated outgroup; however, trees with the same internal branches were always obtained when additional or alternative species of *Terriera* were used as the outgroup. As noted in the methods, *Lophodermium* Group A and *Lophodermium* Group E were each represented by two selected species. All cladograms presented had Group A represented by *L. culmigenum* and *L. gramineum* and Group E by *L. hauturuanum* and *L. unciniae*, but analyses carried out with alternative representatives for

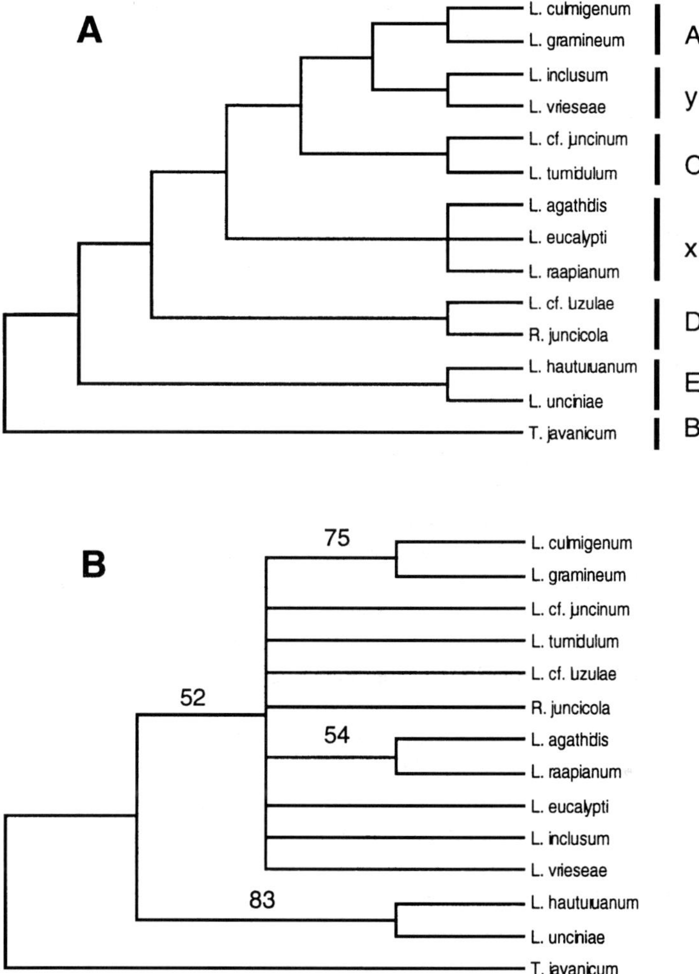

Fig. 15. **A**, strict consensus of the three shortest trees produced by a Branch and Bound analysis of the monocotyledon-inhabiting species of *Lophodermium s. lat.*, with *Terriera javanica* as the outgroup. Each of the groups proposed within *Lophodermium s. lat.* is resolved within the cladogram (marked A, B, C, D, E). Similarly, within clade 'x' a possible relationship had been suggested between *L. agathidis* and the poorly-known *L. raapianum* (see Notes under *L. raapianum*). However, on the kind of morphological evidence used to postulate Groups A–E, no relationship between these two species and the other member of clade 'x', *L. eucalypti*, had been suggested. Similarly the two taxa in clade 'y' had not been linked. **B**, bootstrap analysis of the tree illustrated in **A**.

these two groups resulted in trees with the same branching patterns.

1. *Relationship between the monocotyledon-inhabiting species only.* Three shortest trees were produced, of length 45, homoplasy index 0·306 and rescaled consistency index 0·536. The strict consensus tree is shown in **Fig. 15A**. All the groups recognized within *Lophodermium s. lat.* were

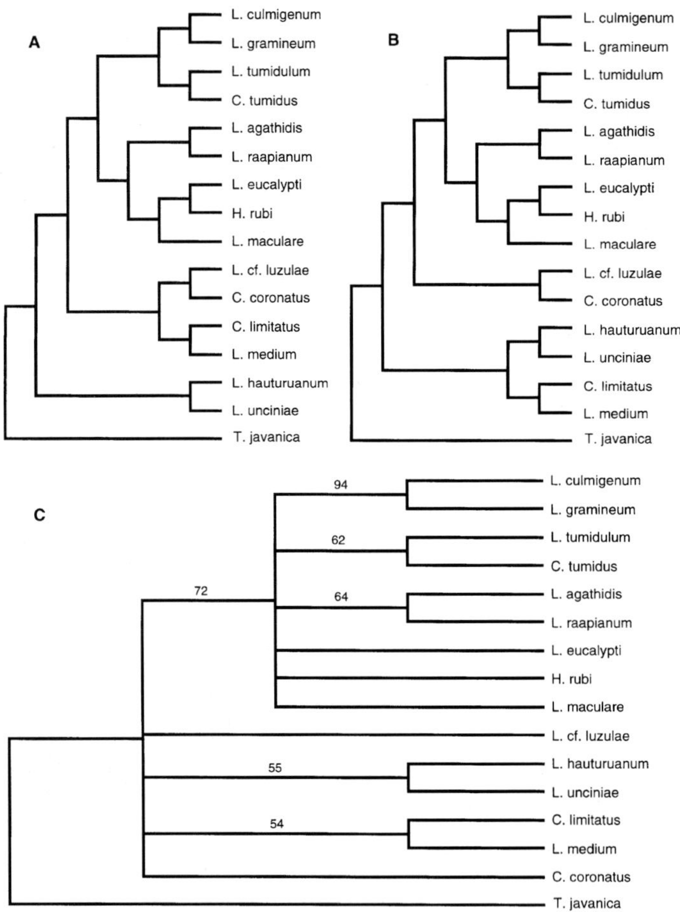

Fig. 16. A, the single shortest tree produced by a Branch and Bound analysis, testing the postulated relationship between the groups recognised amongst the monocotyledon-inhabiting species of *Lophodermium* and species in other genera within the *Rhytismataceae*, with *Terriera javanica* as the outgroup. Most of the postulated relationships are resolved within the tree, except the unexpected position of *C. limitatus* and *L. medium*. **B**, rearrangement of the branches within the tree illustrated in **A**, to test the strength of the clade *L.* cf. *luzulae/C. coronatus/C. limitatus/L. medium*. When the branch is repositioned (using MacClade) to reflect the relationship hypothesised between *C. limitatus/L. medium* and *L. hauturuanum/L. unciniae*, the resulting tree is only one step longer than the shortest tree. **C**, bootstrap analysis of the tree illustrated in **A**.

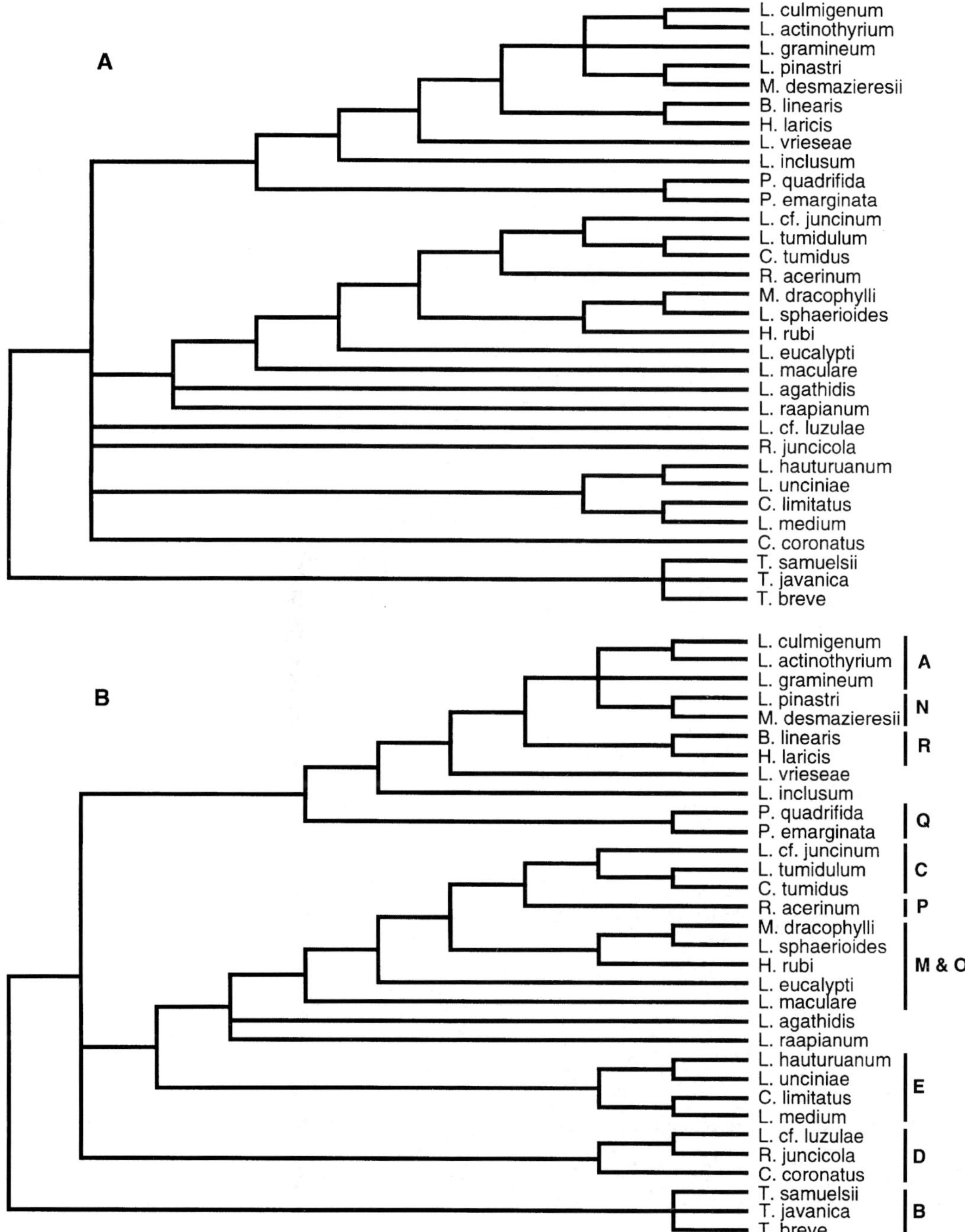

Fig. 17. Strict and 50% majority rule consensus of the 225 trees produced from the heuristic search of leaf-inhabiting *Rhytismataceae*, selected to represent a range of genera and informal groups across the family. **A**, strict consensus tree; **B**, majority rule consensus tree, with hypothesized Groups A–E and M–R indicated.

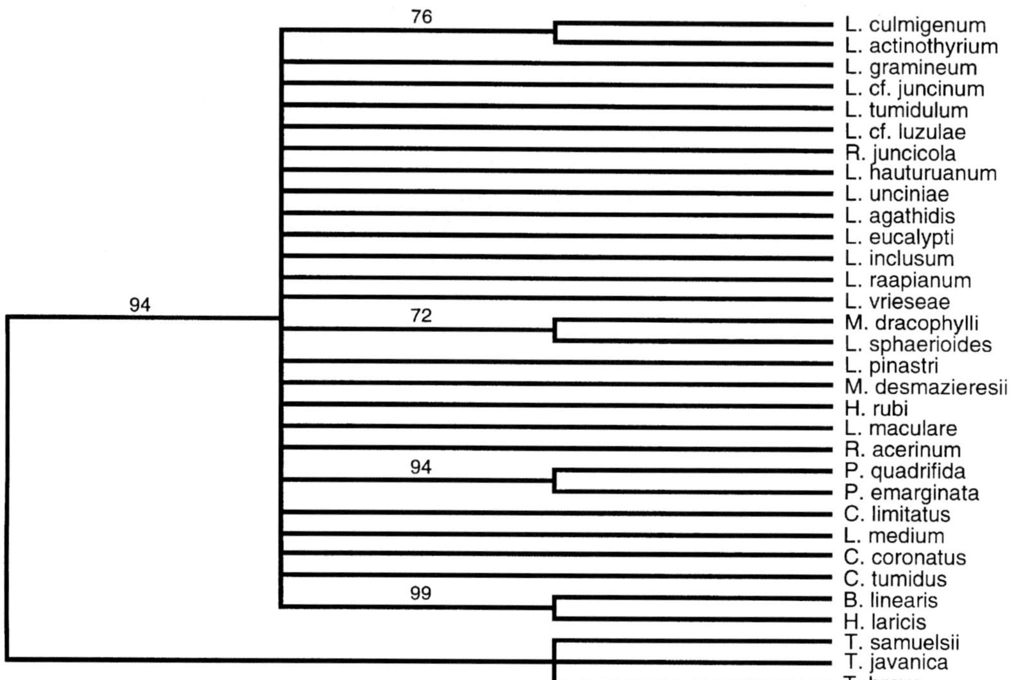

Fig. 18. Bootstrap analysis of the trees illustrated in Fig. 17.

resolved; however, the bootstrap analysis (**Fig. 15B**) showed weak support for almost all the branches in this tree.

2. *Relationship between the monocotyledon-inhabiting Lophodermium s. lat. groups and other species within the Rhytismataceae.* A single shortest tree was produced, of length 52, homoplasy index 0·525 and rescaled consistency index 0·591 (**Fig. 16A**). The relationships postulated between *Lophodermium* Group C and *Coccomyces tumidus*, *Lophodermium* Group D and *C. coronatus* and *L. eucalypti* and *Hypoderma* were all resolved. However, the hypothesized relationship between the clade containing *L. medium*, *C. limitatus* and *Lophodermium* Group E was not supported, *L. medium* and *C. limitatus* clustering with *Lophodermium* Group D. An alternative branching arrangement was tested using MacClade, linking *C. limitatus* and *L. medium* with *Lophodermium* Group E (**Fig. 16B**) and the resulting tree was only a single step longer, indicating only weak support for the arrangement on the shortest tree. The bootstrap analysis (**Fig. 16C**) showed little support for most branches, only *Lophodermium* Group A being well supported.

3. *Relationship between the monocotyledon-inhabiting groups and other species of leaf-inhabiting Rhytismataceae on diverse hosts.* The initial Stepwise Addition search resulted in 201 shortest trees, of length 86, rescaled consistency index 0·403, homoplasy index 0·483. These were used as the source trees for subsequent Branch Swapping using TBR, resulting in 24 additional 86-step trees. The consensus trees from both analyses were identical (**Fig. 17A, B**).

It is interesting to note that all the postulated groups among the monocotyledon-inhabiting species of *Lophodermium* were resolved, as were most of the other postulated monophyletic groups of leaf-inhabiting *Rhytismataceae*. Again, however, a bootstrap analysis (**Fig. 18**) showed little support for any of the internal branches of the tree.

Given the poorly-resolved bootstrap analyses, the results from the cladistic analyses have not been used to propose formal taxonomic groups. However, the results did provide some support for the notion that present generic concepts within the *Rhytismataceae* mean little in a phylogenetic sense (JOHNSTON, 1990*a*). These analyses provided a preliminary hypothesis of relationships among the leaf-inhabiting species of *Lophodermium*. For example, they suggested that *Lophodermium s. str.* (as defined here) is most closely related to a group of *Pinaceae*-inhabiting species. These results also provided a base for choosing a sensible sample of taxa for future molecular studies, almost certainly required if the phylogenetic relationships among these fungi are to be resolved.

KEYS TO ACCEPTED SPECIES

Identification of many of the monocotyledon-inhabiting species of *Lophodermium* can be problematic, with individual characters often somewhat variable, or atypical in some collections or in individual ascomata within a collection (see discussion, p. 23 *ff.*). Because of this, the keys provided below, although workable for the 'ideal' collection, must be used with care. Any identification suggested by these keys should be verified by checking the specimen at hand with the full description.

Two dichotomous keys have been provided to identify the groups recognized within *Lophodermium s. lat.*, and additional keys to species within each of these groups. Of the keys to the groups, one is based on geographical distribution and host preference, the other on morphology. Although the first key, based on host and geographical features, is not definitive, it is quick and easy to use and usually provides the correct answer. The morphologically-based key, on the other hand, requires measurement of many detailed micro-morphological and anatomical features of the specimens.

Computer-based interactive keys are often more user-friendly than the traditional dichotomous key. With such keys the user is able to choose which characters to apply at each step, and they allow some mistakes to be made when scoring individual characters during the identification, while still providing the correct identification. An interactive key to the monocotyledon-inhabiting species of *Lophodermium* is being developed using the DELTA-based programme INTKEY (DALLWITZ *et al.*, 1995). Although only at an early stage of development, this key is available on request from the author.

Key to groups of monocotyledon-inhabiting species within *Lophodermium s. lat.* based on geographical distribution and host preference

Most collections can be placed in a group on the basis of the following geographical distributions and host preferences. Rare exceptions occur and, if the collection does not appear to fit the group or groups suggested by this key (see p. 23), then the morphology-based key should be used.

1 Temperate Northern Hemisphere **2**
Temperate Southern Hemisphere **3**
Tropics .... **Group B (*Terriera*)**

2 (1) On *Poaceae* or *Allium*, *Carex*, *Convallaria* and *Iris* .... **Group A**
On other *Cyperaceae* or *Juncaceae* .... **Group C or D**

3 (1) On *Poaceae* .... **Group A**
On *Cyperaceae* .... **Group A, B or E**

Key to groups of monocotyledon-inhabiting species within *Lophodermium s. lat.* based on morphology

1 Opening slit of ascomata lined on both sides with narrow, flat, black shelf-like areas covering top of hymenium; ascospores lacking gelatinous sheath; ascomatal primordium comprising *textura prismatica*, in vertical section with appearance of cylindrical cells arranged in ± vertical columns .... **Group B (*Terriera*)**

Opening slit of ascomata not lined with black shelf-like areas covering top of hymenium, often with pale lip cells; ascospores with gelatinous sheaths, or gelatinous caps or barbs at apex, base or along sides of spores; ascomatal primordium comprising *textura intricata*, in vertical section with appearance of hyphal cells arranged ± horizontal to host surface **2**

2 (1) Ascomata becoming erumpent from host tissue, *Hysteriales*-like in appearance with upper wall either raised into a narrow ridge along opening slit or sunken along opening slit .... **Group C**

Ascomata remaining covered by host tissue **3**

3 (2) Ascomata circular to irregular in outline, with one or more irregular, sometimes radiate opening slits; never with differentiated cells along future line of opening of unopened ascomata or along edge of opening slit .... **Group D**

Ascomata elliptical to oblong in outline, with a single longitudinal opening slit, sometimes with short side-slits near ends of ascoma; most species with differentiated cells of some kind associated with future line of opening or along edge of opening slit **4**

4 (3) Ascospores with gelatinous cap at either end and loose, gelatinous sheath surrounding entire spore, best observed from fresh collections with spores mounted in water; common on *Poaceae* **5**

Ascospores with small gelatinous caps or barbs at apex, base or along sides of spores, rarely with no gelatinous appendages; rare on *Poaceae* **6**

5 (4) Common on *Poaceae*, also found on *Allium*, *Carex*, *Convallaria* and *Typha*; ascomata subepidermal, with well-developed clypeus in epidermal cells above ascoma; lip cells, when present, almost always branched; upper wall of ascomata with distinct, restricted dark patch near opening slit, two patches in unopened ascomata, one to either side of future line of opening .... **Group A**

Rare on monocotyledons, not known from *Poaceae*; ascomata subcuticular, with clypeus poorly developed; lip cells unbranched; cells of upper wall ± uniformly in pigmentation .... **eucalypti**

6 (4) Ascomata oblong to sublinear in outline; longitudinal opening slit often with short side slits near ends of ascomata; hymenium often remaining exposed, not covered by ascomatal wall even when dry; ascospores usually bent or curved on release, with gelatinous appendages along sides of spores; few periphysoids near opening slit elongating after ascomata open and, together with excipular elements and lip cells, forming persistent layer between hymenium and ascomatal wall; lip cells becoming embedded in dark brown material .... **Group E**

Ascomata elliptical in outline; single opening slit without short side slits; hymenium always covered by upper wall when dry; ascospores ± straight after release, with gelatinous appendages at ends of spores; periphysoids lacking and excipulum remaining poorly developed in open ascomata; lip cells, when present, not embedded in dark brown material **7**

7 (6) Lacking lip cells **8**

Lip cells well developed **9**

8 (7) Ascomata subepidermal, upper wall 30–50 μm thick; asci 5·5–75 μm wide; ascospores 1·5 μm wide; on *Cyperaceae* in New Zealand .... **inclusum**

Ascomata intra-epidermal, upper wall 25–30 μm thick; asci 14–16·5 μm wide, ascospores 2–3 μm wide; on *Bromeliaceae* in South America .... **vrieseae**

9 (7) Asci 135–150 × 9–10 μm; ascospores 80–120 × 1·5–2 μm .... **agathidis**

Asci 95–120 × 5·5–8 μm; ascospores 60–80 × 1·5–2 μm .... **raapianum**

Key to species in *Lophodermium* Group A

1 Lip cells absent, periphysoids usually well developed **2**

Lip cells present, although sometimes macroscopically indistinct, periphysoids usually poorly developed **7**

2 (1) Asci mostly < 10 μm wide **3**

Asci mostly > 10 μm wide **5**

3 (2) Ascospores mostly < 80 μm long, 1·5 μm wide; ascomata developing within discrete lesions on green leaves .... **seriatum**

Ascospores mostly > 80 μm long, 2–2·5 μm wide; ascomata developing on dead leaves **4**

4 (3) Ascomata associated with black conidiomata; ascospores tapering slightly to rounded base; dark cells in lower wall of ascoma angular/globose; lip cells present but very poorly developed and easily missed .... **petriniae**

Ascomata not associated with conidiomata; ascospores tapering to ± acute base; dark cells in lower wall of ascoma irregularly cylindrical .... **sesleriae**

5 (2) Asci subsaccate, often > 16 μm wide; ascospores often < 60 μm long; paraphyses circinate .... **alpinum**

Asci cylindrical to subsaccate-clavate, most < 16 μm wide; ascospores in most collections > 60 μm long; paraphyses undifferentiated to slightly swollen near apex, sometimes becoming tangled when mature **6**

6 (5) Asci strictly cylindrical, 11–13 μm wide; ascospores rarely > 90 μm long, tapering slightly to rounded base .... **gramineum**
Asci subsaccate-clavate, 13–16 μm wide; ascospores rarely < 90 μm long, tapering to ± acute base .... **grandialpinum**

7 (1) Dark-walled cells within lower wall of ascoma angular to cylindrical, when several rows of darkened cells present horizontal section may be required **8**
Dark-walled cells within lower wall of ascoma hyphal, often forming *textura intricata* **21**

8 (7) Lower wall of ascoma pale, in squash mount comprising single row of dark-walled cells across base of ascoma **9**
Lower wall of ascoma dark, in squash mount comprising several rows of dark-walled cells across base of ascoma **19**

9 (8) Asci mostly < 9 μm wide × 105 μm long **10**
Asci mostly > 9 μm wide × 105 μm long **15**

10 (9) On *Typha* or *Allium* **11**
On *Poaceae* **12**

11 (10) On *Typha* .... **typhinum**
On *Allium* .... **alliaceum**

12 (10) Ascospores tapering suddenly in lower part to tail-like base .... **sieglingiae**
Ascospores tapering gradually and slightly to base **13**

13 (12) Asci 75–90 × 6·5–7·5 (–8) μm; ascospores 35–45 × 1·5 μm .... **robergei** (see also **danthoniae**)
Asci 90–105 × 7·5–9 μm; ascospores 45–60 × 1·5 μm **14**

14 (13) Paraphyses bent or circinate at apex .... **actinothyrium**
Paraphyses swollen at apex; from New Zealand (see p. 73) .... *cf.* **actinothyrium**

15 (9) On *Convallaria* or *Iris* **16**
On *Poaceae* **17**

16 (15) On *Convallaria*; most ascospores 50–60 μm long; asci 10–11·5 μm wide .... **herbarum**
On *Iris*; most ascospores > 60 μm long; asci 9–10·5 μm wide .... **iridicolum**

17 (15) Paraphyses swollen at apex, not circinate .... **fusiforme**
Paraphyses circinate or coiling at apex, not swollen **18**

18 (17) Lip cells well developed, unbranched; asci mostly > 130 µm × 12·5–14 µm; not associated with conidiomata .... **nonramosum**
Lip cells poorly developed; asci mostly < 130 µm × 9–10 µm; associated with black conidiomata .... **petriniae**

19 (8) Asci 70–90 × 5·5–7 µm; ascospores 25–35 µm long; periphysoid layer well developed in unopened ascomata; confined to *Carex* .... **caricinum**
Asci > 90 µm × 7 µm; ascospores > 35 µm long; periphysoids poorly developed in unopened ascomata; rarely on *Carex* **20**

20 (19) Asci often > 130 µm × 12–14 µm; ascospores 80–100 × 2·5 µm; lip cells unbranched .... **nonramosum**
Asci 110–125 × 8·5–9·5 µm; ascospores 55–65 × 1·5–2 µm; lip cells branched dichotomously... **nitidum**

21 (7) Ascomata associated with conidiomata; on *Ampelodesmos* or *Phragmites* **22**
Ascomata not associated with conidiomata; on other host plants **23**

22 (21) On *Phragmites*; invaded part of leaf not associated with zone lines; conidiomata macroscopically indistinct, pale, deeply immersed; asci 9–10 µm wide; ascospores 65–85 µm long .... **arundinaceum**
On *Ampelodesmos*; invaded part of leaf associated with zone lines; conidiomata macroscopically distinct, dark, pustulate; asci 7·5–9 µm wide; ascospores 45–60 µm long .... **eximium**

23 (21) Asci 9·5–11 µm wide **24**
Asci 8·5–9·5 µm wide **26**

24 (23) Asci 90–100 µm long; ascospores 45–55 µm long; on *Festuca novae-zelandiae* (see p. 109) .... *cf.* **culmigenum**
Asci > 110 µm long; most ascospores > 60 µm long **25**

25 (24) Asci 115–130 µm long; mycelium in leaf tissue around ascoma hyaline; known from many hosts in Europe and North America .... **culmigenum**
Asci 120–165 µm long; mycelium in leaf tissue around ascoma pink to red; known only on *Cortaderia* from New Zealand .... **rubrum**

26 (23) Asci 95–110 µm long; ascospores 45–55 µm long; known from Greece (see p. 109) .... *cf.* **culmigenum**
Asci 110–125 µm long; ascospores 60–75 µm long; known from Australia, New Zealand and Chile (see p. 109) .... *cf.* **culmigenum**

Key to species in *Lophodermium* Group B (*Terriera*)

Most species placed in this group are known from few small old collections. Although the type material of the species accepted in *Terriera* can be distinguished using the following key, it is not certain whether the features used to separate the species are appropriate for discrimination at the species level.

A. Paraphyses branched near apex, slightly swollen to swollen

B. Ascus apex broadly rounded or with slight taper to ± broadly rounded apex

	asci	**ascospores**
'*L. andropogonis*' (zones)[5]	80–95 × 6·5–7·5 μm	50–60 × 1·5 μm
T. javanica var. *pandani*	90–95 × 6–6·5 μm	50–60 × 1·5 μm

B. Ascus tapering to small, rounded apex

T. javanica (zones, conidiomata)[6]	85–95 × 5·5–7 μm	50–60 × 1·5 μm
T. latiascus	80–95 × 7–8·5 μm	40–50 × 2–2·5 μm
T. fourcroyae	90–110 × 6–6·5 μm	60–70 × 1·5–2 μm
T. pandani	100–120 × 5–6 μm	50–70 × 1–1·5 μm
T. dracaenae	130–160 × 6–7 μm	100 × 2 μm
T. longissima	175–210 × 6–6·5 μm	120–130 × 1 μm

B. Ascus apex subtruncate

T. clithris (zones)	100–120 × 6·5–7 μm	60–80 × 1–1·5 μm

A. Paraphyses unbranched near apex, undifferentiated to slightly swollen

C. Ascus apex broadly rounded or with slight taper to ± broadly rounded apex

T. fuegiana	75–95 × 7–10 μm	60–65 × 1–2·5 μm
T. sacchari	90–100 × 5–6 μm	50–60 × 1·5 μm
T. stevensii (conidiomata)	100–125 × 5–6 μm	60–80 × 1·5–2 μm
T. samuelsii (zones, conidiomata)	125–140 × 7–8 μm	75–90 × 2 μm
T. arundinacea	130–160 × 8–9 μm	90–100 × 2–2·5 μm
'*L. miscanthi*'	130–175 × 7·5–9 μm	100–120 × 2–2·5μm

C. Ascus apex tapering to small rounded apex

T. breve	110–130 × 6–7 μm	60–70 × 1·5–2 μm

C. Ascus apex subtruncate

T. nematoidea	70–80 × 5–6·5 μm	30–35 × 1 μm
T. asteliae	75–105 × 8–10·5 μm	45–70 × 2–2·5 μm

---

[5] 'zones' refers to the presence of narrow black zone lines around the margins of the bleached part of the leaf invaded by the fungus (see p. 47).

[6] 'conidiomata' refers to the presence of conidiomata in association with the ascomata.

Key to species in *Lophodermium* Groups C and D

1 Paraphyses circinate and often branched near apex, not forming epithecium; upper wall of ascoma with broad inner layer of pale, globose cells forming periphysoid-like layer **3**
Paraphyses swollen, unbranched at apex, forming palisade-like or tangled epithecium; upper wall of ascoma with periphysoids comprising short-cylindrical cells embedded in gelatinous matrix **2**

2 (1) Asci 115–130 × 8·5–10 µm; ascospores 50–70 × 1·5 µm, becoming 3-septate .... *cf.* **luzulae**
Asci 115–125 × 11·5–13 µm; ascospores 50–60 × 2–2·5 µm, non-septate .... **Rhytisma juncicola**

3 (1) Ascomata subcuticular; on *Cyperaceae*, *Poaceae* and *Juncus* .... **tumidulum**
Ascomata subepidermal; known only from *Juncus* .... *cf.* **juncinum**

Key to species in *Lophodermium* Group E

1 Asci < 100 µm long, > 7 µm wide; ascospores 45–60 µm long .... **alienum**
Asci > µm long, < 7 µm wide; ascospores > 70 µm long **2**

2 (1) Conidiomata up to 0·2 mm diam.; asci mostly > 125 µm long; ascospores mostly > 85 µm long; conidiogenous cell with sympodial proliferation; on *Uncinia*, rarely *Carex*, *Gahnia* and *Juncus* .... **unciniae**
Conidiomata > 0·4 mm diam.; asci mostly < 125 µm long; ascospores mostly < 85 µm long; conidiogenous cells with percurrent proliferation; on *Gahnia* .... **hauturuanum**

SPECIES DESCRIPTIONS

Lophodermium actinothyrium Fuckel, *Symb. mycol. Nachtr.* **3**: 28 (1875).

Lophodermium arundinaceum var. *actinothyrium* (Fuckel) Rehm, *Rabenh. Krypt-Fl.* Edn 2, **1**(3): 47 (1887).

Hypoderma actinothyrium (Fuckel) Kuntze, *Revis. gen. pl.* **3**(3): 487 (1898).

Hysterium culmigenum var. *abbreviatum* Roberge ex Desm., *Annls Sci. nat.* Sér. 3, **8**: 179 (1847).

Lophodermium arundinaceum var. *culmigenum* Fuckel, *Symb. mycol.*: 257 (1870), *nom. illegit.*, *ICBN* Art. 52.1.

Lophodermium dactylidis Hilitzer, *Věd. Spisy čsl. Akad. zeměd.* **3**: 92 (1929).

Infected areas on dead leaves, slightly paler than surrounding host tissue, not associated with zone lines, containing ascomata, conidiomata not seen. *Ascomata* 0·5–0·8 × 0·3–0·5 mm, broadly elliptical to subovate in outline, ends rounded, often with small apiculus, unopened ascomata with wall pale grey, no obvious preformed line of dehiscence, opened ascomata with wall black, often shiny, longitudinal opening slit, lips initially white, thick and well differentiated macroscopically, although sometimes becoming dark and indistinct as ascomata mature. *Ascomatal insertion* subepidermal. *Covering layer* up to 50–65 µm thick in vertical section, comprising clypeus of dark-walled hyphae within partially broken-down epidermal cells and upper wall of ascomatal stroma. *Upper wall* 40–50 µm thick in vertical section, comprising mostly angular to globose cells with walls slightly and irregularly encrusted with dark brown material, with group of very dark cells adjacent to well-developed lip cells. *Lower wall* 10–15 µm thick in vertical section, comprising two or three rows of angular to globose cells, 6–12 µm diam., lowermost row of cells with walls darkened and slightly thickened, inner rows of cells with walls thin, pale brown to hyaline, in squash mount darkened cells of lower wall short-cylindrical to subglobose, 7–10 µm diam., arranged in ± single layer across base of ascoma. *Paraphyses* 1·5–2 µm diam., hooked to circinate at apex, sometimes also slightly and irregularly swollen. *Asci* (80–) 90–105 (–115) × 8–9 (–9·5) µm, subclavate to subfusoid, tapering gradually to small rounded to subtruncate apex with undifferentiated wall, 8-spored, short basal stalk at maturity with spores confined to upper 70–80 µm. *Ascospores* 45–60 (–65) × 1·5–2 µm (see Notes), tapering gradually to base, apical gelatinous cap 3–4 µm diam., globose to broad subcylindrical, basal gelatinous cap 2–3 × 1·5 µm, gelatinous sheath 4–5 µm thick.

Typification: Germany: NORDRHEIN-WESTFALEN: nr Köln, on *Molinia coerulea*, *s. coll.*, *dat. nec num.* (Fuckel, *Fungi rhenani* no. 2675; **S**!, *lectotypus* of *L. actinothyrium*, selected here; **K**!, *isolectotypus*, selected here). **France:** NORMANDY: dunes at Lyon-sur-Mer (Calvados), on *Ammophila* (as *Calamagrostis*) *arenaria*, *Roberge s. num.* (**K**!, *lectotypus* of *H. culmigenum* var. *abbreviatum*, selected here). **Germany:** RHEINLAND-PFALZ: Nassau, Oestrich, *Fuckel* (Fuckel, *Fungi rhenani* no. 738; **UPS**!; **DAOM**!, type of *L. arundinaceum* var. *culmigenum*). **Czech Republic:** BOHEMIA: Strekov, on *Dactylis glomerata*, 31 Oct. 1927, *A. Hilitzer* (**PRM** 693208!, *lectotypus* of *L. dactylidis*, selected here).

Illustrations: Figs 3A, 19, 20, 25.

Hosts: *Arrhenatherum*, *Bromus*, *Calamagrostis*, *Dactylis*, *Deschampsia*, *Festuca*, *Koeleria*, *Molinia*, *Sesleria* (*Poaceae*).

Distribution: Europe, North America, Pakistan, ?New Zealand (see Notes).

Notes: *Lophodermium actinothyrium* has an ascomatal structure typical of *Lophodermium* Group A and is one of the species constituting the *actinothyrium*-group within Group A. Ascomata

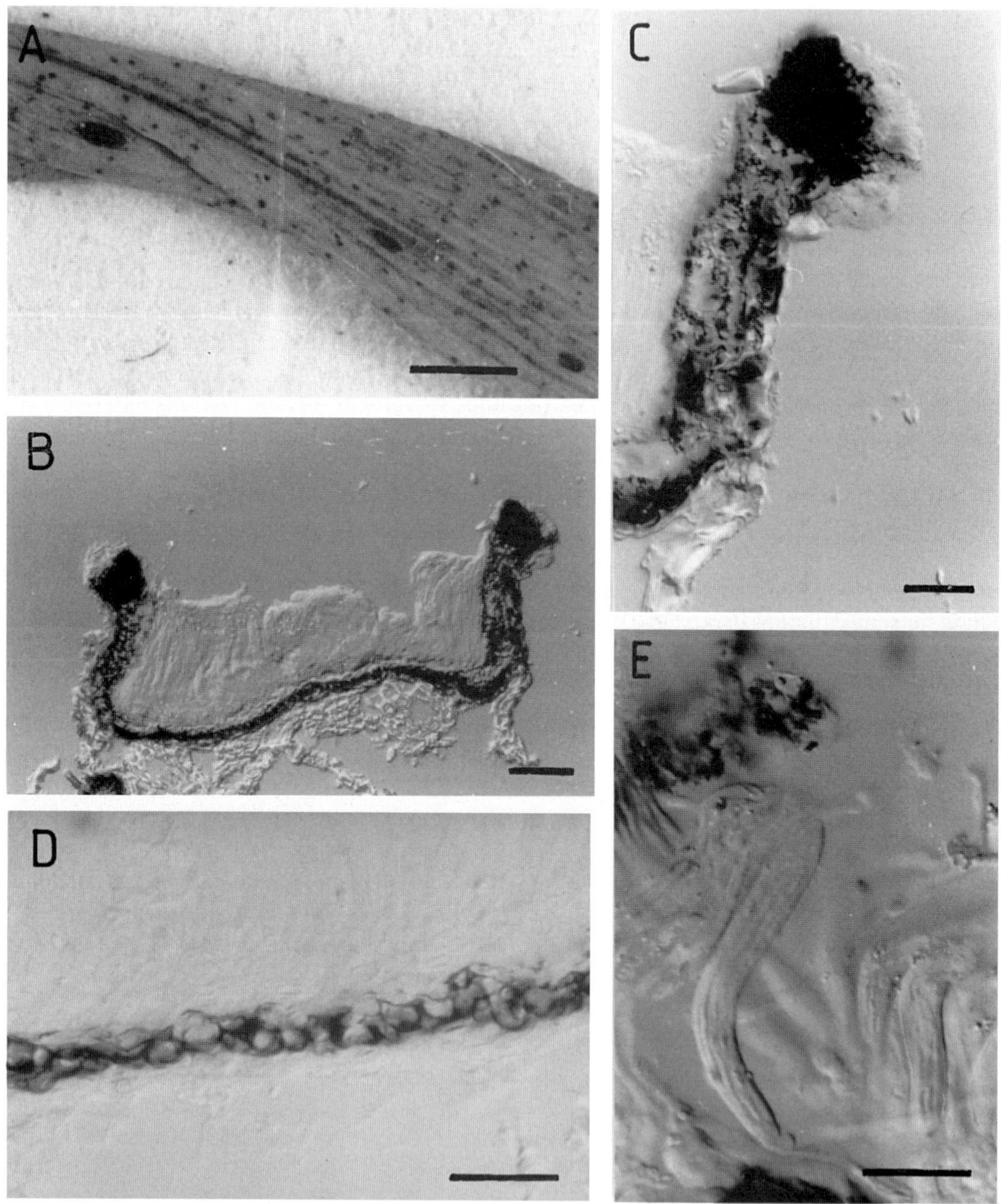

Fig. 19. *Lophodermium actinothyrium* (Fuckel, *Fungi Rhenani* no. 2675, **K**, isolectotype). **A**, ascomata (bar = 1 mm); **B**, ascoma in vertical section (bar = 50 µm); **C**, upper wall of ascoma in vertical section; **D**, lower wall of ascoma in vertical section; **E**, ascus (bars = 20 µm).

have dull black walls with margins quite sharply defined, oblong-elliptical in outline and a small apiculus at the rounded ends. It is characterized microscopically by ascus size (mostly 90–105 × 8–9 μm), ascospore size (mostly 45–60 × 1·5–2 μm), paraphyses hooked but typically not coiling at the apex and the darkened cells of the lower wall being short-cylindrical to subglobose.

Lophodermium actinothyrium is one of the most common species in Europe. The collections from California and Pakistan appear typical in all respects.

Several well-defined species with a similar structure to the lower wall of the ascomata are macroscopically similar to *L. actinothyrium*: *L. sieglingiae* Hilitzer is distinguished by its ascospores narrowing suddenly to a more or less tail-like base, while *L. robergei* has consistently shorter ascospores and narrower asci. For historical and practical reasons, *L. typhinum* has been retained as distinct on the basis of host substratum alone (see Notes under *L. typhinum*). See also Notes under *L. alliaceum* and *L. nitidum*.

Hysterium culmigenum var. *abbreviatum* has been placed in synonymy with *L. actinothyrium*, although the more or less ovate, shiny-walled ascomata of the type collection are somewhat atypical. In addition, the cells of the lower wall of the ascomata are larger than in most collections of *L. actinothyrium*. However, these differences are not regarded as sufficient to treat DESMAZIÈRES's (1847) taxon as distinct. DESMAZIÈRES cited one specimen in the protologue: 'Cette variété a été trouvé par M. Roberge dans les dunes de Lyon-sur-Mer, sur ... *Calamagrostis arenaria*.' Desmazières, *Pl. crypt. N. France* Edn 2, Sér. 2, no. 171 in **K** shares these collecting details, as does a collection there in a handwritten packet and one in **S** (*Herbier Barbey-Boissier* no. 999). All three packets contain what appears to be the same fungus, but the collection in the handwritten packet at Kew is the largest and in the best condition and is here selected as the lectotype.

The collection of Desmazières, *Pl. crypt. N. France* Edn 2, Sér. 2, no. 171 contains two fungi. One, present on all but one of the leaves in the packet, matches DESMAZIÈRES's description of *H. culmigenum* var. *abbreviatum* and is macroscopically the same as the species in the handwritten packet in **K**, but the hymenium has been largely lost. The second species, found on only one of the leaves, matches *L. culmigenum*, both macroscopically and microscopically.

FUCKEL (1870) placed *H. culmigenum* var. *abbreviatum* in synonymy with his *L. arundinaceum* var. *culmigenum* and cited the type specimen of *H. culmigenum* var. *abbreviatum* in the protologue, making his new variety superfluous and thus illegitimate. FUCKEL cited another specimen (Fuckel, *Fungi rhenani* no. 738), which matches *L. actinothyrium* both macroscopically and microscopically. TEHON (1935) incorrectly placed *H. culmigenum* var. *abbreviatum* in synonymy with *L. arundinaceum*.

PRM 693208, selected here as the lectotype for *L. dactylidis*, is the only specimen in **PRM** collected by Hilitzer from *Dactylis*. It had previously been annotated 'holotype' and matches the concept of *L. actinothyrium* adopted here in all respects.

The species limits being applied to taxa in *Lophodermium* Group A are somewhat tentative (see p. 24). Although most of the European grass-inhabiting collections examined fit into one of the species accepted here, the existence of apparent 'intermediates' raises some doubt as to whether the criteria are appropriate. Some *L. actinothyrium*-like collections provide examples. Two collections from *Elymus* (**Sweden:** Bohuslän, Koster-Archipel, Strömstad, on *Elymus arenarius*, 11–13 Jun. 1990, *C. Scheuer s. num.* (**GZU**!); **Scotland:** *s. loc.*, on *Elymus*, 16 Jul. 1991, *D.W. Minter* (**IMI** 351614!)) are typical of *L. actinothyrium* in the structure of the lower wall, but they have larger asci (115–130 × 9–10·5 μm) and ascospores (65–80 × 1·5–2 μm), more characteristic of *L. culmigenum* or *L. nitidum*. Macroscopically, they match typical collections of *L. culmigenum* from *Elymus* (specimens listed under *L. culmigenum*). Two North

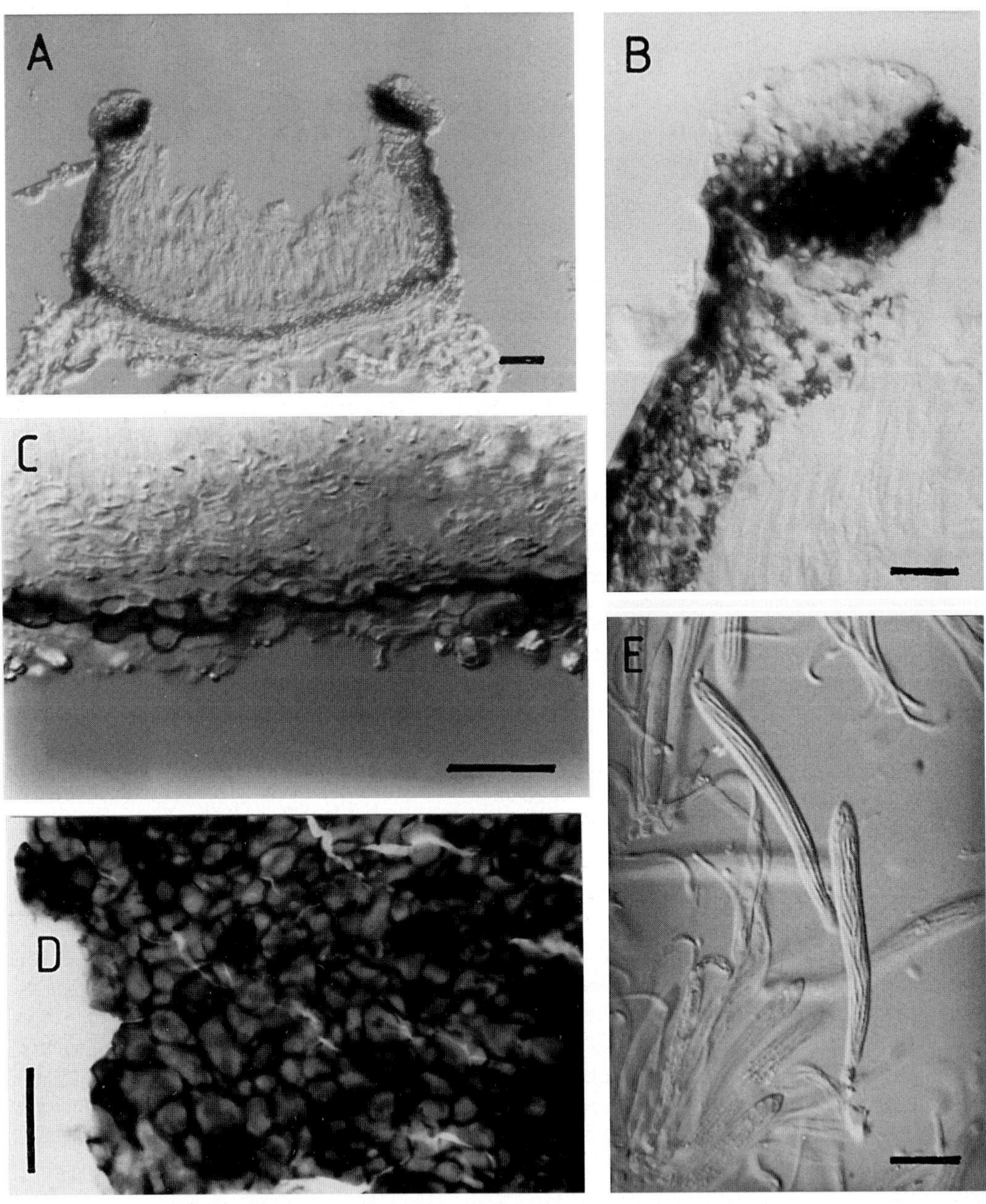

Fig. 20. *Lophodermium actinothyrium* (**A–C**, **PDD** 67186; **D**, **PDD** 67192.; **E**, **PRM** 819851). **A**, ascoma in vertical section (bar = 50 µm); **B**, upper wall of ascoma in vertical section; **C**, lower wall of ascoma in vertical section; **D**, lower wall of ascoma in squash mount; **E**, asci (bars = 20 µm).

American collections (**USA:** UTAH: Weber Co.: Wasatch Mtns E of Ogden, nr Waterfall Canyon, on *Festuca kingii*, 27 Apr. 1979, *C.T. Rogerson* (**NY**!, as *L. arundinaceum*); Wasatch Mtns E of Ogden, Malan's Peak, on *Festuca* (as *Leucopoa*) *kingii*, 6 Jun. 1979, *C.T. Rogerson* (**NY**!, as *L. arundinaceum*)) and a Swiss collection (**Switzerland:** GRAUBÜNDEN: Etan, Alp Laret, on *Trisetum distichophyllum*, 27 Aug. 1984, *K. & L. Holm* 3227c (**PDD** 64953, **ZT**!)) are macroscopically similar to the collections from *Elymus* and have the same lower wall structure, but are distinct in having larger asci (120–135 × 10·5–12 µm) and ascospores (60–70 × 2·5–3 µm) and in either lacking, or having very poorly-developed, lip cells.

Several collections from New Zealand (listed below as *L.* cf. *actinothyrium*) are similar to *L. actinothyrium* in macroscopic appearance, lower wall structure and ascus and ascospore size but have paraphyses slightly swollen rather than circinate at the apex. They may represent a distinct species, but because the genetic significance of small morphological differences such as these is poorly understood, a new species is not proposed here.

Additional specimens examined: ***Lophodermium actinothyrium*****: Austria:** KÄRNTEN: Lavanttal, Saualpe W. St. Andrä, on *Calamagrostis epigeios*, 24 Nov. 1979, *D. Kores s. num.* (**GZU**, as *L. apiculatum*). **Czech Republic:** *s. loc.*, on *Calamagrostis villosa*, 20 Sep. 1934, *A. Hilitzer s. num.* (**PRM** 820250, as *L. apiculatum*); BOHEMIA: Dobříš, Voznice, on *Calamagrostis epigeios*, 6 Sep. 1977, *M. Svrček s. num.* (**PRM** 756744); Těpín, Jílové, on *Calamagrostis epigeios*, 30 Sep. 1954, *M. Svrček s. num.* (**PRM** 756747, as *L. apiculatum*); Vrábsko pr. Čimelice, ad ripam piscinae Nerestec, on *Deschampsia caespitosa*, 5 Aug. 1977, *M. Svrček s. num.* (**PRM** 819847, 819851); Klánovice prope Pragam, in silvis 'Vidrholec', on *Molinia coerulea*, 26 May 1973, *M. Svrček s. num.* (**PRM** 734499); Praha, Thrěskýles, on *Calamagrostis epigeios*, 27 May 1948, *M. Svrček s. num.* (**PRM** 756751, as *L. apiculatum*); Pragam, on *Dactylis glomerata*, 25 May 1947, *M. Svrček s. num.* (**PRM** 824322); Karlštejn, Velká hora, on *Sesleria coerulea*, 22 May 1960, *M. Svrček s. num.* (**PRM** 620299); MORAVIA: Útěchov, Bílovice, on *Calamagrostis epigeios*, 24 Jun. 1973, *M. Svrček s. num.* (**PRM** 734438, as *L. apiculatum*). **France:** NORMANDY: Lion-sur-Mer (Calvados), on *Ammophila* (as *Calamagrostis*) *arenaria*, in autumn, *s. coll.* (Desmazières, *Pl. crypt. N. France* Edn 2, Sér. 2, no. 171; **K**). **Germany:** Mecklenburg, on *Molinia coerulea*, 11 Aug. 1909, *H. Sydow s. num.* (Sydow, *Mycotheca germanica* no. 897; **PDD** 51983); BERLIN: on *Molinia coerulea*, Aug. 1886, *P. Sydow s. num.* (Sydow, *Mycotheca marchica* no. 1170; **NY**; **HBG**, as *L. arundinaceum* var. *actinothyrium*); SAXONY: Leipzig, on *Calamagrostis sylvatica*, Jun. 1875, *G. Winter s. num.* (de Thümen, *Mycotheca universalis* no. 471; **K**, **NY**, as *L. arundinaceum* var. *apiculatum*); Frankfurt, on grass, *Bagge s. num.* (Rabenhorst, *Fungi europaei* no. 1226; **NY**, as *L. arundinaceum* var. *apiculatum*). **Hungary:** Kövestetó, on *Calamagrostis epigeios*, 14 Oct. 1958, *S. Tóth & A. Vass s. num.* (**DAOM** 150896, as *L. arundinaceum* f. *apiculatum*). **Latvia:** Adarzi, on *Molinia coerulea*, 8 Sep. 1929, *F. Smarods s. num.* (**K**). **Pakistan:** Shogran, on grass, 28 Jul. 1956, *S. Ahmad* 14816 (**K**, as *Lophodermium* sp.). **Switzerland:** AARGAU: Geissberg, Besserstein, on *Arrhenatherum*, 15 May 1980, *L. Petrini* 25 (**PDD** 67185, **ZT**); Besserstein, on *Sesleria coerulea*, 15 May 1980, *L. Petrini* 23 (**PDD** 67192, **ZT**); GRAUBÜNDEN: Bergün, Val Tuors, on *Sesleria coerulea*, 14 Jul. 1980, *L. Petrini* 116 (**PDD** 67186, **ZT**); Bergün, Val Tuors, on *Dactylis glomerata*, 14 Jul. 1980, *L. Petrini* 117B (**PDD** 67187, **ZT**); *idem loc., hosp. et dat., L. Petrini* 118B (**PDD** 67188, **ZT**); SCHAFFHAUSEN: Thayngen, Kerzenstiibli, on *Bromus erectus*, 31 May 1980, *L. Petrini* 30 (**PDD** 67189, **ZT**); Hemmental-Guetbuck, on *Bromus erectus*, 8 Jun. 1980, *L. Petrini* 31 (**PDD** 67190, **ZT**); TICINO: San Gottardo, on *Festuca*, 21 Jul. 1980, *L. Petrini* 188B (**PDD** 67191, **ZT**). **USA:** CALIFORNIA: Marin Co.: Tiburon, on *Calamagrostis ophitidis*, 25 May 1966, *J.T. Howell s. num.* (**K**).

***Lophodermium* cf. *actinothyrium*: New Zealand:** AUCKLAND: Waitakere Ranges, Karamatura Stream, on *Cynodon dactylon*, 2 Mar. 1984, *P.R. Johnston* R389 (**PDD** 45304); CHATHAM IS: on *Hierochloe*, *E.H.C. McKenzie* 146 (**PDD** 49369); KERMADEC IS: Raoul I, Denham Bay, on *Poa polyphylla*, 21 Sep. 1988, *E.H.C. McKenzie s. num.* (**PDD** 54562); STEWART I: Lee Bay, on *Bromus*, 4 May 1984, *P.R. Johnston* R480 (**PDD** 49365); TAUPO: Tongariro National Park, Mahuia campsite, on *Chionochloa rubra*, 20 Mar. 1984, *P.R. Johnston* R616 (**PDD** 49367); WAIKATO: Mt Pirongia, Te Tahi Rd, on grass, 19 Nov. 1984, *P.R. Johnston* R603 (**PDD** 49361); WELLINGTON: Island Bay, on *Puccinella fasciculata*, 26 Nov. 1989, *E.H.C. McKenzie s. num.* (**PDD** 56701).

Lophodermium agathidis Minter & Hettige, *N.Z. Jl Bot.* **21**: 39 (1983).

Infected areas on dead leaves, slightly paler than surrounding host tissue, not associated with zone lines, containing scattered ascomata. *Ascomata* 1–2 × 0·4–0·5 mm, elliptical to oblong-elliptical

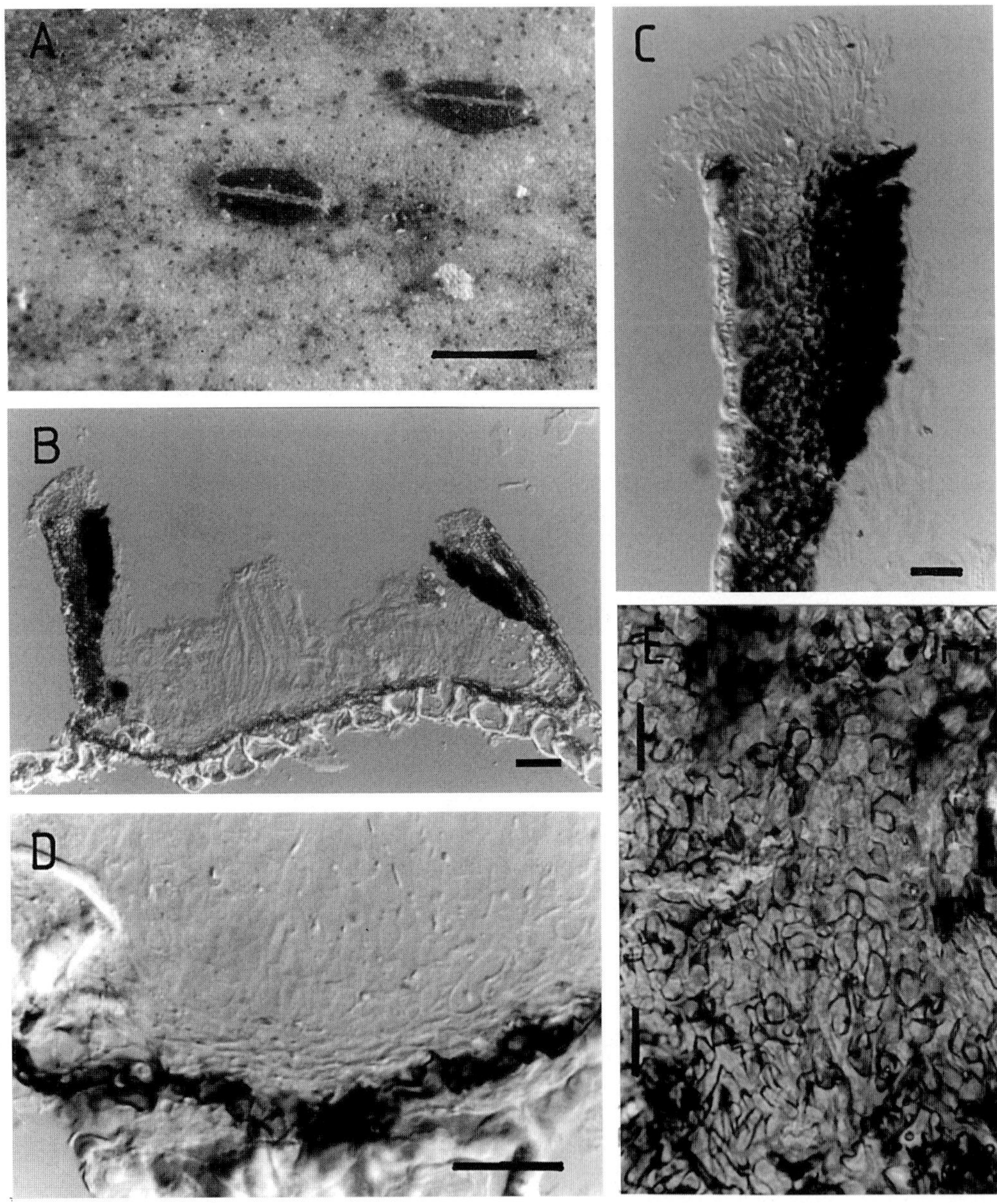

Fig. 21. *Lophodermium agathidis* (**PDD** 46708). **A**, ascomata (bar = 1 mm); **B**, ascoma in vertical section (bar = 50 µm); **C**, upper wall of ascoma in vertical section; **D**, lower wall of ascoma in vertical section; **E**, lower wall of ascoma in squash mount (bars = 20 µm).

in outline, unopened ascomata with wall black, usually with broad pale zone along future line of opening, open ascomata with wall black, single, longitudinal opening slit with bright orange-yellow lip cells. *Ascomatal insertion* subepidermal, rarely developing within upper layers of hypodermis. *Covering layer* of unopened ascomata (at stage where cavity developing within ascomatal primordium, but organized hymenium not yet evident) 65 µm thick, comprising ± intact epidermal cells and ascomatal wall, in opened ascomata covering layer up to 70 µm thick, comprising epidermal cells packed with dark-walled hyphae of clypeus and ascomatal wall. *Upper wall* in unopened ascomata 30 µm thick in vertical section, comprising mostly angular to globose cells, 3–5 µm diam., cell walls slightly and irregularly encrusted, becoming darker and thicker-walled as ascomata mature, inner part of ascomatal wall becoming very dark but, in outer part of wall, group of cells along future line of opening remain pale and thin-walled, ascomatal wall lined with poorly-developed layer of periphysoids; in open ascomata upper wall up to 50 µm thick, comprising angular cells, 5–7 µm diam., cells of inner part of wall with very thickly encrusted walls, often with cellular structure completely obscured, cells in outer part of wall with paler, irregularly encrusted walls; lip cells unbranched, forming well-developed palisade along edge of opening slit. *Lower wall* darkening before upper wall, comprising one or two rows of angular to globose, dark, slightly thick-walled cells, 5–8 µm diam., with scattered narrow-cylindrical cells present across lower side of wall in older ascomata; in vertical section lower wall 10–15 µm thick, comprising one or two rows of angular to globose cells, only the lowermost row with darkened walls. *Reduced excipulum* of paraphysis-like elements around edge of hymenium, differentiated from paraphyses by being more closely septate and not intermixed with asci. *Paraphyses* 1·5–2 µm diam., slightly and irregularly swollen to 3–3·5 µm at apex. *Asci* 135–150 × 9–10 µm, subclavate, tapering to rounded apex with undifferentiated wall, 8-spored, empty basal stalk at maturity with spores confined to upper 110–130 µm, ascus development sequential. *Ascospores* 80–120 × (1·5–) 2 µm, straight when released, tapering slightly to base, (0–) 1-septate, cup-like gelatinous appendage at base of spore.

Typification: New Zealand: AUCKLAND: Waitakere Ranges, Oratia, Kelly's Bush, on *Agathis australis*, 10 Nov. 1979, *G. Hettige s. num.* (**IMI** 259066!).

Illustrations: Figs 21, 22.

Hosts: Numerous monocotyledonous and dicotyledonous plants, the former including *Xeronema* (*Phormiaceae*) and an unidentified species of *Poaceae*.

Distribution: New Zealand, tropical Asia, tropical America.

Notes: *Lophodermium agathidis* occurs on a wide range of hosts which, unusually for a species of *Lophodermium*, includes both monocots and dicots, although most specimens in herbarium collections are on various dicotyledons. It is a distinctive species with large ascomata, a broad pale zone along the future line of opening, bright orange-yellow lip cells and a small, cup-like gelatinous appendage at the base of the ascospores. The grass-inhabiting collection from Guadeloupe is typical in all respects, including the ascospore appendage.

This species appears to be common and widely distributed throughout tropical Asia and tropical America, with collections on dicotyledons examined from India, Malaysia (**IMI**) and Taiwan (**TNS**). It may be conspecific with *L. clusiae* (Lév.) P.R. Johnst., the latter being regarded as distinct primarily because of its bright red lip cells and geographical distribution (JOHNSTON, 1989*a*). Collections typical of *L. agathidis* in all respects are now known to be widespread in tropical regions, including America, and the lip cells of at least some collections of *L. agathidis* from New Zealand turn red in 3% KOH, the red pigment diffusing out into the KOH.

The relationship of *L. agathidis* and *L. clusiae* to other species of *Lophodermium* is unknown. *Lophodermium raapianum* may have a similar ascomatal structure, although only a single, old

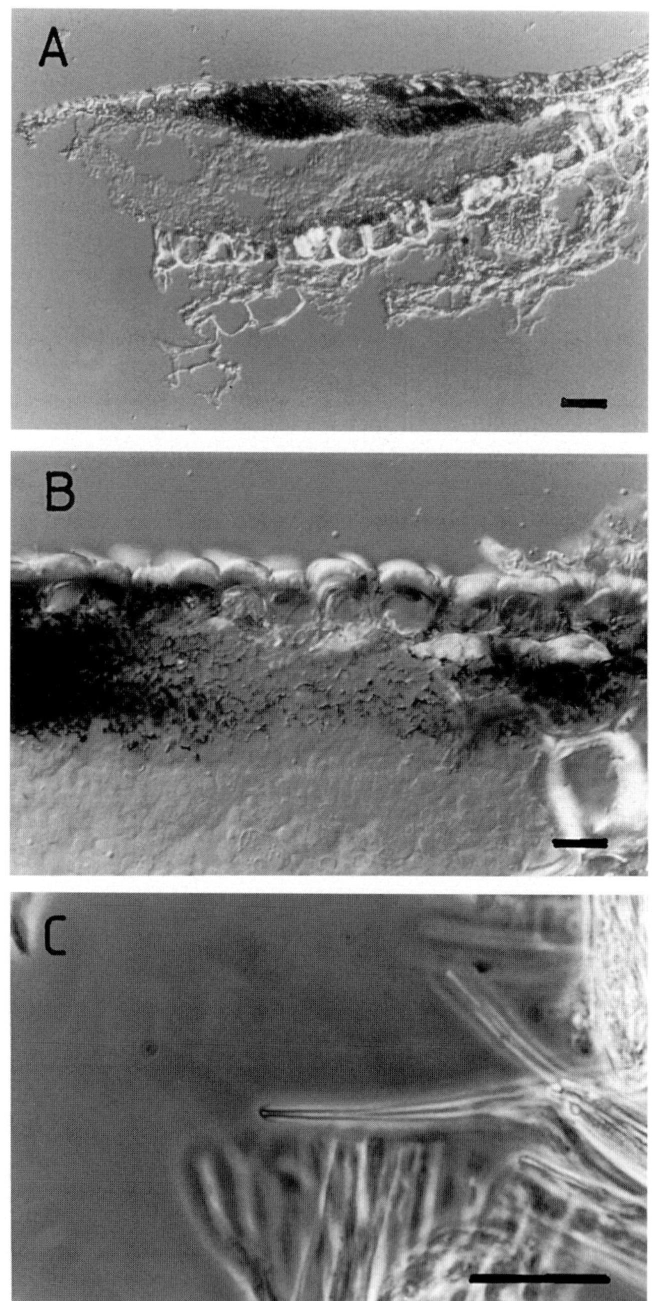

Fig. 22. *Lophodermium agathidis* (**PDD** 46708). **A**, unopened ascoma in vertical section (bar = 50 μm); **B**, detail of paler zone along future line of opening in upper wall of unopened ascoma; **C**, released ascospore in water, showing basal, cup-like gelatinous appendage (bars = 20 μm).

specimen is available; see Notes under *L. raapianum.*

Additional specimens examined: Guadeloupe: Matouba, Victor Hugo Trail, on grass, 9 Jan. 1974, *D.H. Pfister* P1166 (**FH**). **New Zealand:** NORTHLAND: Poor Knights Is: Aorangi, Tatua Peak, on *Xeronema callistemon*, 28 Aug. 1984, *R.E. Beever s. num.* (**PDD** 46704); Poor Knights Is: Pumeto Valley, Western Cliffs, on *Xeronema callistemon*, 2 Sep. 1984, *R.E. Beever s. num.* (**PDD** 46708).

Lophodermium airarum Hilitzer, *Věd. Spisy čsl. Akad. zeměd.* **3**: 149 (1929).

Typification: 'In caulibus, vaginis et foliis Deschampsiae flexuosae haud rar in valle rivuli Kačák prope Březnice' (HILITZER, 1929).

Host: *Deschampsia* (*Poaceae*).

Distribution: Europe.

Notes: HILITZER (1929) regarded *L. airarum* as a recombination of FRIES's '*Hysterium culmigenum* a. *airarum*'. However, despite this name being cited by a number of authors, including HÖHNEL (1906), TEHON (1935) and TERRIER (1942), it was never used by FRIES; 'Airarum' was mentioned in a list of host plants for *H. culmigenum.* The only infraspecific taxon published by FRIES (1823) under this combination was *H. culmigenum* var. *gramineum.*

HILITZER (1929: 149) cited no individual specimen as the type, giving only locality notes under the Latin description. There is no material at **PRM** collected by Hilitzer on *Deschampsia* from localities matching those cited in the protologue. However, HILITZER also listed Sydow, *Mycotheca germanica* no. 1599 under this name. The hymenium of the material examined from this collection was in poor condition, but other features of the ascomata, including the characteristic structure of the lower wall, were typical of *L. culmigenum.*

Specimen examined: Germany: POTSDAM: Brandenburg, Sophienstadt bei Ruhlsdorf, Kreis Nieder-Barnim, on *Deschampsia* (as *Aira*) *flexuosa*, 10 Aug. 1920, *H. Sydow & P. Sydow s. num.* (Sydow, *Mycotheca germanica* no. 1599; **PDD**, as *L. culmigenum*).

Lophodermium alienum P.R. Johnst., **sp. nov.**

Etymology: Referring to the geographical locality of the type specimen, outside the primarily tropical and Southern Hemisphere range for species in *Lophodermium* Group E and the closely related *Coccomyces leptosporus*-like species.

Ascomata oblonga vel sublinearia subepidermalia, cellulae prope aperturam ascocarpi parietem superiorem vestientes longicylindricae. *Apices paraphysium* simplices, implicati. *Asci* clavati vel subclavati, 85–95 × 8–8·5 µm. *Ascosporae* 45–50 × 1 µm.

Infected areas on partially decomposed leaves, somewhat blackened compared with surrounding host tissue, not associated with zone lines, containing scattered ascomata, conidiomata not seen. *Ascomata* oblong-elliptical to sublinear in outline, ends rounded, wall black, somewhat irregular opening slit with no obvious lip cells, hymenium often partly exposed, even when ascomata dry. *Ascomatal insertion* subepidermal. *Covering layer* of ascomata 30–40 µm thick, comprising clypeus of dark-walled hyphae within epidermal cells, and upper wall of ascomatal stroma. *Upper wall* comprising several layers of thick- and dark-walled cells, 4–6 µm diam., lip cells lacking or very poorly developed, few periphysoids near ascomatal opening persistent, becoming elongate.

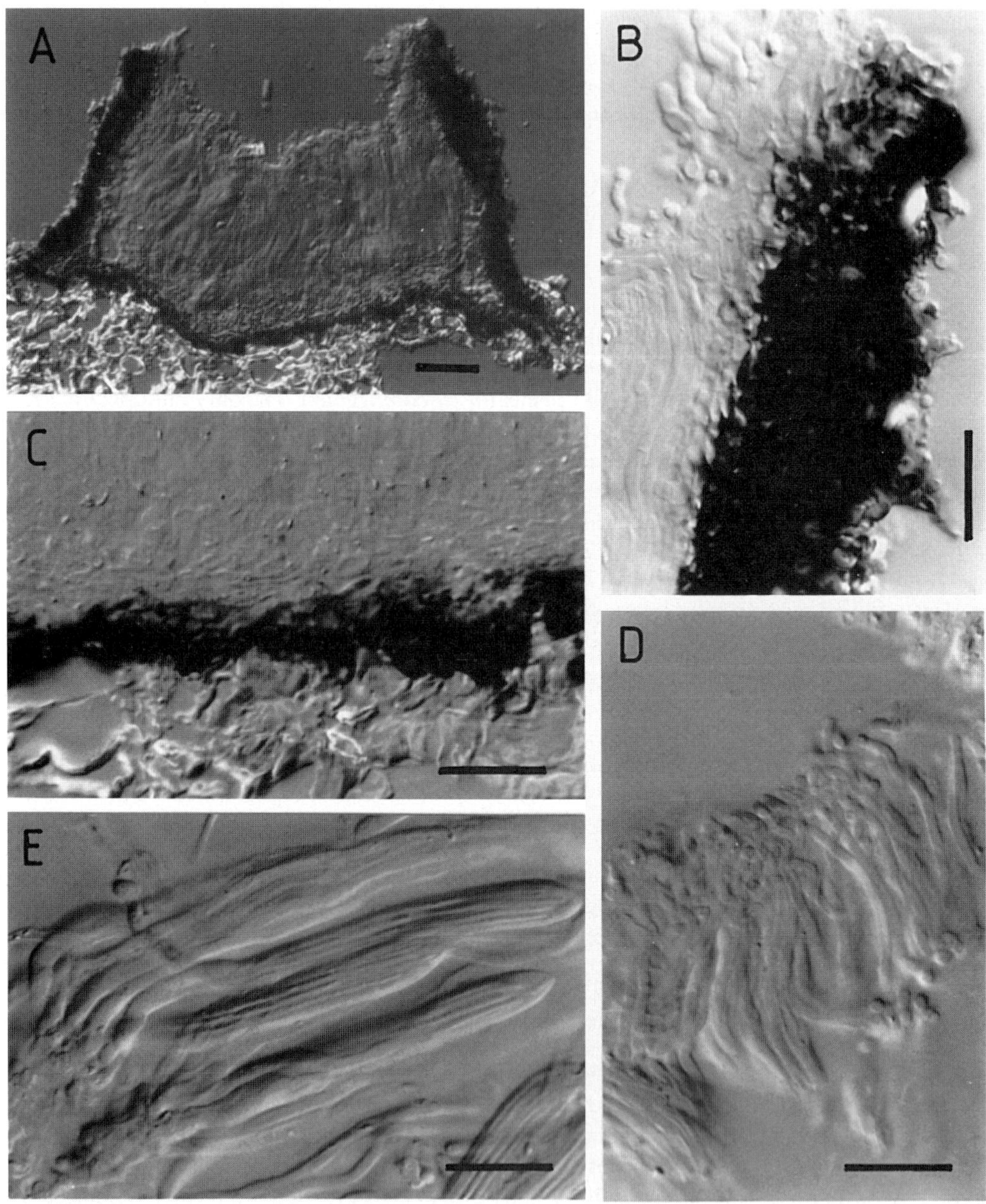

Fig. 23. *Lophodermium alienum* (**CUP** 8625). **A**, ascoma in vertical section (bar = 50 µm); **B**, detail of upper wall of ascoma in vertical section; **C**, detail of lower wall of ascoma in vertical section; **D**, apex of paraphyses; **E**, asci (bars = 20 µm).

Reduced excipulum forming distinct layer around hymenium, comprising closely septate paraphysis-like elements, partially mixed with the persistent periphysoids. *Lower wall* 10–15 µm thick in vertical section, comprising three or four rows of small, angular, thick- and dark-walled cells. *Paraphyses* 1·5–2 µm diam., unbranched but tangled to form dense epithecium extending 15–20 µm beyond asci. *Asci* 85–95 × 8–8·5 µm, clavate to subclavate, tapering to small, rounded apex, 8-spored, short basal stalk with spores confined to upper 65–70 µm. *Ascospores* not seen released, 45–50 × 1 µm, with small, gelatinous apical cap.

Typification: USA: NEW YORK: Newfield, on *Andropogon*, 11 Oct. 1879, *s. coll.* (**CUP** 8625!, *holotypus*).

Host: *Andropogon* (*Poaceae*).

Distribution: North America.

Illustrations: Figs 14D, 23.

Notes: *Lophodermium alienum* is typical of *Lophodermium* Group E (see p. 43). Although collected from New York, in temperate North America, the host is typically tropical to subtropical in distribution. Likewise, the group within *Lophodermium* to which this species belongs is otherwise not known from the Northern Hemisphere. It is probable that the fungus was transported to New York along with its host.

Lophodermium andropogonis Tehon, also described from *Andropogon* and with similar sized asci and ascospores, is a different species, typical of the genus *Terriera*.

Lophodermium alliaceum Feltgen, *Vorst. Pilzfl. Grossh. Luxembourg Nachtr.* **4**: 34 (1905).

Infected areas on dead leaves, slightly paler than surrounding host tissue, not associated with zone lines, containing scattered ascomata, conidiomata not seen. *Ascomata* 0·4–0·6 × 0·2–0·3 mm, elliptical in outline, ends ± acute, wall black, lips macroscopically indistinct. *Ascomatal insertion* not clearly determined because host leaf tissue largely decomposed. *Covering layer* 40 µm thick in vertical section, comprising mostly upper wall of ascomatal stroma, dark-walled hyphae, characteristic of clypeus in subepidermal species, present below the cuticle. *Upper wall* mostly of pale brown, angular to globose cells with slightly and irregularly encrusted walls, but with group of dark cells with very thickly encrusted walls adjacent to poorly-developed lip cells. *Lower wall c.* 10 µm thick in vertical section, apparently comprising two or three rows of large, angular to globose cells, only the lowermost row of cells with darkened walls, in squash mount darkened cells of lower wall angular to short-cylindrical, 4–5 µm diam. *Paraphyses* 1·5–2 µm diam., circinate and coiling at apex. *Asci* 85–95 × 7–8 µm, subclavate to subfusoid, tapering gradually to small, rounded apex with undifferentiated wall, 8-spored, spores extending ± to base of ascus. *Ascospores* not seen released, 50–60 × 1·5 µm, gelatinous caps and sheaths not observed.

Typification: Luxembourg: Grossherzogthums, Höbenhof-Park, on *Allium oleraceum*, 14 May 1903, *J. Feltgen s. num.* (**LUX**!, holotype; **FH**!, isotype).

Illustrations: Figs 24, 26.

Host: *Allium* (*Alliaceae*).

Distribution: Luxembourg.

Notes: No *Lophodermium* ascomata were found on the type material at **LUX**, although a few stromatic structures macroscopically similar to immature ascomata of *Lophodermium* were present. However, drawings attached to the type specimen appear to show mature *Lophodermium* ascomata with lip cells. A few *Lophodermium* ascomata typical in structure of Group A are present on isotype material in **FH**.

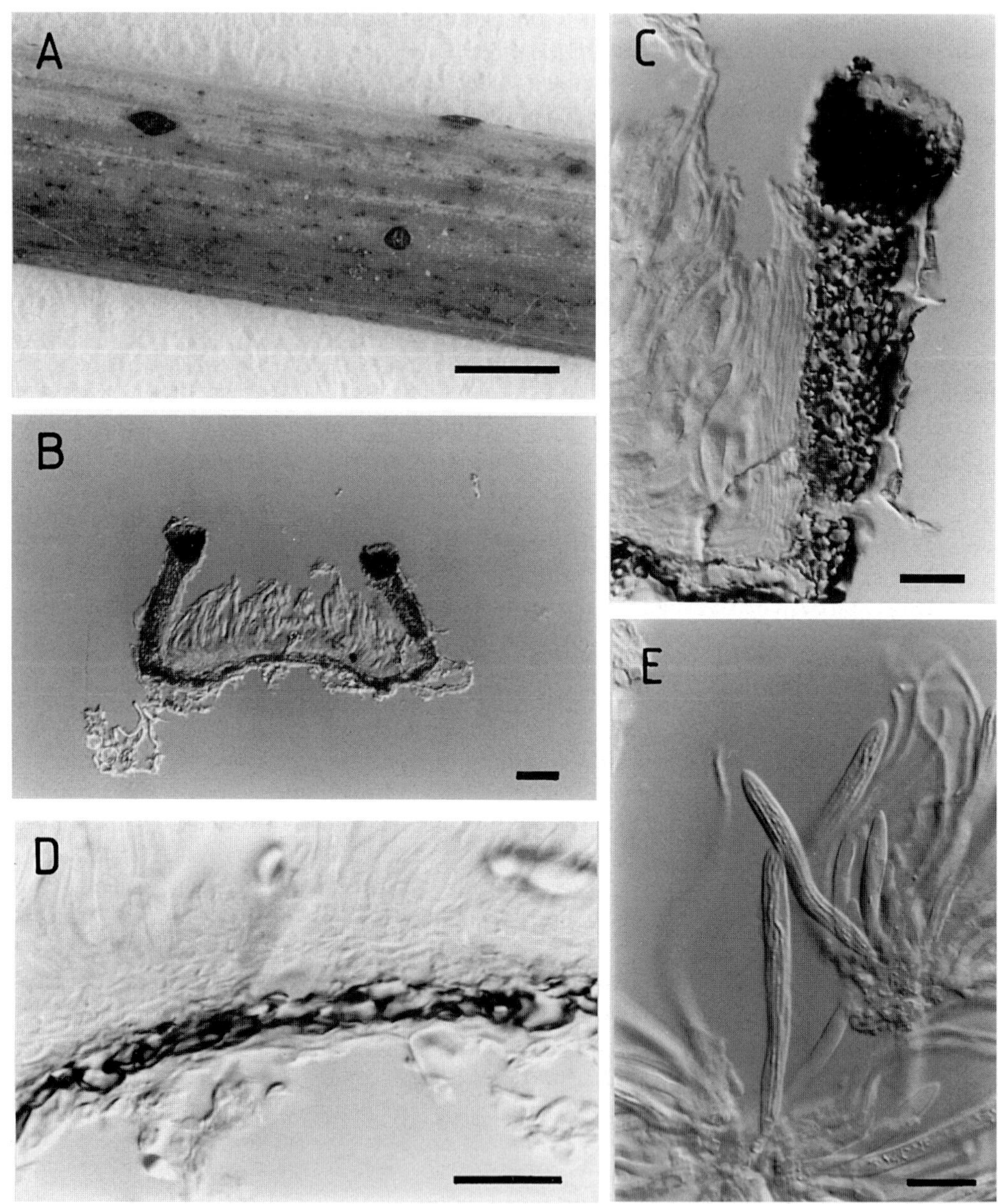

Fig. 24. *Lophodermium alliaceum* (isotype, **FH**). **A**, ascomata (bar = 1 mm); **B**, ascoma in vertical section (bar = 50 µm); **C**, upper wall of ascoma in vertical section; **D**, lower wall of ascoma in vertical section; **E**, asci (bars = 20 µm).

Lophodermium alliaceum has an ascomatal structure typical of *Lophodermium* Group A and within this group is similar to the species in the *actinothyrium*-group. Ascus and ascospore sizes appear similar to those of *L. actinothyrium*, but although the asci appear to be a little narrower than in most collections of *L. actinothyrium*, this feature must be used with care. It was regularly observed in the grass-inhabiting species of *Lophodermium* that a short basal stalk develops before ascospore release, and until this occurs the asci continue to increase in both length and width. Ascospores of the *L. alliaceum* ascomata examined extended to the base of the asci, suggesting that the ascomata were still slightly immature. In addition, hymenial elements from old herbarium specimens often appear a little smaller than those from more recent collections. The darkened cells of the lower wall of *L. alliaceum* are cylindrical rather than more or less globose, as in *L. actinothyrium*, and squash mounts seem to show additional layers of narrow-cylindrical cells across at least part of the base of the wall.

More collections from *Allium* must be examined before the validity of this species can be assessed properly but, for now, it is retained as distinct primarily on the basis of host difference.

Lophodermium alliaceum was placed in synonymy with *L. herbarum* by HÖHNEL (1906) and this was followed by TEHON (1935). However, *L. herbarum* is restricted to *Convallaria* and differs in ascus and ascospore size.

Lophodermium alpigenum Rehm ex E. Müll., *Beitr. Krypt.-Fl. Schweiz* **15**(1): 15 (1977), *nom. inval.*, *ICBN* Arts 36.1, 37.1, 41.3.

Notes: '*Lophodermium alpigenum* Rehm' cited by MÜLLER (1977) is presumably a misprint for *L. alpinum*.

Lophodermium alpinum (Rehm) Weese, *Mitt. bot. Inst. tech. Hochsch. Wien* **10**: 80 (1933).
Lophodermium arundinaceum var. *alpinum* Rehm, *Ber. naturhist. Ver. Augsburg* **26**: 80 (1881).
Lophodermium alpinum Rehm ex Tehon, *Illinois biol. Monogr.* **13**(4): 61 (1935), *nom. inval.*, *ICBN* Art. 34.
Dermascia alpina (Rehm) Tehon, *Illinois biol. Monogr.* **13**(4): 61 (1935), *nom. inval.*, *ICBN* Art. 43.
Ophiobolus festucae Tracy & Earle, in GREENE, *Pl. baker.* **1**: 34 (1901), *fide* FARR *et al.* (1989: 771).
Lophodermium magellanicum (Speg.) J. Walker, *Mycotaxon* **11**: 46 (1980).
Linospora magellanica Speg., *Boln Acad. nac. Cienc. Córdoba* **27**: 378 (1924).

Infected areas on dead leaves, slightly paler than surrounding host tissue, not associated with zone lines, containing scattered ascomata, conidiomata not seen. *Ascomata* 0·4–0·6 (–0·8) × 0·2–0·4 mm, ± ovate, oblong-elliptical or rhomboid in outline, wall grey to black, unopened ascomata often with pale zone along future line of opening (this apparently depending at least partially on host), single longitudinal opening slit with no lip cells along edge, but pale remains of preformed line often visible along sides of slit in recently opened ascomata. *Ascomatal insertion* subepidermal. *Covering layer* up to 60 μm thick in vertical section, comprising dark-walled hyphae forming clypeus within epidermal cells, and upper wall of ascomatal stroma. *Upper wall* 25–40 μm thick, with two layers of ± equal thickness, outer layer of two to four rows of angular

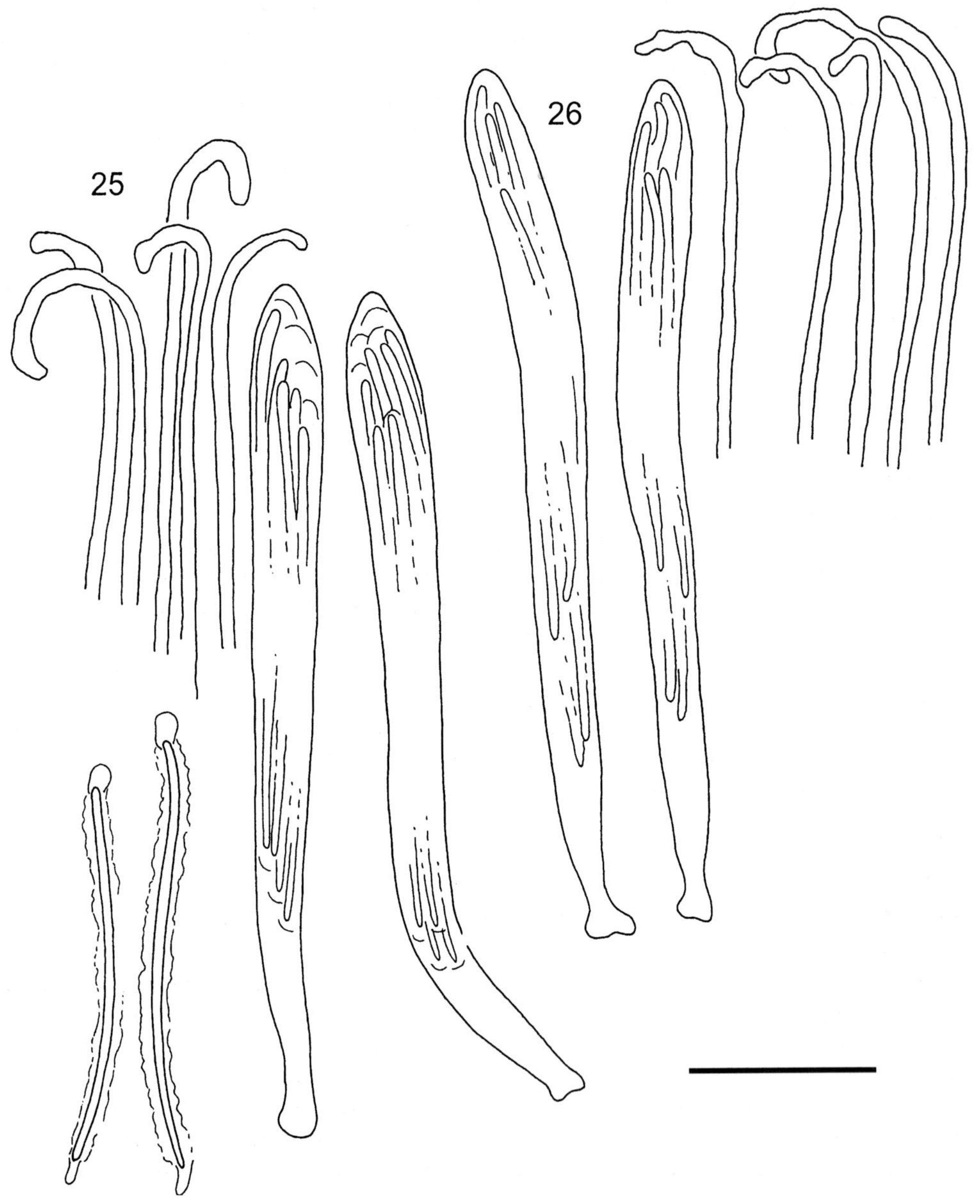

Figs 25–26. 25. *Lophodermium actinothyrium*. Asci and apex of paraphyses (**PRM** 819847) and ascospores released into water (**PRM** 819851). **26.** *Lophodermium alliaceum* (isotype, **FH**). Asci and apex of paraphyses (bar = 20 μm).

to globose cells, 6–8 μm diam., with slightly and irregularly encrusted walls, but with patch of very dark-walled cells adjacent to opening slit, inner layer comprising hyaline, thin-walled, ± monilioid periphysoids, sometimes lost after ascomata open, lip cells absent. *Lower wall* up to 25 μm thick in vertical section, comprising two (or three) rows of large, globose cells, the lowermost row, or its lowermost wall, darkened and slightly thickened, in squash mount lower wall with one or two rows of darkened, angular to globose cells, 10–15 μm diam., across base of ascoma. *Paraphyses* 1·5–2 μm diam., circinate. *Asci* 70–125 (–140) × 15·5–22 μm, broad-fusoid to subsaccate, tapering to subtruncate apex and suddenly to foot-like base, wall undifferentiated at apex, 8-spored, spores spaced widely within ascus, extending to base of ascus. *Ascospores* (30) 40–80 (–90) × 2·5–3·5 (–4) μm, apical gelatinous cap globose, 7–10 μm diam., basal gelatinous cap globose to broad cylindrical, from 2 × 5 μm to 5–8 μm diam., gelatinous sheath 6–15 μm wide.

Typification: Austria: TIROL: Mittagskogel bei Mittelberg, on *Sesleria disticha*, Aug. 1875, *H. Rehm s. num.* (Rehm, *Ascomyceten* no. 319; **K**!, *lectotypus* of *L. arundinaceum* var. *alpinum*, selected here). **Chile:** Isla Capitán Aracena, Sholl Bay, on *Festuca purpurascens*, 13 Jan. 1924, *C. Spegazzini s. num.* (**LPS** 881!, holotype of *Linospora magellanica*).

Illustrations: Figs 2C, D, 5A, 27, 29.

Hosts: *Agrostis, Anthoxanthum, Calamagrostis, Dactylis, Elymus, Koeleria, Nardus, Festuca, Pentoschistis, Phleum, Poa, Sesleria, Trisetum* (*Poaceae*); *Carex* (*Cyperaceae*).

Distribution: North America, Europe, Turkey, India, central Africa, temperate South America.

Notes: *Lophodermium alpinum* is typical of Group A in ascomatal development and structure and within the group is one of several species making up the *alpinum*-group (see p. 33). Among these species *L. alpinum* is distinguished by its very distinctive broad, saccate asci, as well as the shape and size of the ascospores and paraphyses, but species characteristic of this group differ in lacking lip cells.

A wide range of ascus and ascospore sizes has been accepted for this species. However, the range present within individual collections is usually quite restricted, most ascospores, for example, falling within a range of *c.* 15 μm in length. This range may be from 75–90 μm, or from 40–55 μm, depending on the collection being examined. Collections with longer ascospores tend to have broad-cylindrical asci, while in those with shorter ascospores the asci tend to be subsaccate. As the variation in ascus and ascospore size does not appear to correlate with any other feature, it has not been considered significant at the species level.

This species is quite distinctive macroscopically, with smallish ascomata, more or less ovate or rhomboid in outline, walls often grey rather than black and unopened ascomata with a paler zone along the future line of opening. Although lip cells are lacking, the paler zone along the future line of opening is often persistent and, macroscopically, may be mistaken for lip cells. Macroscopically, it is similar to *L. gramineum* and, in those collections with longer ascospores and broad-cylindrical asci, ascus shape can also appear similar to that of *L. gramineum*. However, the two species are easily distinguished by the difference in shape of paraphyses.

Lophodermium alpinum was originally described as having ascomata variable in orientation. This is true of the type specimen, but in most other collections they are regularly oriented with respect to the longitudinal veins on the host leaf. This difference may simply be due to physical characteristics of the different host leaves.

TERRIER (1942) considered that the original description of this species was based on characters combined from two distinct fungi, both contained within Rehm, *Ascomyceten* no. 319. However,

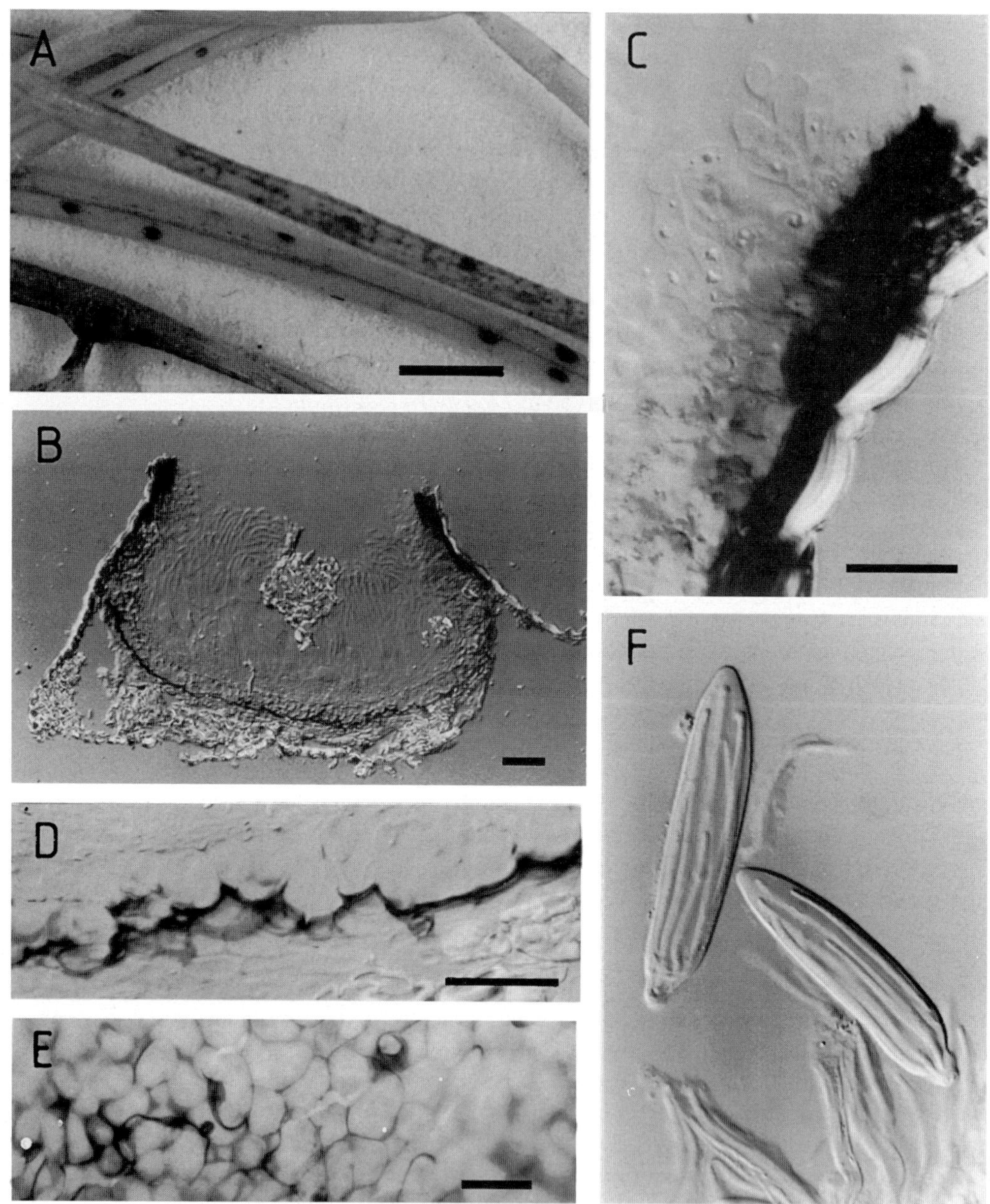

Fig. 27. *Lophodermium alpinum*. **A**, ascomata (Rehm, *Ascomyceten* no. 319, **K**) (bar = 1 mm); **B**, ascoma in vertical section (**PDD** 59526) (bar = 50 μm); **C**, upper wall of ascoma in vertical section (**PDD** 59526); **D**, lower wall of ascoma in vertical section (**PDD** 59526); **E**, lower wall of ascoma in squash mount (**PDD** 59606); **F**, asci (Rehm, *Ascomyceten* no. 319, **K**) (bars = 20 μm).

he cited a Rehm collection from *Nardus*, so was presumably working with Rehm, *Ascomyceten* no. 319b rather than the type specimen no. 319. Collections of Rehm, *Ascomyceten* no. 319b examined (**Switzerland:** Gotthard Pass, on *Nardus stricta*, Sep. 1892, *H. Rehm s. num.* (**K**, **DAOM**)) match *L. nitidum*. They can be distinguished from *L. alpinum* by their darker ascomata, well-developed lips, narrower asci and longer, narrower ascospores, as well as by the appearance of the ascomata in vertical section.

Additional specimens examined: Algeria: Miliane, 1400–1500 m, on *Koeleria splendens*, *H. Trabut s. num.* (**K**, as *L. koeleriae*). **Austria:** SALZBURG: on *Sesleria disticha*, Jul. 1918, *G. Wiessl-Mayendorf s. num.* (Weese, *Eumycetes selecti exsiccati* no. 636, **FH**, **K**); STEIERMARK: Wölzer Tauern, on *Festuca pulchella*, 5 Aug. 1985, *C. Scheuer* 807 (**GZU**); Koralpe, on *Sesleria varia*, 2 Sep. 1991, *A. Nograsek & W. Pongratz s. num.* (**GZU**, as *L. festucae*); Schladminger Tauern, Kleinsölk, on *Sesleria varia*, 5 Jul. 1988, *C. Scheuer* 1852 (**GZU**); TIROL: Ötzaler Alpen, Obergurgl, nr Rotmoostal, on *Carex atrata*, 23 Aug. 1991, *C. Scheuer et al. s. num.* (**GZU**). **Canada:** NEW BRUNSWICK: Kouchibouguac National Park, on *Elymus mollis*, 24 Jul. 1978, *K. Egger & R. A. Shoemaker s. num.* (**CUP** 58902, ex **DAOM** 169470). **England:** ESSEX: *s. loc.*, on *Festuca rubra* subsp. *arenaria*, 13 Jun. 1912, *B. Britton s. num.* (**K**). **Greenland:** Disks I, on *Festuca ovina*, 2 Aug. 1896, *Cornell Peary Expedition* no. 14974 (**CUP**); Jensen's Nunatakker, Amphibolitxyggen, on *Festuca brachyphila*, 16 Aug. 1967, *L. Ryvarden s. num.* (**K**). **India:** PUNJAB: Lahul, Bailing Nulla, 3650 m, on *Trisetum spicatum*, 11 Sep. 1930, *W. Koelz* 1340 (**DAOM** 123911a); HIMACHAL PRADESH: on *Poa*, 14 Sep. 1983, *P. Cannon & D.W. Minter s. num.* (**IMI** 285207 & 285208, as *L. gramineum*). **Kenya:** Mt Elgon, on *Pentoschistis brassica*, 7 Mar. 1954, *A. Bogden* 3939 (**K**); Mt Kenya, on *Agrostis volkensis*, 7 Aug. 1948, *O. Hedberg* 1849 (**K**). **Russia:** SOVIET FAR EAST: Magadan obl.: Bilibinsky distr.: River Hisiý, on *Calamagrostis purpurascens*, 11 Aug. 1981, *T. Plieva & V. Petrovsky s. num.* (**K**). **Scotland:** E INVERNESSHIRE: Dalwhinnie, Allt Coire Chuirn, on *Festuca vivipora*, 24 Jul. 1948, *N.Y. Sandwith s. num.* (**K**). **South Georgia:** *s. loc.*, on *Festuca erectus*, 18 Jan. 1883, *Will s. num.* (**K**). **Sweden:** UPPLAND: Västerboten: Lövånger Parish: Avan, seashore, on *Elymus arenarius*, 29 Jul. 1946, *L. Holm* 330 (*Fungi suecici praesertim upsaliensis* no. 3487, **K**). **Switzerland:** GRAUBÜNDEN: Sufers, Bächliställe, on *Sesleria coerulea*, 20 Aug. 1981, *L. Petrini* 148 (**PDD** 59528, **ZT**); Sufers, on *Sesleria coerulea*, 20 Aug. 1981, *L. Petrini* 147 (**PDD** 59530, **ZT**); Albula Pass, on *Sesleria coerulea*, 14 Jul. 1980, *L. Petrini* 127 (**PDD** 59521, **ZT**); Davos Dorf, on *Poa alpina*, 11 Jun. 1980, *L. Petrini* 43 (**PDD** 59623, **ZT**); Ofenpass, Murtarol, on *Nardus stricta*, 4 Oct. 1986, *L. Petrini* 132.1 (**PDD** 59537, **ZT**); Ofenpass, on *Sesleria coerulea*, 3 Sep. 1984, *L. & O. Petrini* 130 (**PDD** 59525, **ZT**); Ofenpass, Murtarol, on *Sesleria coerulea*, 3 Sep. 1984, *L. & O. Petrini* 131 (**PDD** 59534, **ZT**); Unterengadin, Prui-Laret, on *Carex*, 4 Sep. 1986, *O. Petrini* 136 (**PDD** 59580, **ZT**); Prui-Laret, on *Phleum hirsutum*, 27 Aug. 1984, *O. Petrini* 137 (**PDD** 59594, **ZT**); *idem loc.*, on *Nardus stricta* 135.1, 27 Aug. 1984, *O. Petrini s. num.* (**PDD** 59538, **ZT**); Unterengadin, Val Tavru, on *Sesleria coerulea*, 4 Sep. 1984, *L. & O. Petrini* 145 (**PDD** 59533, **ZT**); Oberengadin, Zuoz, Crasta, on *Festuca*, 15 Jul. 1980, *L. Petrini* 112 (**PDD** 59617, **ZT**); Zuoz, San Batrumieu, on *Festuca*, 15 Jul. 1980, *L. Petrini* 102 (**PDD** 59609, **ZT**); *idem loc.*, on *Festuca*, 15 Jul. 1980, *L. Petrini* 103.1 (**PDD** 59608, **ZT**); *idem loc.*, on *Festuca rubra*, 15 Jul. 1980, *L. Petrini* 107 (**PDD** 59607, **ZT**); *idem loc.*, on *Festuca rubra*, 15 Jul. 1980, *L. Petrini* 101.1 (**PDD** 59610, **ZT**); *idem loc.*, on *Poa*, 15 Jul. 1980, *L. Petrini* 105.2 (**PDD** 59622, **ZT**); Bergun, Val Tuors, on *Nardus stricta*, 14 Jul. 1980, *L. Petrini* 52.2 (**PDD** 59536, **ZT**); *idem loc.*, on *Sesleria coerulea*, 29 Aug. 1980, *O. Petrini* 115 (**PDD** 59599, **ZT**); Bergun, Murtigal dal Crap Alv., on *Sesleria disticha*, 30 Aug. 1984, *O. & L. Petrini* 123 (**PDD** 59598, **ZT**); TICINO: Alpe di Piora, on *Dactylis glomerata*, 26 Jul. 1980, *L. Petrini* 212b (**PDD** 59522, **ZT**); *idem loc.*, on *Festuca rubra*, 25 Jul. 1980, *L. Petrini* 210 (**PDD** 59602, **ZT**); *idem loc.*, on *Nardus stricta*, 23 Jul. 1980, *L. Petrini* 54 (**PDD** 59545, **ZT**); *idem loc.*, on *Sesleria coerulea*, 23 Jul. 1980, *L. Petrini* 208 (**PDD** 59531, **ZT**); *idem loc.*, on *Sesleria coerulea*, 23 Jul. 1980, *L. Petrini* 199 (**PDD** 59535, **ZT**); Alpe di Piora, Cadagno, on *Anthoxanthum odoratum*, 27 Jul. 1980, *L. Petrini s. num.* (**PDD** 60252, **ZT**); *idem loc.*, on *Festuca ovina*, 27 Jul. 1980, *L. Petrini* 195 (**PDD** 59603, **ZT**); *idem loc.*, on *Festuca varia*, 27 Jul. 1980, *L. Petrini* 196 (**PDD** 59601, **ZT**); *idem loc.*, on *Sesleria coerulea*, 23 Jul. 1980, *L. Petrini* 202 (**PDD** 59523, **ZT**); Alpe di Piora, Lago Ritorn, on *Festuca rubra*, 23 Jul. 1980, *L. Petrini* 191 (**PDD** 59611, **ZT**); *idem loc.*, on *Sesleria coerulea*, 23 Jul. 1980, *L. Petrini* 206 (**PDD** 59532, **ZT**); Alpe di Piora, Scopella, on *Nardus stricta*, 23 Jul. 1980, *L. Petrini* 205 (**PDD** 59544, **ZT**); Alpe di Ruvina, on *Nardus stricta*, 22 Jul. 1980, *L. Petrini* 167.1 (**PDD** 59541, **ZT**); Leventina, Alpe Campolungo, Fontana, on *Festuca*, 24 Jul. 1980, *L. Petrini* 164 (**PDD** 59616, **ZT**); Alpe Campolungo, Venett, on *Sesleria coerulea*, 24 Jul. 1980, *L. Petrini* 165 (**PDD** 59524, **ZT**); Alpe Campolungo, on *Nardus stricta*, 24 Jul. 1980, *L. Petrini* 163 (**PDD** 59540, **ZT**); Airolo, Nante, on *Festuca*, 22 Jul. 1980, *L. Petrini* 179.1 (**PDD** 59615, **ZT**); *idem loc.*, on *Festuca*, 22 Jul. 1980, *L. Petrini* 180 (**PDD** 59614, **ZT**); Airolo, Sasso della Boggia, on *Nardus stricta*, 22 Jul. 1980, *L. Petrini* 168 (**PDD** 59542, **ZT**); *idem loc.*, on *Sesleria coerulea*, 22 Jul. 1980, *L. Petrini* 169 (**PDD** 59529, **ZT**); San Gottardo, on *Festuca*, 21 Jul. 1980, *L. Petrini* 186 (**PDD** 59605, **ZT**); *idem loc.*, on *Festuca*, 21 Jul. 1980, *L. Petrini* 188.1 (**PDD** 59612, **ZT**); *idem loc.*, on *Festuca rubra*, 21 Jul. 1980, *L. Petrini* 189 (**PDD** 59604, **ZT**); San Gottardo, Motta Bartola, on *Festuca*, 21 Jul. 1980, *L. Petrini* 184 (**PDD** 59606, **ZT**); *idem loc.*, on *Festuca varia*, 21 Jul. 1980, *L. Petrini* 185 (**PDD** 59613, **ZT**); Val Bedretto, Bolle di Paltano, on *Nardus stricta*, 21 Jul. 1980, *L. Petrini* 161.1 (**PDD** 59539, **ZT**); WALLIS: Lotschental, Falferalp, on *Festuca violacea*,

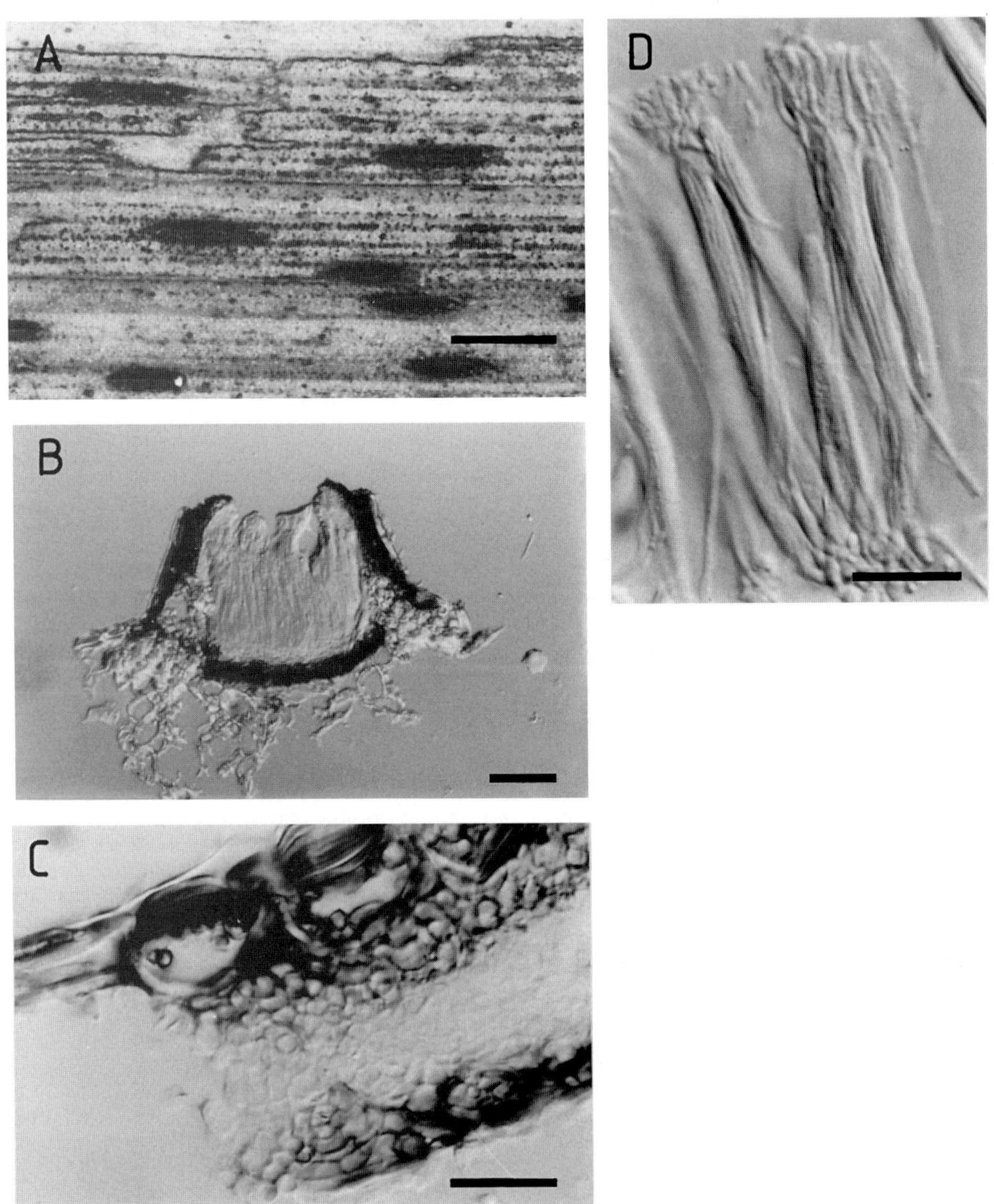

Fig. 28. *Lophodermium andropogonis* (**ILL** 7571). **A**, ascomata (bar = 1 mm); **B**, ascoma in vertical section (bar = 50 μm); **C**, margin of very immature ascoma in vertical section, showing palisadic structure of primordium; **D**, paraphyses and asci (bars = 20 μm).

26 Jul. 1979, *L. & O. Petrini* 2 (**PDD** 59600, **ZT**); Falferalp, on *Poa alpina*, 8 Jul. 1979, *L. & O. Petrini* 9 (**PDD** 59621, **ZT**); Lotschental, Gletschertor, on *Poa alpina*, 8 Jul. 1979, *L. & O. Petrini* 6 (**PDD** 59624, **ZT**); St Gallen, Speer, on *Sesleria coerulea*, Aug. 1981, *L. Petrini* 215 (**PDD** 59526, **ZT**). **Tanzania:** *s. loc.*, on *Pentoschistis minor*, *E. Stuart Watt s. num.* (**K**). **Turkey:** Kap, Sendor Bulak, Gralik, Büjük, Agri Dagi, on *Nardus stricta*, 20 Jul. 1966, *Wavis s. num.* (**K**); Kastramouni Prov.: Igaz Dağ, on *Sesleria argentea*, 28 Aug. 1967, *Davis* 38385 (**K**). **Uganda:** Mt Elgon, on *Pentoschistis minor*, 5 Dec. 1967, *O. Hedberg* 4487 (**K**). **USA:** ALASKA: Anchorage, Portage Glacier, on *Elymus mollis*, 16 Aug. 1956, *J.G. Dickson s. num.* (**DAR** 15271 ex Cryptogamic Herbarium, University of Wisconsin); UTAH: Weber Co.: Wasatch Mts, Taylors Creek, on *Poa glaucifolia*, 19 Apr. 1981, *C.T. Rogerson s. num.* (**NY**).

Lophodermium ambiguum Speg., *Boln Acad. nac. Cienc. Córdoba* **27**: 380 (1924).

Typification: 'Sobre los culmos y hojas muertas de *Poa fuegiana*, en el playa de Scholl bay' (SPEGAZZINI, 1924).

Host: *Poa fuegiana* (*Poaceae*).

Distribution: Chile (Cape Horn).

Notes: SPEGAZZINI (1924) labelled this species '*Lophodermium ?ambigua*'. The type material, not in **LPS**, is presumably lost.

Lophodermium ampelodesmi Ces. ex Tehon, *Illinois biol. Monogr.* **13**(4): 48 (1935), *nom. inval.*, *ICBN* Art. 34.1.

Notes: TEHON (1935: 48) cited '*Lophodermium Ampelodesmi* de Cesati, in literature' as a synonym of *Lophodermium eximium* Ces. TEHON (1935: 116) stated 'The *Lophodermium* on *Ampelodesmus tenax*, to which this name is applied in literature, is properly called *Lophodermium eximium* de Cesati'. However, no reference to *L. ampelodesmi* has been found before TEHON (1935).

Lophodermium andropogonis Tehon, *Illinois biol. Monogr.* **13**(4): 41 (1935), *nom. inval.*, *ICBN* Art. 36.1.

Infected areas on dead leaves, slightly darker than surrounding host tissue, associated with faint, incomplete, brown zone lines, containing scattered ascomata, conidiomata not seen. *Ascomata* 0·5–1·5 × 0·2 mm, narrow-cylindrical to sublinear in outline, ends rounded, wall black, with broad, pale zone along future line of opening, single longitudinal opening slit with flat, shelf-like area along either side. *Ascomatal insertion* subepidermal, with clypeus of dark-walled hyphae within epidermis. *Ascomatal structure* characteristic of *Terriera* (see p. 37), but palisadic stroma becoming disorganized at early stage in ascomatal development, arrangement of this layer ± lost in opened ascomata, but characteristic very dark, narrow extension of upper wall present across top of hymenium. *Paraphyses* 1–1·5 µm diam., becoming branched several times and swollen to 2·5–3·5 µm near apex, embedded in thick, slightly brownish gel, forming compact epithecium 20 µm above asci. *Asci* 80–95 × 6·5–7·5 µm, cylindrical with short basal stalk, apex ± broadly rounded, wall slightly thickened at apex with tiny, broad central pore, 8-spored. *Ascospores* not seen released from type material, 50–60 × 1–1·5 µm, tapering slightly to base, non-septate, with no gelatinous sheath (released ascospores seen from *Korf*, 22 Jan. 1980, gently sigmoid, tapering slightly to both ends, 60–70 × 1·5 µm, no sheath).

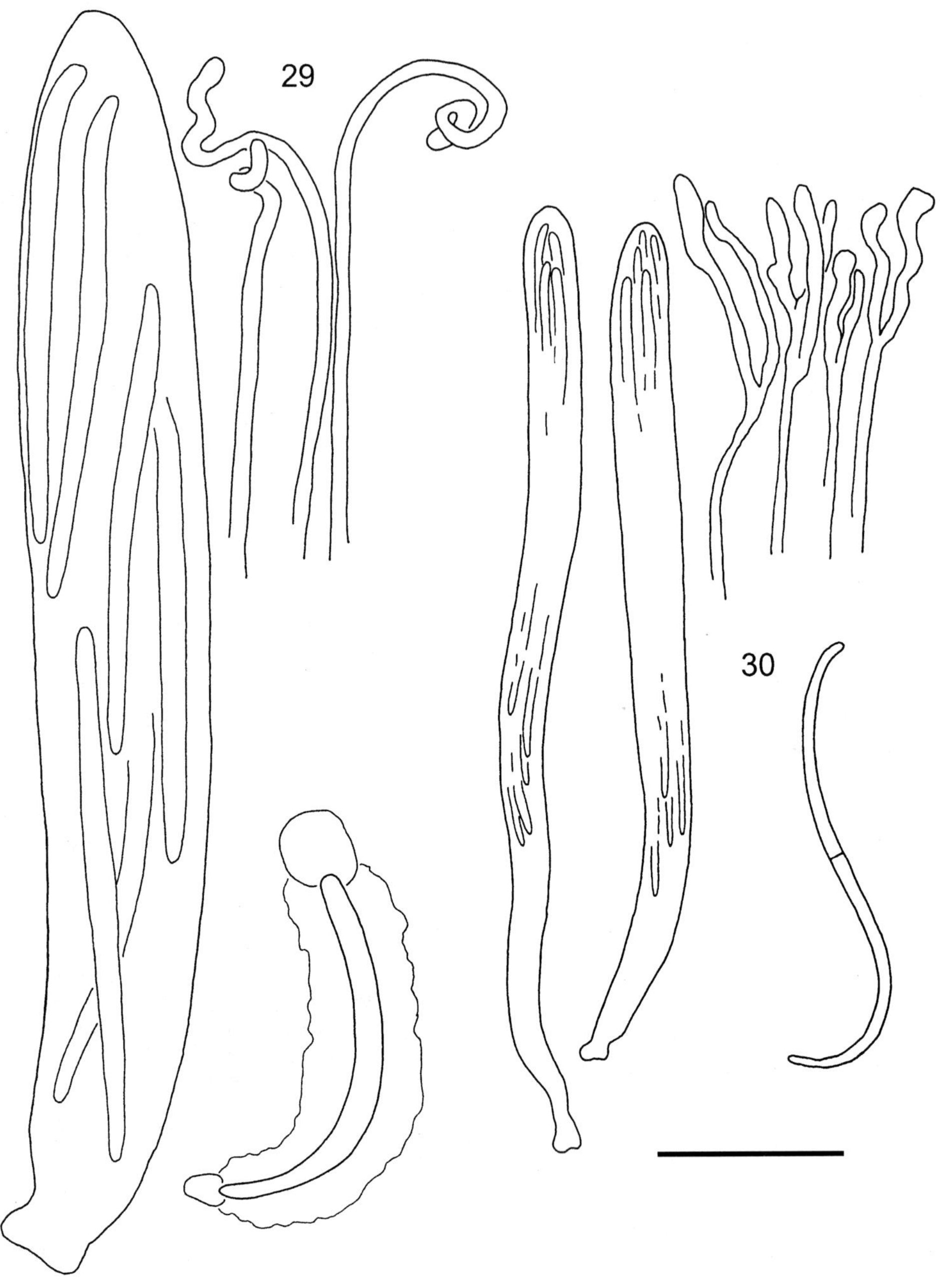

Figs 29–30. 29. *Lophodermium alpinum* (*Scheuer*, 5 Jul. 1988, **GZU**). Asci, apex of paraphyses and ascospore released in water. **30.** *Lophodermium andropogonis.* Asci and apex of paraphyses (**ILL** 7571) and released ascospore (*Korf et al.*, 22 Jan. 1980, **NY**) (bar = 20 μm).

Typification: Costa Rica: Peralta, on *Andropogon bicornis*, 13 Jul. 1923, *F.L. Stevens* 463 (**ILL** 7571!, **FH**!, collections on which the invalid *L. andropogonis* was based).

Host: *Andropogon* (*Poaceae*).

Distribution: Costa Rica.

Illustrations: Figs 28, 30.

Notes: *Lophodermium andropogonis* is macroscopically and microscopically typical of the genus *Terriera*, although in the *Stevens* 463 collection the characteristic arrangement of the cells in the ascomatal stroma can be seen only at early stages of development.

Because *L. andropogonis* was invalidly published the name needs validating in *Terriera.* Although two other collections were examined which appear to match *Stevens* 463 (**French Guiana:** Saul, toward Mt Galbao, on panicoid grass, 24–28 Jan. 1986, *G.J. Samuels* 3340 (**NY**). **Bermuda:** HAMILTON PARISH: Walsingham Nature Trust, on wild ginger, 22 Jan. 1980, *R.P. Korf et al. s. num.* (**NY**)), the taxon is not validated here because of doubt over appropriate species limits in *Terriera* (see p. 36). *Lophodermium andropogonis* is very similar to *T. javanica* var. *pandani* (Penz. & Sacc.) P.R. Johnst., with asci, ascospores and paraphyses comparable in size and shape, although the latter taxon lacks the zone lines of *L. andropogonis.*

Duplicaria antarctica (Speg.) P.R. Johnst., **comb. nov.**

Lophodermium antarcticum Speg., *Boln Acad. nac. Cienc. Córdoba* **11**: 249 (1887).

Hypoderma antarcticum (Speg.) Kuntze, *Revis. gen. pl.* **3**(3): 487 (1898).

Lophodermina antarctica (Speg.) Tehon, *Illinois biol. Monogr.* **13**(4): 87 (1935).

Infected areas on dead leaves, slightly paler than surrounding host tissue, sometimes associated with poorly-differentiated zone lines, containing groups of ascomata and structures resembling conidiomata of *Rhytismataceae*. *Ascomata* 0·8–1·2 × 0·5–0·7 mm, broadly elliptical to ± ovate in outline, wall uniformly shiny black, single longitudinal opening slit, with no differentiated cells associated. *Ascomatal insertion* subcuticular. *Covering layer* up to 45 μm thick in vertical section, comprising host cuticle and upper wall of ascomatal stroma. *Upper wall* in unopened ascomata (at stage where paraphyses starting to elongate but asci not yet visible) up to 45 μm thick in vertical section, comprising mostly cells with dark, thick walls but inner half of wall in central part of ascoma with hyaline, thin-walled cells forming loose tissue. *Lower wall* 10 μm thick, comprising two or three layers of pale brown, thin-walled, cylindrical cells. *Paraphyses* 1–1·5 μm diam., undifferentiated at apex. *Asci* 100–125 × 18–22 μm, saccate, tapering to broad-truncate apex, some with tiny apical pore, 8-spored, spores extending to base, ascus development sequential. *Ascospores* 45–70 × 3·5–4 μm, bifusiform, constricted to 1–1·5 μm at centre, tapering to base, surrounded by thick gelatinous sheath. *Conidiomata*-like structures 0·2 mm diam., round in outline, pustulate, walls pale to dark brown. *Conidiogenous cells* and *conidia* not seen.

Typification: Argentina: TIERRA DEL FUEGO: Staten I, on *Rostkovia grandiflora*, Mar. 1882, *C. Spegazzini s. num.* (**LPS** 1010!, holotype of *L. antarcticum*).

Host: *Rostkovia grandiflora* (*Juncaceae*).

Distribution: Tierra del Fuego.

Illustrations: Figs 31, 34.

Notes: The type collection is in poor condition; although the macroscopic appearance of the fungus present can be seen and some intact unopened ascomata are present, no hymenial or conidiogenous elements remain. A specimen collected by Dusén (**BPI**) is from the same host and

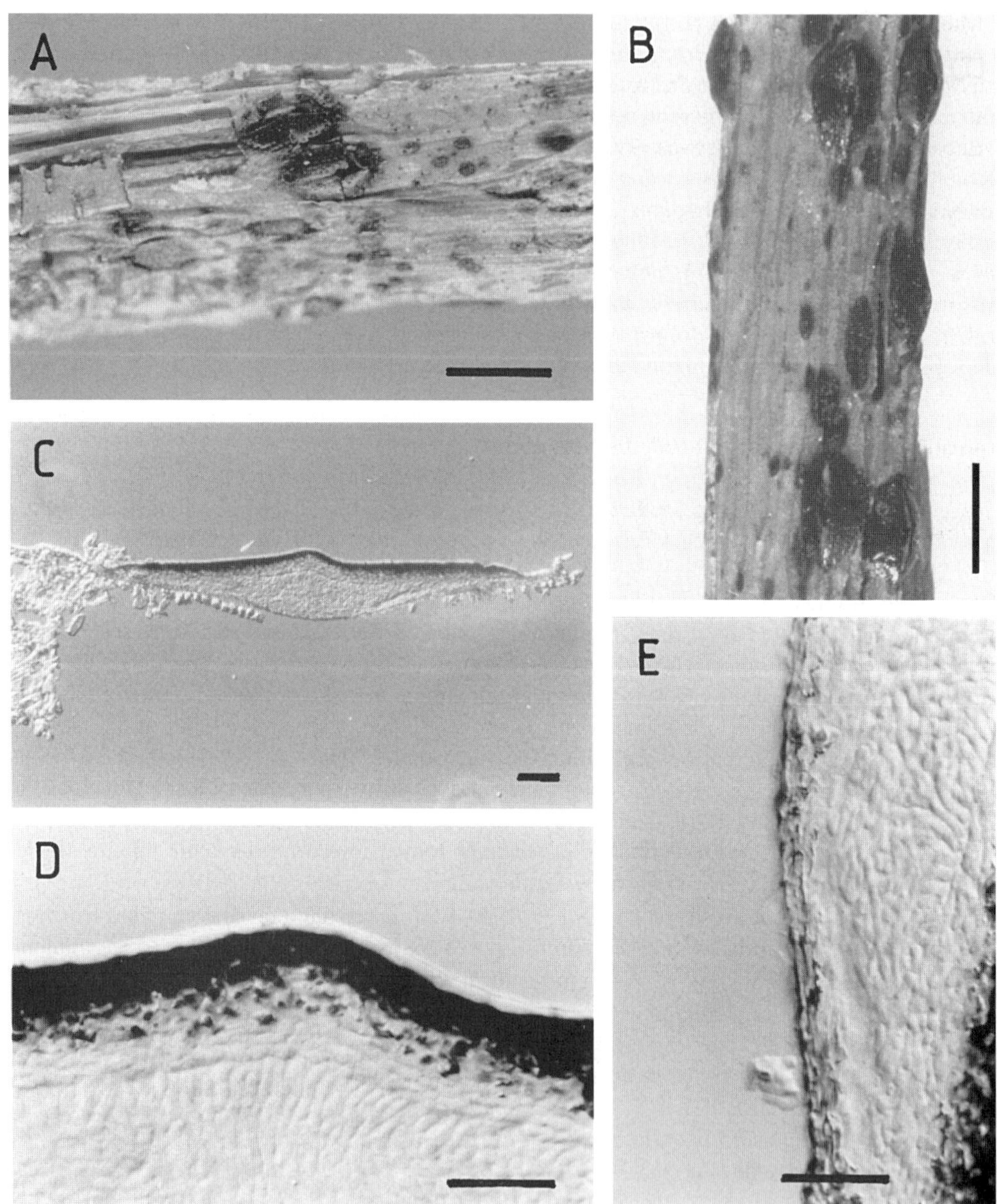

Fig. 31. *Duplicaria antarctica* (**A**, **C–E**, holotype, **LPS**; **B**, *Dusén s. num.*, **BPI**). **A**, ascomata; **B**, ascomata (bars = 1 mm); **C**, unopened ascoma in vertical section (bar = 50 µm); **D**, upper wall of unopened ascoma in vertical section; **E**, lower wall of ascoma in vertical section (bars = 20 µm).

the same general locality. Two distinct species of *Rhytismataceae* are present on this collection: one is macroscopically indistinguishable from the type specimen of *D. antarctica*; the second matches *Terriera fuegiana*. The former is distinguished by its larger, more ovate ascomata and by being associated with numerous conidiomata.

Although SPEGAZZINI's (1887) description of *D. antarctica* cited ascospores as filiform and 1 µm wide, with asci as 66–80 × 8–10 µm, his description otherwise matches observations made here with respect to macroscopic appearance of the ascomata and shape of the asci. Also, he clearly illustrated broad asci with subtruncate apexes in drawings on the herbarium packet of the type. For several species, the ascus sizes cited by SPEGAZZINI are smaller than observed in this study (e.g. *T. fuegiana*). As most of the ascomata remaining on the type collection are immature, SPEGAZZINI could easily have missed the distinctive ascospore shape of this species.

Duplicaria antarctica differs from *D. acuminata*, known from the Northern Hemisphere on species of *Juncus* and *Carex* (POWELL, 1973), in having shorter asci and longer ascospores and no doubt represents a closely related species. Both *D. antarctica* and *D. acuminata* are very similar in ascomatal structure to other *Juncaceae*-inhabiting species, discussed as *Lophodermium* Group D (see p. 42).

Additional specimen examined: Argentina: TIERRA DEL FUEGO: Isla Desolarion, Puerto Angusto, on *Rostkovia* (as *Marsippospermum*), 1896, *P. Dusén s. num.* (**BPI**, *p.p.*).

Lophodermium apiculatum (Wormsk. ex Fr.) De Not., *G. Bot. ital.* **2**(2): 48 (1847).
Hysterium apiculatum Wormsk. ex Fr., *Syst. mycol.* **2**: 593 (1823).
Lophodermium arundinaceum var. *apiculatum* (Wormsk. ex Fr.) Duby, *Mém. Soc. Phys. Hist. nat. Genève* **16**: 59 (1862).
Hypoderma apiculatum (Wormsk. ex Fr.) Kuntze, *Revis. gen. pl.* **3**(3): 487 (1898).
Lophodermina apiculata (Wormsk. ex Fr.) Tehon, *Illinois biol. Monogr.* **13**(4): 88 (1935).

Typification: 'Wormsk. Mscr. in foliis graminis sub Pinubus pygmaeis in Kamtschatka habitantium' (FRIES, 1823).

Host: *Poaceae*.

Distribution: Soviet Far East (Kamchatka).

Notes: There is no material in **UPS** matching the protologue of *L. apiculatum* as cited by FRIES. Collections labelled as *L. apiculatum* or *L. arundinaceum* var. *apiculatum* in various exsiccati represent several different species, for example, *L. gramineum* (Rehm, *Ascomyceten* no. 775; **K**!; *Fungi suecici praesertim upsaliensis* no. 3485; **K**!), *L. robergei* (Petrak, *Flora Bohemiae et Moraviae exsiccata* no. 217; **K**!; **FH**!), *L. actinothyrium* (Thümen, *Mycotheca universalis* no. 417; **K**!) and *L. sesleriae* Hilitzer (Libert, *Plantae cryptogamicae quas in Arduenna collegit* no. 73; **K**!). This supports the finding reported by TERRIER (1942) that collections with apiculate ascomata differed in ascomatal structure, some matching his concept of *L. arundinaceum*, others matching *L. alpinum* in structure of the lower wall of the ascomata.

TERRIER considered that it was not possible to define species of *Lophodermium* from grasses on their macroscopic appearance, with ascomatal shape reflecting the nature of the host substratum rather than indicating the relationship of the fungus. This is true to some extent, and uncritical use of macroscopic appearance as a taxonomic character has probably been the reason for the wide range of species misidentified as *L. apiculatum* by various authors. However, although it is not possible to identify definitively any species from macroscopic appearance alone, results here show that ascomatal shape is often a useful guide to species identification among the monocotyledon-inhabiting species of *Lophodermium s. str.* (compare, for example, *L. alpinum*, *L. culmigenum* and *L. nitidum*).

FRIES (1823) did not record the host genus in his original description. HILITZER (1929), MÜLLER (1977) and FARR *et al.* (1989) restricted the application of the name *L. apiculatum* to those collections from *Calamagrostis*; TEHON (1935), however, considered it to occur on *Calamagrostis* and *Molinia*. TERRIER (1942) noted that it was not possible to define grass-inhabiting species of *Lophodermium* on the basis of host. It is certainly true in this instance, three of the species from the exsiccati noted above being known from *Calamagrostis* (*L. actinothyrium*, *L. gramineum and L. sesleriae*) and two from *Molinia* (*L. actinothyrium* and *L. sieglingiae*).

Since no type material is available for *L. apiculatum*, and because it has been used by various authors for groups defined solely by host substratum or macroscopic appearance of the ascomata, TERRIER (1942), who considered it best to avoid use of this name, is followed here.

Terriera arundinacea (Penz. & Sacc.) P.R. Johnst., **comb. nov.**
Clithris arundinacea Penz. & Sacc., *Malpighia* **15**: 222 (1902).

Infected areas on partially decomposed leaves, not obviously associated with bleached areas or zone lines, containing groups of ascomata, conidiomata not seen. *Ascomata* 0·7–2 × 0·4 mm, in type specimen oriented conspicuously perpendicular to direction of main leaf veins, oblong to sublinear (rarely branched) in outline, unopened ascomata with wall very pale grey to ± concolorous with surrounding host tissue, with faint but broad paler zone along future line of opening, open ascomata with wall concolorous with surrounding host tissue or pale grey, edge of longitudinal opening slit marked with narrow black line. *Ascomatal insertion* subepidermal. *Ascomatal structure* typical of *Terriera* (see p. 37). *Paraphyses* 1·5–2 µm diam., irregularly swollen and tangled at apex, unbranched, forming distinct epithecium above asci. *Asci* 130–160 × 8–9 µm, ± cylindrical, tapering slightly to ± broadly rounded apex with undifferentiated wall, 8-spored, basal stalk slightly narrower than upper portion with spores confined to upper 100 µm of ascus. *Ascospores* not seen released, 90–100 × 2–2·5 µm, tapering slightly to base, with no gelatinous sheath.

Typification: Indonesia: JAVA: Bogor Botanical Gardens, on *Bambusa*, 21 Dec. 1896, *O. Penzig* (**PAD**!, lectotype of *C. arundinacea*; **W**!, isolectotype).

Host: *Bambusa* (*Bambusaceae*).

Distribution: Java.

Illustrations: Fig. 32.

Notes: *Terriera arundinacea* is distinctive within the genus because of its large asci and ascospores. The type specimen is very similar to the invalid *Lophodermium miscanthi* in ascus and paraphysis shape, and ascus and ascospore size. However, the black shelf-like ridge along the ascomatal opening typical of *Terriera* is poorly developed in the *L. miscanthi* specimen, possibly owing to the very robust nature of the host tissue covering the ascoma.

HÖHNEL (1917*a*) compared *T. arundinacea* to *Lophodermium*, but did not make the required combination.

Additional specimens examined: Indonesia: JAVA: Buitenzorg, on *Bambusa*, 1907–1908, *von Höhnel* (**FH**, von Höhnel collection sheet 7843, as *Lophodermium arundinaceum*).

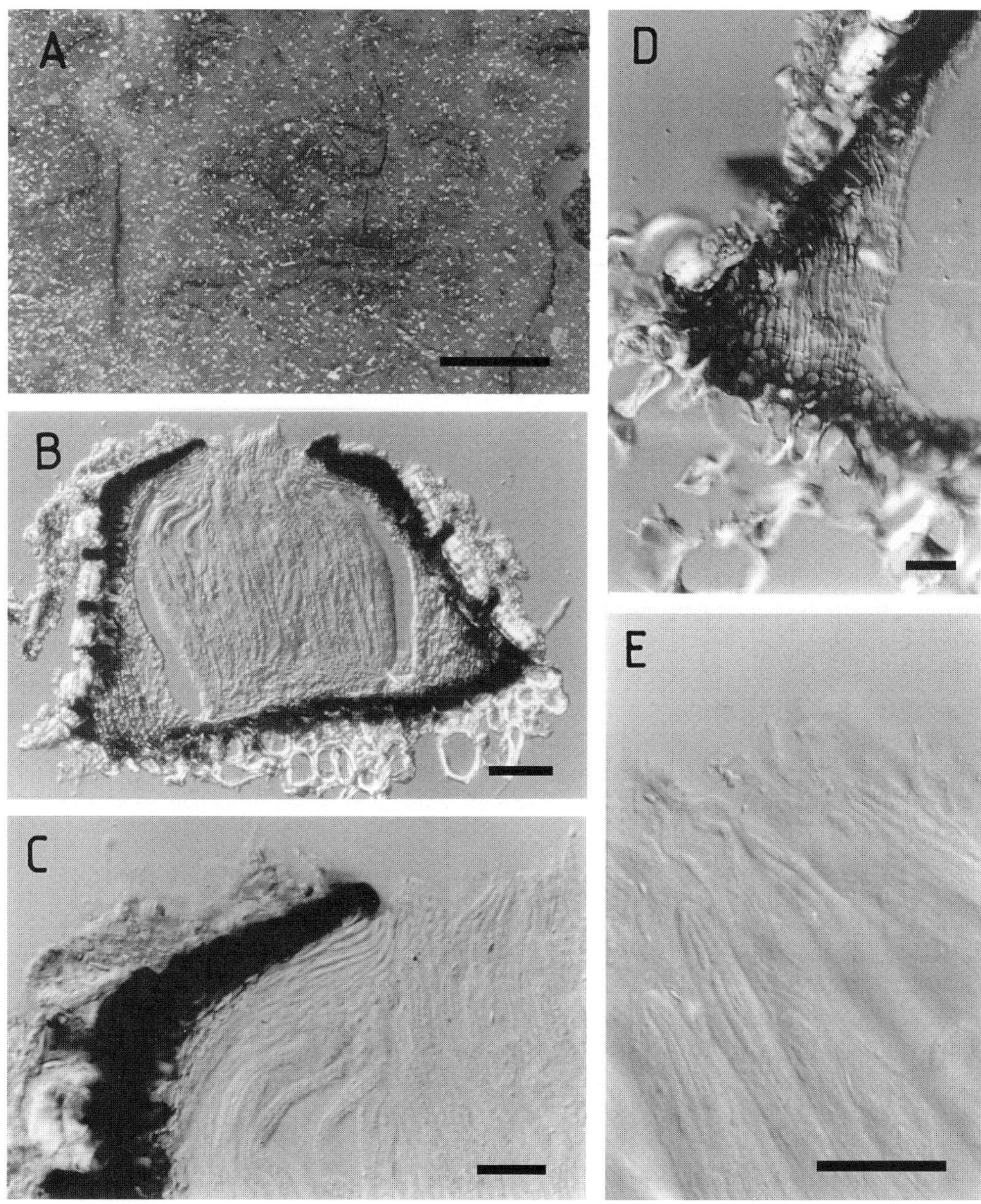

Fig. 32. *Terriera arundinacea* (lectotype, **PAD**). **A**, ascomata (bar = 1 mm); **B**, ascoma in vertical section (bar = 50 µm); **C**, detail of margin of opening slit in vertical section; **D**, detail of lower corner of ascoma in vertical section; **E**, apex of asci and paraphyses (bars = 20 µm).

Lophodermium arundinaceum (Schrad.) Chevall., *Fl. gén. env. Paris* **1**: 435 (1826).
Hysterium arundinaceum Schrad., *J. Bot.* (Schrader) **2**: 68 (1799).
Xyloma arundinis (Schrad.) Rebent., *Prodr. fl. neomarch.*: 342 (1804).
Hypoderma arundinaceum (Schrad.) DC., in LAMARCK & CANDOLLE, *Fl. franç.* **2**: 305 (1805).
Lophodermium arundinaceum var. *vulgare* Fuckel, *Symb. mycol.*: 256 (1870).

Infected areas on dead leaves and leaf sheaths, slightly paler than surrounding host tissue, not associated with zone lines, containing groups of ascomata, most of these associated with conidiomata. *Ascomata* 0·6–1 × 0·3–0·5 mm, oblong-elliptical in outline, ends rounded, unopened ascomata with wall pale grey-brown, with an indistinct preformed line of dehiscence, open ascomata with wall darker, sometimes ± black, opening slit lined with narrow creamy-white lips, although in older ascomata lips may be dark and macroscopically indistinct. *Ascomatal insertion* subepidermal or intrahypodermal. *Covering layer* up to 50–100 µm thick in vertical section, comprising clypeus of dark-walled hyphae within epidermal cells and upper wall of ascomatal stroma. *Upper wall* often with partially decomposed host fibre and hypodermal cells embedded, comprising mostly brown to pale brown cells, 3–5 µm diam., with irregularly encrusted walls, but with group of very dark cells adjacent to lip cells. *Lower wall* 15–30 µm thick in vertical section, comprising four to six rows of angular to subcylindrical cells, 3–6 µm diam., with walls brown to dark brown, in squash mount darkened cells of lower wall narrow-cylindrical, 3–5 µm diam., forming broad layer of *textura intricata* across base of ascoma. *Paraphyses* 1·5 µm diam., circinate at apex. *Asci* 110–130 × (8·5–) 9–10 (–10·5) µm, subfusoid to subclavate, tapering gradually to small, rounded apex with undifferentiated wall, 8-spored, spores initially extending ± to base, with basal stalk developing near maturity. *Ascospores* (55–) 60–80 (–90) × (1·5–) 2 µm, tapering slightly and gradually to base, non-septate, apical gelatinous cap broad-cylindrical, sometimes slightly curved, 4·5–5 × 2·5–3 µm, basal gelatinous cap cylindrical, 3–4 × 1·5 µm, gelatinous sheath 4·5–5 µm wide. *Conidiomata* deeply immersed, subglobose, appearing round from above, very pale grey, scattered, not numerous. *Conidiogenous cells* solitary, cylindrical, tapering slightly to apex, 15–20 × 2·5–3 µm, proliferation usually percurrent, rarely sympodial. *Conidia* 3–4·5 × 1 µm, oblong-elliptical with rounded ends, sterile elements (trichogynes) filiform, up to 65 µm long in group near centre of conidiomata, sometimes branched near base, often slightly swollen at apex.

Typification: *S. loc., hosp., dat. nec coll.* (Mougeot and Nestler, *Stirpes cryptogamae vogeso-rhenanae* no. 655; **K**!, neotype; **IMI**, isoneotype, *fide* CANNON & MINTER, 1983).

Host: *Phragmites* (*Poaceae*).

Distribution: Europe, North America.

Illustrations: Figs 33, 35.

Notes: *Lophodermium arundinaceum* has an ascomatal structure typical of the *arundinaceum*-group in *Lophodermium* Group A (see p. 33), where it is distinguished by the presence of conidiomata and by host, being confined to *Phragmites*. There is some variation in the pigmentation of ascomata macroscopically, which relates to the amount of fibre cell tissue embedded in the upper wall.

REHM (1887) cited *Leptostromella hysterioides* var. *graminicola* De Not. as the anamorph of *L. arundinaceum*. GROVE (1937) agreed, but raised the status of this taxon to species as *Leptostromella graminicola* (De Not.) Grove. However, these authors were almost certainly referring to a distinct fungus, since *L. graminicola* was described from *Molinia*. REHM, GROVE and DE NOTARIS all described the conidia of *Leptostromella graminicola* as curved and *c.* 18 µm

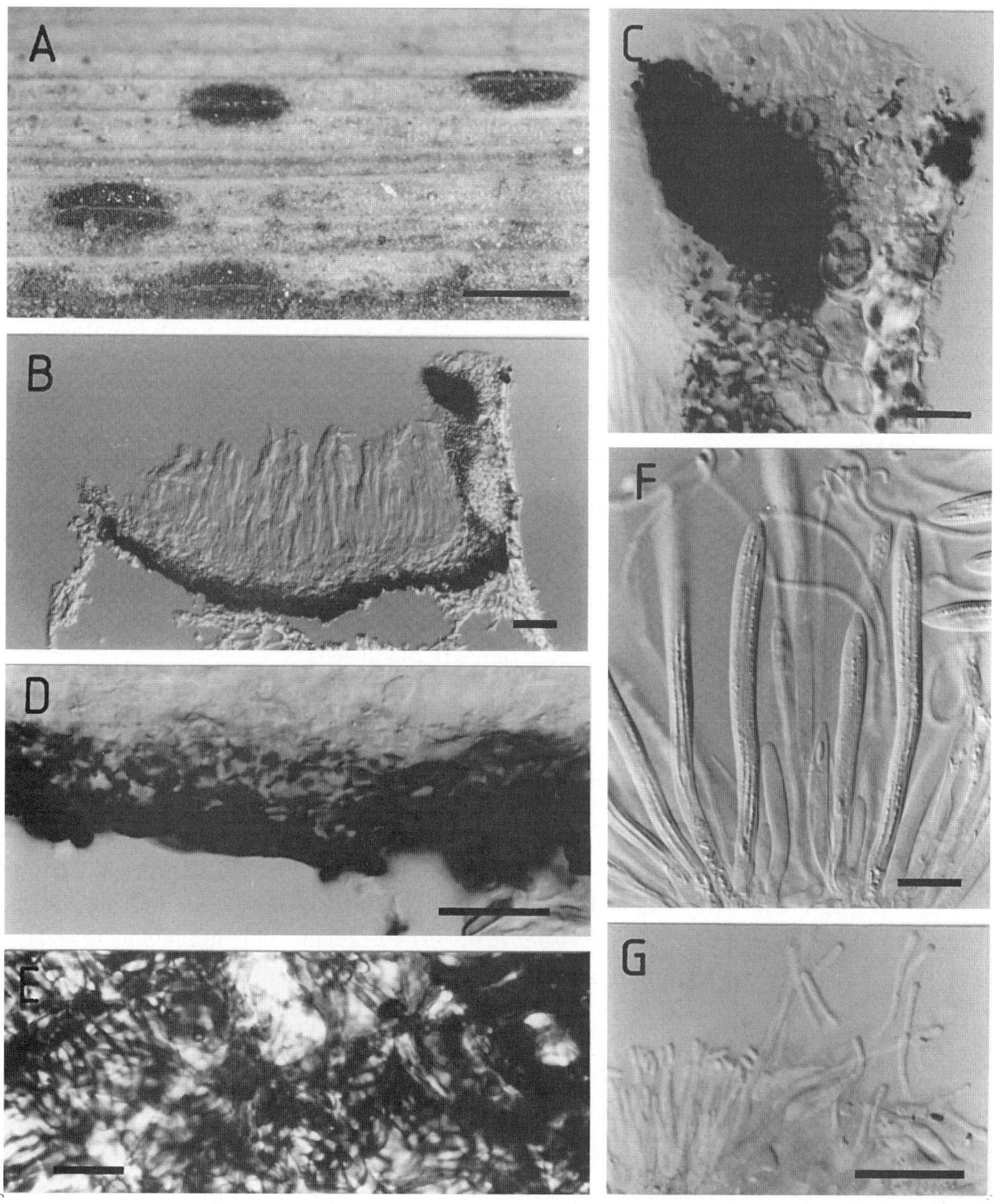

Fig. 33. *Lophodermium arundinaceum* (**A**, **PRM** 802642; **B–D**, **PDD** 59498; **E–G**, **PDD** 59492). **A**, ascomata (bar = 1 mm); **B**, ascoma in vertical section (bar = 50 µm); **C**, upper wall of ascoma in vertical section; **D**, lower wall of ascoma in vertical section; **E**, lower wall of ascoma in squash mount; **F**, asci and paraphyses; **G**, conidiogenous cells, conidia and trichogynes (bars = 20 µm).

long, much larger than for any *Leptostroma*-like species which has been considered definitely as an anamorph of *Lophodermium*.

FUCKEL (1870) cited *Hysterium arundinaceum* Schrad. as the basionym of his variety *L. arundinaceum* var. *vulgare*, hence the type specimen of *H. arundinaceum* (also cited by FUCKEL) must be the type of *L. arundinaceum* var. *vulgare*. A second specimen cited by FUCKEL (1870), *Fungi rhenani* no. 737, is also typical of *L. arundinaceum*. REHM (1881) placed *L. arundinaceum* var. *vulgare* in synonymy with his *L. arundinaceum* var. *secalis*, but this second variety is a synonym of *L. culmigenum*.

Additional specimens examined: Czech Republic: BOHEMIA: Sudoměř, Škaredý, 9 Jun. 1977, *M. Svrček & J. Kubička s. num.* (**PRM** 819550); Ražice, Řežabinec, on *Phragmites communis*, 2 Jun. 1974, *M. Svrček s. num.* (**PRM** 756749); *idem loc.*, on *Phragmites communis*, 21 May 1976, *M. Svrček & J. Kubička s. num.* (**PRM** 802642). **Germany:** BADEN-WÜRTTEMBERG, Rastatt, on *Phragmites communis*, May 1876, *J. Schroeter* 159 (**PAD**); BAYERN: Unterfranken, Gerolzhofen, Kleinrheinfeld, on *Phragmites communis*, Mar. 1912, *A. Vill s. num.* (Sydow, *Mycotheca germanica* no. 1172; **PDD** 55990); *s. loc., hosp., dat. nec coll.* (Fuckel, *Fungi rhenani* no. 737; **K**); SAXONY: Leipzig, on *Phragmites communis*, Apr. 1874, *Winter s. num.* (Thümen, *Mycotheca universalis* no. 77; **NY**). **Hungary:** Comit. Vas, Ikervár, Herpenyő, on *Phragmites communis*, 25 Apr. 1980, *S. Tóth s. num.* (**DAOM** 188376). **Switzerland:** AARGAU: Reusstal, Unterlunkofen, on *Phragmites communis*, 15 May 1980, *P. Crivelli* 22 (**PDD** 59493, **ZT**); SCHAFFHAUSEN: Thayungen, Moos, on *Phragmites communis*, 5 May 1980, *L. Petrini* 14 (**PDD** 59495, **ZT**); *idem loc.*, on *Phragmites*, 5 May 1980, *L. Petrini* 15 (**PDD** 59492, **ZT**); Thayungen, Talweiher, on *Phragmites communis*, 4 May 1980, *L. Petrini* 12 (**PDD** 59498, **ZT**); *idem loc.*, on *Phragmites communis*, 4 May 1980, *L. Petrini* 11 (**PDD** 59497, **ZT**); *idem loc.*, on *Phragmites communis*, 5 May 1980, *L. Petrini* 13 (**PDD** 59494, **ZT**); THURGAU: Etzwiber Ried, on *Phragmites*, 11 May 1981, *P. Crivelli* 214 (**PDD** 59499, **ZT**); ZÜRICH: Weiach, on *Phragmites*, 10 May 1979, *O. Petrini* 5 (**PDD** 59500, **ZT**); *idem loc.*, on *Phragmites communis*, 14 Apr. 1979, *L. Petrini* 10 (**PDD** 59496, **ZT**).

Terriera asteliae (P.R. Johnst.) P.R. Johnst., **comb. nov.**
Lophodermium asteliae P.R. Johnst., *N.Z. Jl Bot.* **27**: 248 (1989).

Infected areas on dead leaves, paler than surrounding leaf tissue, not associated with zone lines, containing scattered ascomata, conidiomata not seen. *Ascomata* 0·5–0·8 × 0·3–0·4 mm, elliptical to oblong in outline, ends rounded, wall black with poorly-defined paler zone along future line of opening, single, longitudinal opening slit with black flattened zone along either side. *Ascomatal insertion* subepidermal. *Ascomatal structure* typical of *Terriera* (see p. 37). *Paraphyses* 1–2 µm diam., ± undifferentiated at apex, extending 15–20 µm beyond asci. *Asci* 75–105 × 8–10·5 µm, cylindrical, apex broadly-truncate with undifferentiated wall, 8-spored, short, broad basal stalk with spores extending almost to base. *Ascospores* 45–70 × 2–2·5 µm, tapering slightly to ends and slightly constricted near centre, non- or 1-septate, gently curved on release, with no gelatinous sheath.

Typification: New Zealand: NORTHLAND: Waima Forest, Waiotemarama Bush Walk, on *Astelia trinervia*, 21 Oct. 1987, *P.R. Johnston* R743 (**PDD** 48000!, holotype).

Host: *Astelia* (*Asteliaceae*).

Distribution: New Zealand.

Illustrations: See JOHNSTON (1989*b*, figs 1, 20, as *Lophodermium asteliae*).

Notes: *Terriera asteliae* is characterized within the genus by its broad asci with a broad-truncate apex. *Terriera fuegiana*, described on *Rostkovia* from Tierra del Fuego, has similar-sized asci and ascospores. Although the only material of *T. fuegiana* available is in poor condition, the asci appear to be broadly rounded at the apex, and illustrations by SPEGAZZINI included in the packet clearly show asci of this shape. Structures resembling conidiomata of *Rhytismataceae* were associated with the ascomata, but conidiogenous cells and conidia were not seen.

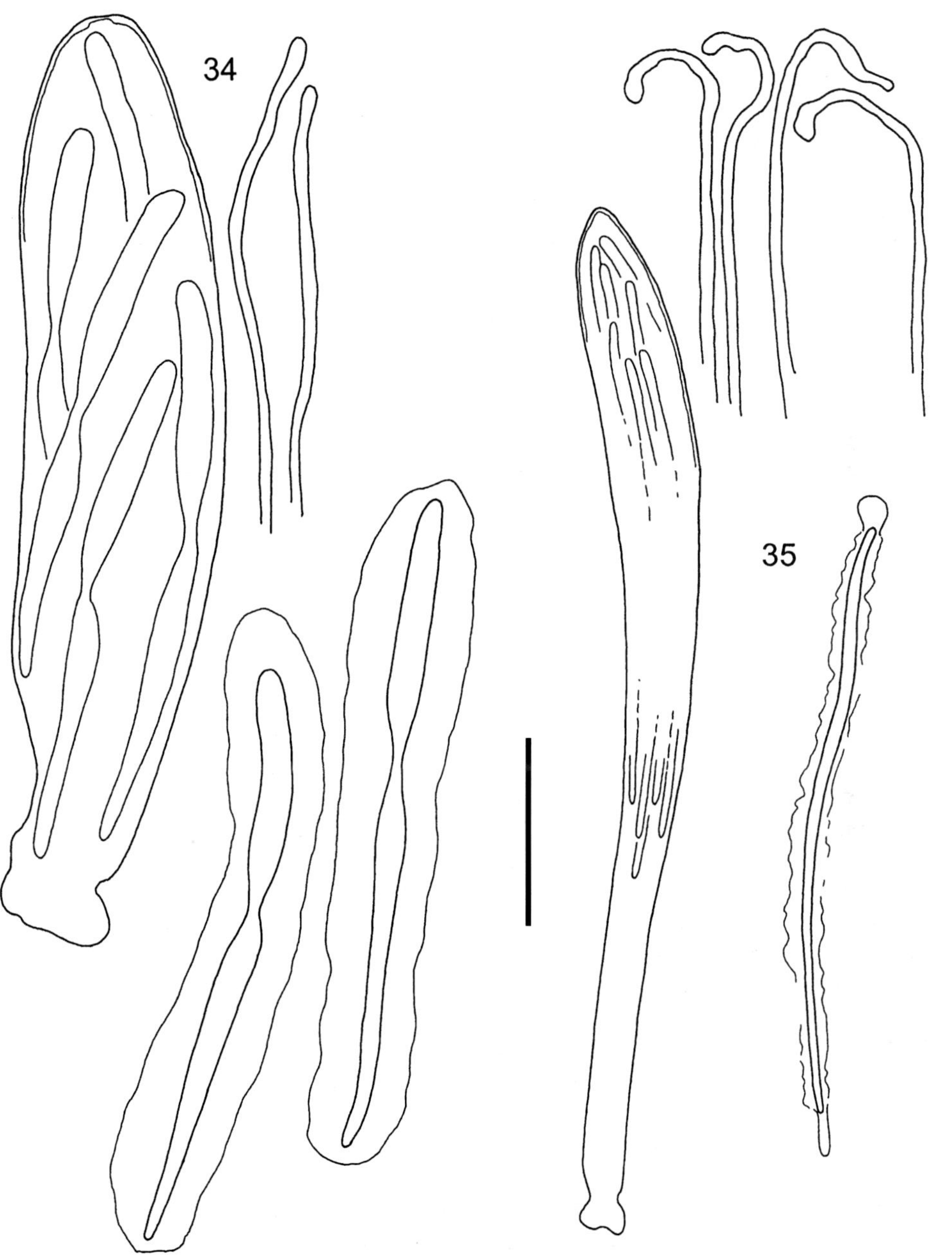

Figs 34–35. 34. *Duplicaria antarctica* (holotype, **LPS**). Asci, apex of paraphyses and released ascospores. **35.** *Lophodermium arundinaceum* (**DAOM** 188376). Asci, apex of paraphyses and ascospores released in water (bar = 20 µm).

Additional specimen examined: New Zealand: AUCKLAND: Waitakere Ranges, Centennial Track, on *Astelia solandri*, 12 Oct. 1982, *P.R. Johnston* R102 (**PDD** 43267).

Terriera breve (Berk.) P.R. Johnst., **comb. nov.**
Lophodermium breve (Berk.) De Not., *G. Bot. ital.* **2**(2): 47 (1847).
Hysterium breve Berk., in HOOKER, *Fl. antarct.*: 174 (1845).
Hypoderma breve (Berk.) Kuntze, *Revis. gen. pl.* **3**(3): 487 (1898).

Infected areas on dead leaves, no paler than surrounding leaf tissue, not associated with zone lines, containing scattered ascomata, conidiomata not seen. *Ascomata* on *Uncinia* 0·6–1·2 × 0·2–0·3 mm, oblong-elliptical in outline, ends rounded, on *Gahnia* 1–5 × 0·2–0·4 mm, often sublinear, in unopened ascomata wall black with paler zone along future line of opening, in opened ascomata wall black, oriented at high angle to surrounding host tissue, single, longitudinal opening slit with narrow, black, shelf-like zone along either side. *Ascomatal insertion* subepidermal. *Ascomatal structure* typical of *Terriera* (see p. 37). *Paraphyses* 1–2 µm diam., undifferentiated to slightly swollen at apex, unbranched, embedded in thick gel, extending as compact epithecium 10–15 µm beyond asci. *Asci* 110–135 (–160) × 6–7 µm, cylindrical, tapering gradually to rounded apex with slightly thickened wall, 8-spored, narrow basal stalk with spores extending 80–90 µm from ascus apex. *Ascospores* (55–) 60–75 × 1·5–2 µm, tapering slightly to ends, non- or 1-septate, gently curved or sigmoid on release, with no gelatinous sheath.

Typification: New Zealand: Campbell I, on *Uncinia hookeri*, *s. dat.*, *coll. nec num.* (**K**!, holotype).

Hosts: *Carex*, *Uncinia*, *Gahnia* (*Cyperaceae*).

Distribution: New Zealand.

Illustrations: Figs 2B, 11D, E, 12, 38.

Notes: *Terriera breve* is distinguished from the other monocotyledon-inhabiting New Zealand species of *Terriera* by its undifferentiated paraphyses and ascus and ascospore size.

Asci from a collection on *Gahnia* (**PDD** 37127) are slightly longer (130–160 µm, *vs* 110–135 µm) than the other collections examined. However, in all other respects the fungi could not be distinguished.

DE NOTARIS (1847) reported what he considered to be this species from Italy. DUBY (1862) placed it in synonymy with *Lophodermium arundinaceum* var. *gramineum* (Fr.) Duby, and this was accepted by TEHON (1935). However, this synonymy is incorrect, the ascomatal structure of *T. breve* differing from that found in all European monocotyledon-inhabiting species of *Lophodermium*.

Additional specimens examined: New Zealand: AUCKLAND: Marguerite Track, on *Gahnia*, 15 Oct. 1975, *G.J. Samuels s. num.* (**PDD** 47493); Waitakere Ranges, Karamatura Stream, on *Gahnia*, 3 Sep. 1976, *G.J. & C. Samuels s. num.* (**PDD** 37127); COROMANDEL: Little Barrier I, Shag Track, on *Uncinia*, 15 Jun. 1984, *P.R. Johnston* LB29 (**PDD** 46121); Port Charles, between wharf and Big Sandy Bay, on *Carex*, 29 Dec. 1993, *P.R. Johnston s. num.* (**PDD** 62551).

Lophodermium calami Henn. & E. Nyman, in HENNINGS, *Monsunia* **1**: 170 (1900).

Typification: 'Java. Hort. Bogor. auf Blattschieden von *Calamus*. 12. Aug. 1898 (E. Nyman.)' (HENNINGS, 1900).

Host: *Calamus* (*Arecaceae*).

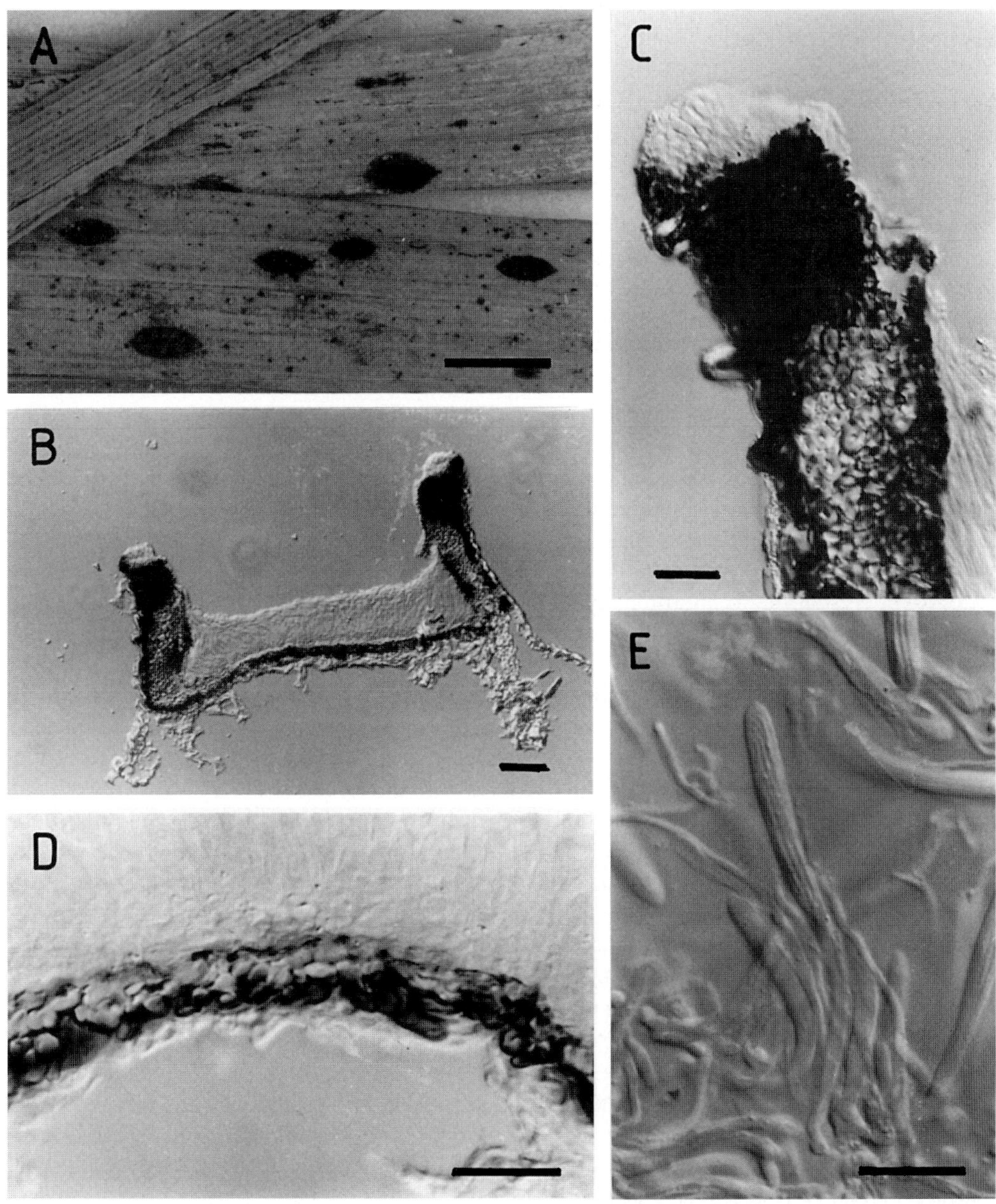

Fig. 36. *Lophodermium caricinum* (Desmazières, *Pl. crypt. N. France* no. 168, **K**). **A**, ascomata (bar = 1 mm); **B**, ascoma in vertical section (bar = 50 μm); **C**, upper wall of ascoma in vertical section; **D**, lower wall of ascoma in vertical section; **E**, asci (bars = 20 μm).

Distribution: Java.

Notes: The type specimen was not located, but geographical location and host suggest that this will almost certainly be a species of *Terriera*.

Lophodermium caricinum (Roberge ex Desm.) Duby, *Mém. Soc. Phys. Hist. nat. Genève* **16**: 59 (1862).

Hysterium caricinum Roberge ex Desm., *Annls Sci. nat.* Sér. 3, **8**: 180 (1847).

Hypoderma caricinum (Roberge ex Desm.) Kuntze, *Revis. gen. pl.* **3**(3): 487 (1898).

Lophodermium arundinaceum var. *caricinum* (Roberge ex Desm.) Rehm, *Rabenh. Krypt-Fl.* Edn 2, **1**(3): 47 (1887).

Lophodermellina caricina (Roberge ex Desm.) Höhn., *Annls mycol.* **15**: 312 (1917).

Dermascia caricina (Roberge ex Desm.) Tehon, *Illinois biol. Monogr.* **13**(4): 62 (1935), *nom. inval.*, *ICBN* Art. 43.1.

Aporia neglecta Duby, *Mém. Soc. Phys. Hist. nat. Genève* **16**: 63 (1862) (*fide* REHM, 1896; TEHON, 1935).

Infected areas on dead leaves, paler than surrounding host tissue, not associated with zone lines, containing scattered ascomata, with conidiomata in some collections. *Ascomata* 0·5–1 × 0·3–0·4 mm, elliptical to oblong-elliptical in outline, ends ± acute, unopened ascomata with wall black, with indistinct paler zone along future line of opening, opened ascomata with wall black, margin sharply defined, single, longitudinal opening slit with indistinct lips. *Ascomatal insertion* subepidermal. *Covering layer* at early stage of development (paraphyses starting to elongate but asci not yet present) up to 65 µm thick in vertical section, clypeus of dark brown, thick-walled hyphae, 3–4 µm diam., within host epidermal cells, and upper wall of ascomatal stroma. *Upper wall* in unopened ascomata without differentiated asci, up to 50 µm thick, outer two-thirds comprising angular cells with pale brown, irregularly encrusted walls, inner third comprising loose, tangled, hyaline, cylindrical cells forming prominent and characteristic layer of periphysoids, later in development (ascospores starting to differentiate within asci) upper wall up to 60 µm thick, mostly of angular to globose cells with pale brown walls, but with two almost confluent patches of very thick-walled, dark brown cells, one to either side of centre of ascoma, periphysoid layer *c.* 25 µm thick (in opened ascomata periphysoids lost). *Lower wall* 15 µm thick in vertical section, in unopened ascomata only lowermost row of cells with darkened walls, with one or more additional rows of irregularly shaped cells with thickened and darkened walls developing across base of maturing ascomata, in squash mount darkened cells in lower wall angular/globose to cylindrical, the angular/globose cells comprising single layer covered with three (or four) layers of ± cylindrical cells, 4–6 µm diam. *Paraphyses* 1·5–2 µm diam., circinate and coiling. *Asci* 70–90 × 5·5–7 µm, subclavate, tapering suddenly to small rounded apex with undifferentiated wall, 8-spored, long basal stalk with spores confined to upper 30–50 µm. *Ascospores* 25–35 × 1–1·5 µm, tapering slightly to ends, apical gelatinous cap globose, 2–3 µm diam., basal gelatinous cap cylindrical, 1·5–2 × 1–1·5 µm, gelatinous sheath poorly developed. *Conidiomata* round in outline, 0·2 mm diam., pale grey with narrow darker line around margin, subepidermal, in vertical section lower wall comprising one to three rows of angular cells, 5–8 µm diam., the lowermost row with walls slightly thickened and darkened. Upper wall lacking, but some dark material with no cellular structure deposited against covering host tissue near centre of conidioma. *Conidiogenous cells* solitary, cylindrical, 8–12 × 2 µm, proliferation sympodial,

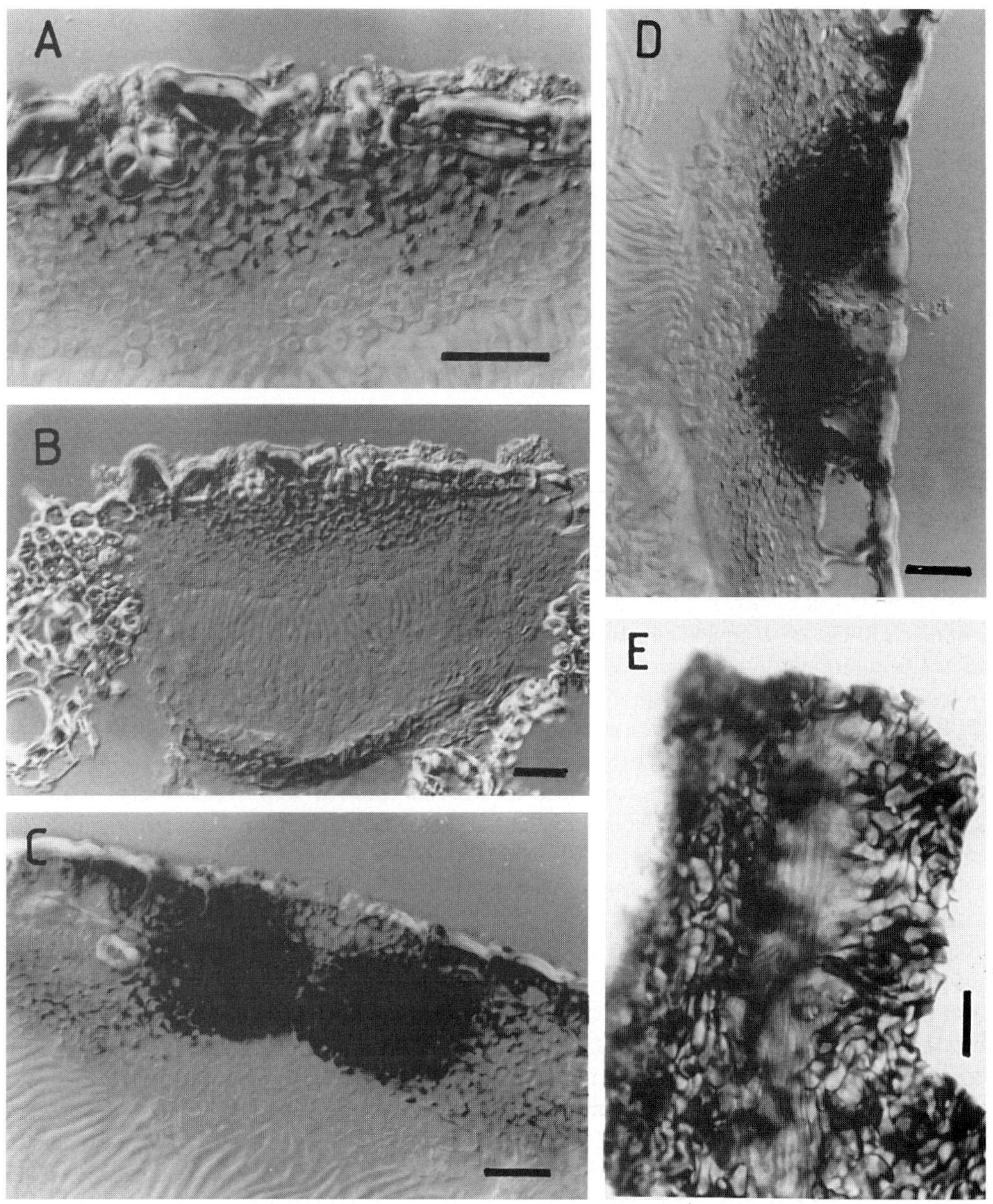

Fig. 37. *Lophodermium caricinum* (**A–C**, **E**, **PDD** 59588; **D**, *Roberge*, **K**). **A**, upper wall of ascoma at early stage of development (paraphyses starting to elongate but asci not present) in vertical section, showing broad layer of periphysoids (bar = 20 µm); **B**, unopened ascoma in vertical section, at later stage (bar = 50 µm); **C**, upper wall of unopened ascoma in vertical section, ascospores starting to differentiate; **D**, upper wall of unopened ascoma; **E**, lower wall of ascoma in squash mount (bars = 20 µm).

lining lower wall of conidioma as palisade. *Conidia* 3–4 × 1 μm, oblong-elliptical, non-septate, hyaline.

Typification: France: *s. loc.*, 'sur plusieurs *Carex*, au printemps' (Desmazières, *Plantes cryptogames de France* Edn 2, Sér. 2, no. 168; **K**!, *lectotypus*, selected here).

Host: *Carex* (*Cyperaceae*).

Distribution: Europe, ?North America (see Notes).

Illustrations: Figs 36, 37, 39.

Notes: *Lophodermium caricinum* has an ascomatal structure typical of *Lophodermium* Group A, but does not fit easily into any of the recognized subgroups (see p. 33). Characteristically the upper wall of unopened ascomata is lined with a broad layer of periphysoids, very similar to that seen in species of the *alpinum*-group, but which differ in lacking lip cells.

Lophodermium caricinum is distinguished microscopically from other species of *Lophodermium* (including those occurring on *Carex*, which in Europe comprise *L. alpinum*, *L. nitidum* and *L. fusiforme*) by having narrow asci, and short ascospores confined to the upper half of the ascus.

No type specimen was cited clearly by DESMAZIÈRES (1847), but he mentioned one collection from *Carex glauca* made by Roberge. A specimen examined from **K** is annotated '*Hysterium caricinum* Roberge (Defn. 2 Serie 168), feuille seche in divers *Carex*. Caen. Roberge'; a smallish collection, it is mostly immature. Desmazières, *Plantes cryptogames de France* Edn 2, Sér. 2, no. 168 (**K**) contains the same species as the Roberge collection and is selected here as the lectotype.

Collections typical of *L. caricinum* are not known from outside Europe. However, a collection from California (on *Carex*, *s. coll*, 2201; **K**, as *L. caricinum*) has asci (75–100 × 6–7·5 μm) and ascospores (30–50 × 1·5 μm) of similar size to the European collections, and ascospores likewise confined to the upper half of the ascus; however, unopened ascomata lack the broad layer of periphysoids. All specimens examined from Europe have the well-developed periphysoids described above. More collections from North America need to be examined in order to evaluate the significance of this apparent difference. *Lophodermium caricinum* has been reported from the Himalayas (MÜLLER, 1977), but CANNON & MINTER (1986) expressed doubts over the authenticity of these records. A collection examined from Kashmir (**India:** KASHMIR: Sonarmarg, Sind River, Srinigar, on *Carex setigera*, Jul.–Aug. 1928, *F.G. Dickason* 25; **DAOM** 124007, as *L. caricinum*) is certainly a distinct species with very large asci (120–140 × 12–14 μm) and ascospores (65–80 × 2–2·5 μm).

FUCKEL (1870: 256) reported *Leptostroma caricinum* Fr. as the anamorph of *Lophodermium caricinum*. Although structures resembling conidiomata of *Rhytismataceae* were seen in several collections of *L. caricinum* (conidia and conidiogenous cells were present in only one of the collections examined: Roumeguère, *Fungi gallici exsiccati* no. 7142), *Leptostroma caricinum* is a distinct species. Although microscopically typical of *Leptostroma*, conidiomata on both the type and a second specimen examined are not associated with ascomata of *Lophodermium*. Conidiomata are large (0·4–0·6 × 0·25–0·3 mm), black, broadly elliptical with what appears to be an irregular, slit-like opening and are subcuticular, the covering layer comprising dense black material less than 5 μm thick. The lower wall comprises two or three rows of angular cells with thin, very pale brown walls, on which the conidiogenous cells are arranged in palisade-like layer. Scattered dark-walled hyphae invade some of the ± intact epidermal cells beneath the conidioma. Conidiogenous cells are 7–10 × 1·5–2 μm, cylindrical, with sympodial proliferation and two conidia often held at the apex. Conidia are 3–5 × 0·8–1 μm, narrow-cylindrical with rounded ends, straight and non-septate.

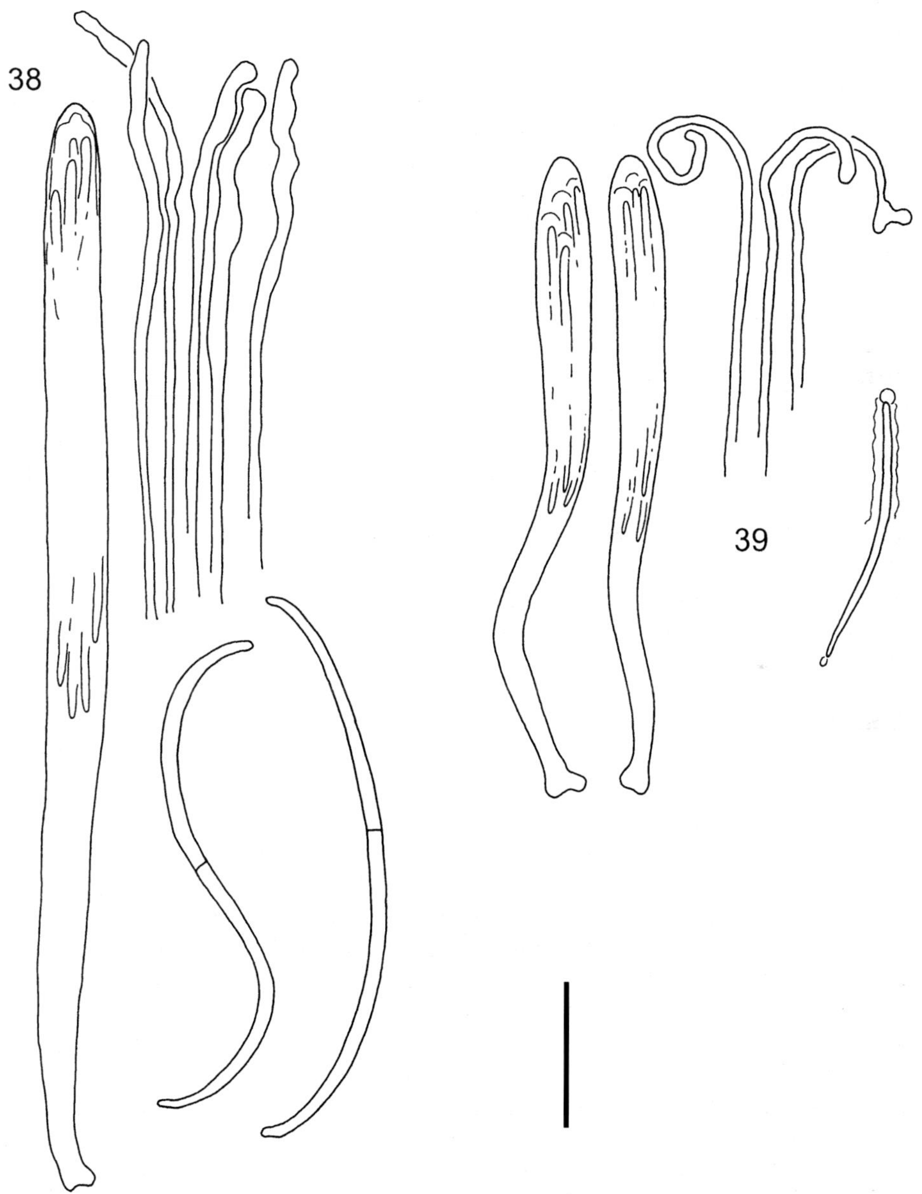

Figs 38–39. 38. *Terriera breve* (**PDD** 62551). Asci, apex of paraphyses and ascospores. **39.** *Lophodermium caricinum*. Asci, apex of paraphyses (**PDD** 59586) and ascospore in water (**PDD** 59588) (bar = 20 µm).

Additional specimens examined: ***Lophodermium caricinum*****: England:** DEVON: Barnstaple, Braunton Burrows, on *Carex arenaria*, 19 Sep. 1988, *C. Scheuer* 358 (**GZU**). **France:** NORMANDY: Caen, 'feuille seche in divers *Carex*', *Roberge s. num.* (**K**); Caen, Parc de Lébiseq, on *Carex glauca*, 9 Apr. 1843, *Roberge s. num.* (**FH**); on *Carex glauca*, Jul. 1897, *F. Fautrey s. num.* (Roumeguère, *Fungi gallici exsiccati* no. 7142; **NY**). **Germany:** BAYERN: Unterfranken, Würzburg, on *Carex flacca*, Sep. 1981, *C. Scheuer s. num.* (**GZU**). **Italy:** Coegliano, on *Carex digitatis*, Jan. 1878, *Spegazzini s. num.* (Saccardo, *Mycotheca Veneta* no. 1280; **K**). **Switzerland:** SCHAFFHAUSEN: Randen, Guetbuck, on *Carex flacca*, 8 Jun. 1980, *L. Petrini* 32 (**PDD** 59587, **ZT**); Hemmental-Guetbuck, on *Carex flacca*, 8 Jun. 1980, *L. Petrini* 33. (**PDD** 59588, **ZT**); [*loc. illeg.*], on *Carex flacca*, 1872, [?]*Tabloavonsky* (**HBG**); VAUD: Yverdon, Montaigny, on *Carex flacca*, 17 Jun. 1980, *L. Petrini* 46 (**PDD** 59586, **ZT**).

Leptostroma caricinum**: Sweden:** *s. loc.* (Fries, *Scleromyceti Sueciae* no. 176; **UPS**, holotype of *Leptostroma caricinum*); Alsike, Rickebasta träsk, on *Eriophorum latifolium*, 9 Aug. 1923, *J.A. Nannfeldt s. num.* (**UPS**, as *Lophodermium eriophori*).

Lophodermium castellani Graniti, *Nuovo G. Bot. ital.* n.s. **59**: 40 (1952).

Typification: 'Su esemplari di *Sesleria disticha* Pers., da me raccolti in un pascolo tra il Corno del Renon e la Sella dei tre sentieri (Altopiano del Renon, prov. di Bolzano) a m. 2150 s.m., il 22 luglio 1947, e di *Sesleria caerulea* Ard. raccolti nello stresso mese presso Collalbro (Altopiano del Renon) a m. 1100 s.m.' (GRANITI, 1952).

Host: *Sesleria* (*Poaceae*).

Distribution: Italy.

Notes: The type specimen could not be located, but GRANITI's description of *L. castellani* is very close to that of *L. alpinum*.

Terriera clithris (Starbäck) P.R. Johnst., **comb. nov.**
Lophodermium clithris Starbäck, *Bih. K. svenska VetenskAkad. Handl.* **25**(1): 16 (1899).

Infected areas on dead host leaves, often slightly darker than surrounding host tissue, associated with faint, but ± complete, broad, grey zone lines, containing scattered ascomata, conidiomata not seen. *Ascomata* 1·5–4 × 0·3 mm, cylindrical to linear in outline, wall pale grey to ± concolorous with surrounding host tissue, edge of ascoma marked by narrow dark line, longitudinal opening slit lined on either side lined with black, shelf-like zone. *Ascomatal insertion* subepidermal and beneath bundles of host fibre cells. *Ascomatal structure* typical of *Terriera* (see p. 37), subhymenium very well developed, comprising layer of narrow, hyaline, cylindrical, thin-walled cells, 60 µm thick, between darkened lower wall and hymenium. *Paraphyses* 1–1·5 µm diam., increasing suddenly to 2–3·5 µm diam. at clavate to knob-like apex, often branched in upper 20–30 µm, with apices embedded in colourless gel forming dense epithecium. *Asci* 110–120 × 6·5–7 µm, cylindrical to subfusoid, tapering slightly to broad subtruncate apex with undifferentiated wall, 8-spored, short basal stalk at maturity with spores extending ± to base. *Ascospores* not seen released, 60–80 × 1–1·5 µm, tapering slightly to ends, with no obvious gelatinous sheath.

Typification: Brazil: RIO GRANDE DO SUL: Silviera Marino, Santa Maria, on grass ('Taquara'), 1 Mar. 1893, *G.A. Malme s. num.* (**S**!, holotype).

Host: Unidentified monocotyledon.

Distribution: Brazil.

Illustration: Fig. 40.

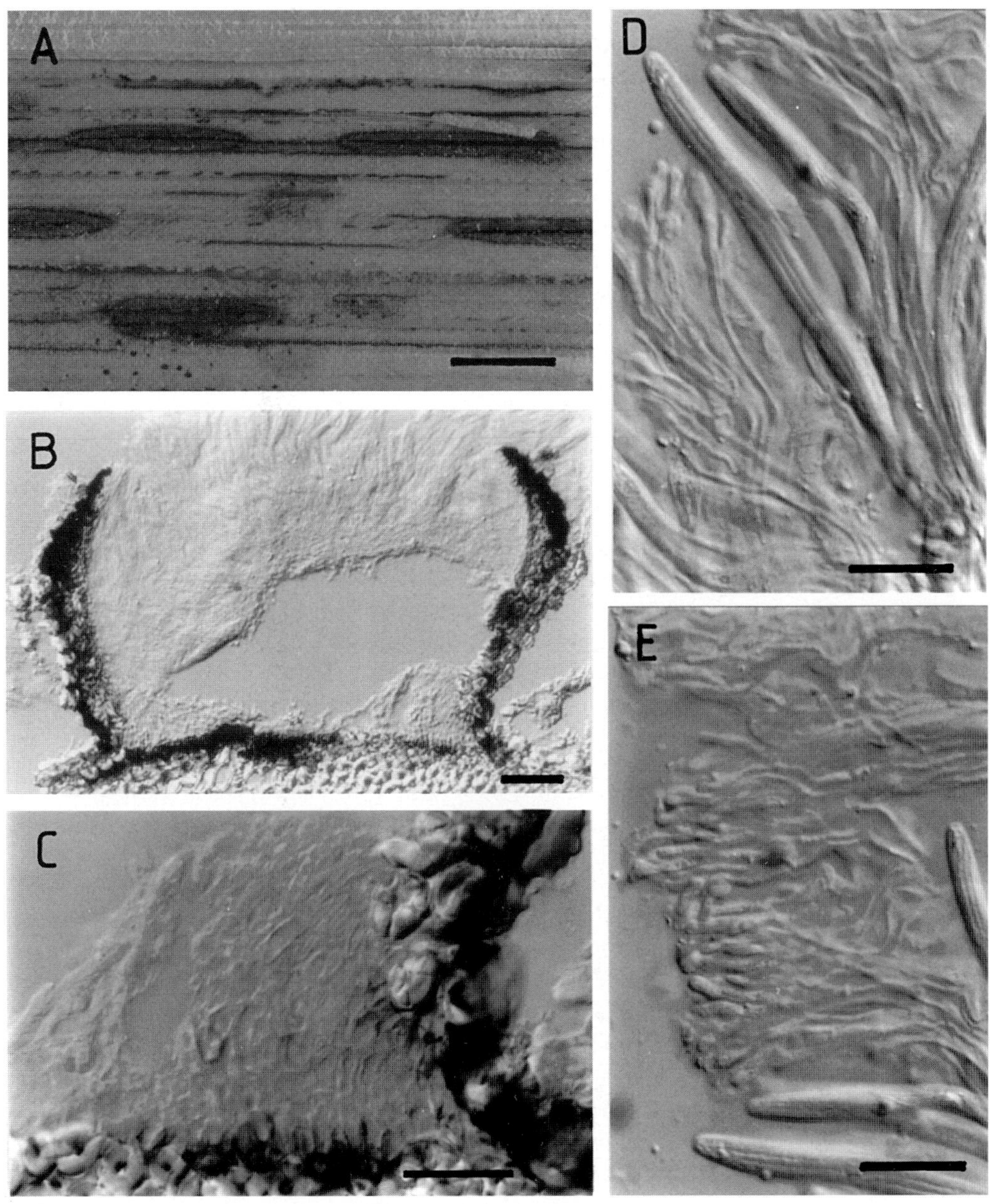

Fig. 40. *Terriera clithris* (holotype, **S**). **A**, ascomata (bar = 1 mm); **B**, ascoma in vertical section (bar = 50 μm); **C**, lower corner of ascoma in vertical section, showing lower wall and palisadic structure of remaining tissue of ascomatal initial; **D**, ascus; **E**, apex of paraphyses and asci (bars = 20 μm).

Notes: *Terriera clithris* is characterized by its branched paraphyses, subtruncate ascus apex and the pale walled ascomata associated with poorly-developed zone lines.

The host leaf is very tough, possibly *Cyperaceae* rather than *Poaceae*.

Lophodermium culmigenum (Fr.) De Not., *G. Bot. ital.* **2**(2): 46 (1847).
Hysterium culmigenum Fr., *Observ. mycol.* **2**: 355 (1818).
Hypoderma culmigenum (Fr.) Kuntze, *Revis. gen. pl.* **3**(3): 487 (1898).
Lophodermellina culmigena (Fr.) Höhn., *Annls mycol.* **15**: 313 (1917).
Lophodermina culmigena (Fr.) Tehon, *Illinois biol. Monogr.* **13**(4): 93 (1935).
Lophodermium arundinaceum var. *secalis* Rehm, *Hedwigia* **20**: 42 (1881).
Lophodermium arundinaceum var. *tritici* Roum., *Revue mycol.* **14**: 173 (1892), *nom. inval.*, *ICBN* Art. 32.1(c).
Lophodermium secalis Hilitzer, *Věd. Spisy čsl. Akad. zeměd.* **3**: 96 (1929).
Lophodermium phlei Tehon, *Illinois biol. Monogr.* **13**(4): 54 (1935), *nom. inval.*, *ICBN* Art. 36.1.
Lophodermellina tritici Tehon, *Illinois biol. Monogr.* **13**(4): 83 (1935), *nom. inval.*, *ICBN* Art. 36.1.

Infected areas on dead leaves and flowering stems, often slightly paler than surrounding host tissue, not associated with zone lines, containing scattered ascomata, conidiomata not seen. *Ascomata* 0·7–1·2 (–1·5) × 0·3–0·4 mm, elliptical, ends rounded, apparently deeply immersed with margins poorly defined, wall usually dark grey to black, but in some collections pale grey with narrow dark line at margin, single longitudinal opening slit with poorly to well-developed lips. *Ascomatal insertion* intrahypodermal. *Covering layer* up to 80 µm thick in vertical section, comprising dark-walled hyphae of clypeus within epidermal cells and upper wall of ascomatal stroma. *Upper wall* mostly of angular to globose cells, 5–8 µm diam., with walls slightly and irregularly encrusted with dark brown material, but with group of very dark cells adjacent to lip cells, commonly some cells in upper, widest part of ascomatal wall becoming cylindrical, with encrusting material confined to ends of cell walls, partially broken-down hypodermal and/or fibre cells of host usually embedded within wall. *Lower wall* 15–25 µm thick in vertical section, comprising several rows of irregularly shaped, globose or cylindrical cells, 3–5 µm diam., with thick, dark brown walls, in squash mounts darkened cells of lower wall long, 3–4 µm diam., forming *textura intricata*-like tissue across base of ascoma, although lower wall sometimes patchy in development across ascoma. *Paraphyses* 1·5–2 µm diam., circinate, typically bent rather than coiling, often also irregularly swollen near tip. *Asci* (90–) 115–130 (–150) × (7·5–) 9·5–10·5 (–11) µm (see Notes below), subclavate, tapering gradually to small, truncate apex with undifferentiated wall, 8-spored, basal stalk developing at maturity with spores confined to upper 85–95 µm. *Ascospores* (45–) 60–75 (–95) × 1·5–2 (–2·5) µm (see Notes below), straight when released, tapering gradually to base, apical gelatinous cap globose, 3·5–5 µm diam., basal gelatinous cap cylindrical, 4–6 × 1·5–2 µm, gelatinous sheath *c.* 5 µm diam.

Typification: Sweden: *s. loc.*, on *Secale cereale*, *s. dat.*, *E.M. Fries s. num.* (Fries, *Scleromyceti sueciae* no. 97; **UPS**!, lectotype of *H. culmigenum*). **Germany:** BERLIN: Zehlendorf, on *Secale cereale*, *E. Loew & W. Zopf s. num.* (*Mycotheca marchica* no. 25; **UPS**!, *lectotypus* of *L. arundinaceum* var. *secalis*, selected here). **Czech Republic:** BOHEMIA: Klínec, Praha, on *Secale cereale*, 5 May 1928, *A. Hilitzer s. num.* (**PRM** 705172, *lectotypus* of *L. secalis*, selected here). **USA:** *s. loc.*, on *Phleum pratense*, Jan. 1878, *s. coll.* (Ellis, *North American Fungi* no. 465,

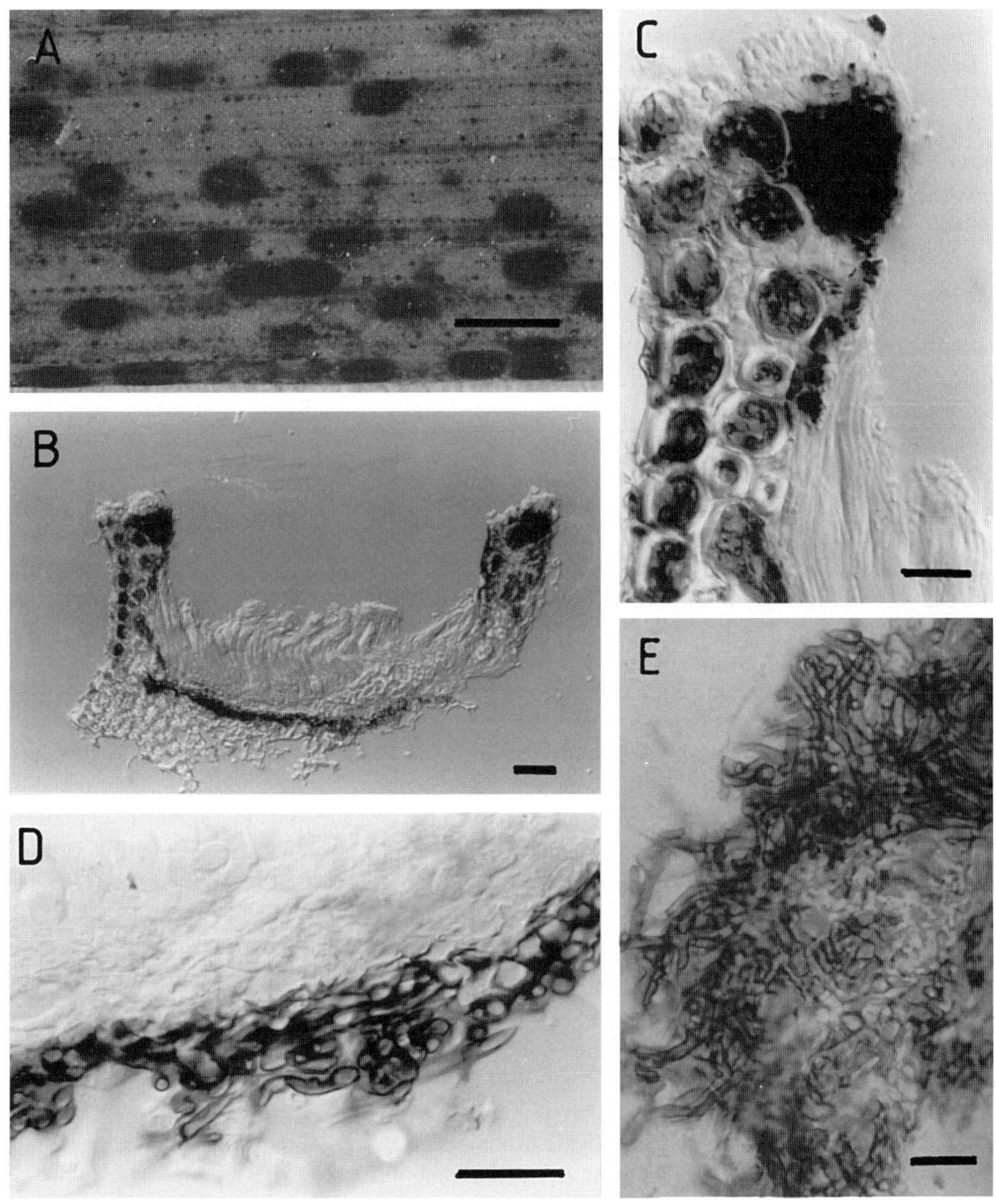

Fig. 41. *Lophodermium culmigenum* (Fries, *Scleromyceti Sueciae* no. 97, **UPS**). **A**, ascomata (bar = 1 mm); **B**, ascoma in vertical section (bar = 50 μm); **C**, upper wall of ascoma in vertical section; **D**, lower wall of ascoma in vertical section; **E**, lower wall of ascoma in squash mount (bars = 20 μm).

specimen on which the invalid name *L. phlei* was based, **K**!; **PDD**!). **France:** *s. loc.*, 'En masse sur vieux chaume de *Triticum hibernum*. F. Fautrey.', *s. dat.* (Roumeguère, *Fungi gallici exsiccati* no. 6144, specimen on which the invalid names *L. arundinaceum* var. *tritici* and *Lophodermellina tritici* were based, **UPS**!).

Hosts: *Agropyron*, *Agrostis*, *Aira*, *Ammophila*, *Dactylis*, *Deschampsia*, *Elymus*, *Festuca*, *Holcus*, *Phleum*, *Poa*, *Secale*, *Triticum* (*Poaceae*); *Typha* (*Typhaceae*); *Juncus* (*Juncaceae*).

Distribution: Europe, North America, Chile, Australia, New Zealand.

Illustrations: Figs 7A, 9E, F, 41, 45.

Notes: *Lophodermium culmigenum* has an ascomatal structure typical of *Lophodermium* Group A, within which it is one of the species making up the *arundinaceum*-group (see p. 33). Its ascomata, characteristically elliptical with a poorly-defined margin, develop among the host hypodermal cells, with partially broken-down hypodermal or fibre cells usually embedded within the upper wall.

Other species in the *arundinaceum*-group include *L. arundinaceum*, *L. eximium* and *L. rubrum*. The former is restricted to *Phragmites* and can be distinguished only by the presence of conidiomata. It may be difficult to distinguish in small collections (where conidiomata may be missing by chance) or in collections in which the host substratum is not known. *Lophodermium eximium* is known from a few collections on *Ampelodesmos*, and is distinguished by its pale, lenticular conidiomata. *Lophodermium rubrum* is known from only one collection, on *Cortaderia* from New Zealand, and is distinguished primarily by the orange-red hyphae within the host tissue around the ascomata.

Lip cell development varies greatly and in some collections lip cells may be almost lacking (e.g. *Parmelee* 5977b, **GZU**), though with all other features, both macroscopic and microscopic, typical of this species. The extent of development of the lower ascomatal wall also varies, often being patchy across a single ascoma.

A wide range of variation is noted in ascus length and width and in ascospore length in the above description. Most of this variation reflects what appear to be regional differences as noted below. Some of these groups may be found to deserve separate taxonomic status in the future, but additional collections, or new characters, are required before such a decision can be made (see p. 33).

European and North American collections mostly have asci 115–130 × 9·5–10·5 (–11) µm and ascospores 60–75 × 1·5–2 µm. A few collections from Europe have very long asci (130–150 µm) and ascospores (80–95 µm), but are otherwise typical (e.g. *Crivelli* 21, **PDD** 65111; *Scheuer s. num.*, 10 Jul. 1990, **GZU**).

Australian, Chilean and New Zealand collections from hosts other than *Festuca novae-zelandiae* have narrower asci (110–125 × (7·5–) 8·5–9·5 µm) than those from Europe, but ascospores of similar size.

New Zealand collections from Festuca novae-zelandiae have short, broad asci (90–100 × 10–11 µm) and short ascospores (45–55 × 2–2·5 µm) (**Fig. 42**).

Three collections from Greece have asci 95–110 (–115) × 8·5–9·5 µm and ascospores 45–55 (–60) × 1·5–2 µm, smaller than in all other European collections (**Fig. 43**).

Lophodermium nitidum has similar sized asci and ascospores to *L. culmigenum* and likewise has a lower wall comprising several rows of dark-walled cells. It differs macroscopically in having smaller ascomata with sharply defined margins and, although the lower wall comprises several rows of dark-walled cells, these are cylindrical to short-cylindrical, rather than hyphal. The difference between *L. culmigenum* and *L. nitidum* is also partly reflected in host substratum. In

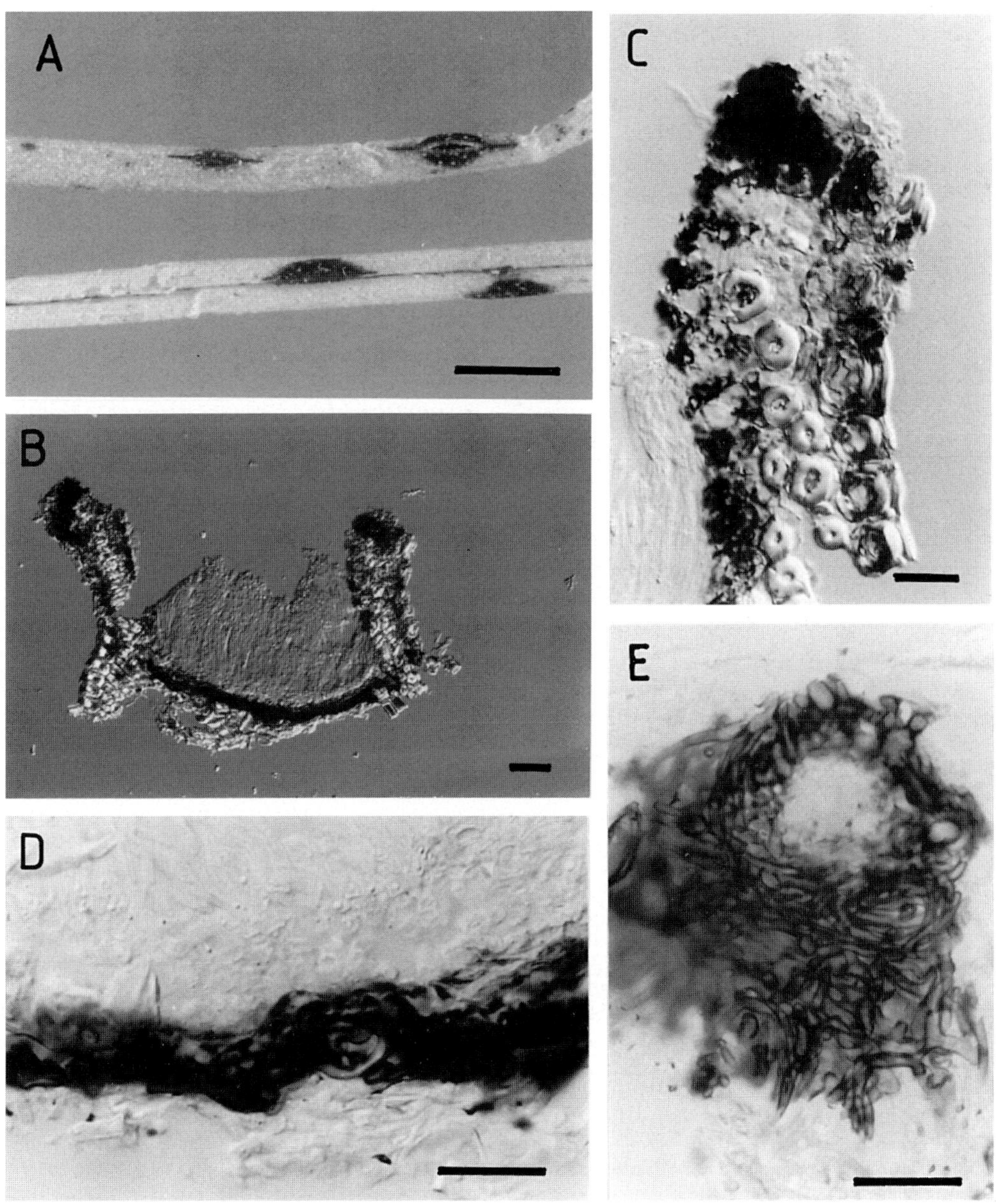

Fig. 42. *Lophodermium* cf. *culmigenum*, collection from *Festuca novae-zelandiae* from New Zealand (**PDD** 48482). **A**, ascomata (bar = 1 mm); **B**, ascoma in vertical section (bar = 50 μm); **C**, upper wall of ascoma in vertical section; **D**, lower wall of ascoma in vertical section; **E**, lower wall of ascoma in horizontal section (bars = 20 μm).

Europe, at least, *L. culmigenum* is most commonly found on members of the tribe *Triticae*, whereas *L. nitidum* is most common on *Avenae* and *Poeae*.

TEHON (1935) cited HÖHNEL (1917*a*) as the author combining the epithet in *Lophodermina*; HÖHNEL, however, considered that *Hysterium culmigenum* Fr. should be a *Lophodermellina* (p. 313), although he did not formally make the combination. TEHON (1935: 93) cited the combination *Lophodermium culmigenum* (Fr.) P. Karst., but it is unclear whether KARSTEN (1872) intended to make such a combination, which in any case is later than that of DE NOTARIS. FARR *et al.* (1989) cited the name *Lophodermium arundinaceum* var. *culmigenum* (Fr.) P. Karst., but the publication relating to this could not be identified.

Two specimens were listed in the protologue of *L. arundinaceum* var. *secalis*: Sydow, *Mycotheca marchica* no. 25 and Rehm, *Ascomyceten* no. 2557. Both match *L. culmigenum* macroscopically, in vertical section and microscopically. Rehm, *Ascomyceten* no. 580, labelled *L. arundinaceum* f. *secalis*, is also this species.

HILITZER (1929) did not designate a type specimen for *L. secalis*, but cited 'In foliis, vaginis et culmis Secalis cerealis prope Klinec'. Two specimens examined from **PRM**, labelled as syntypes on the herbarium packets, have collecting details matching this citation; neither can be distinguished from *L. culmigenum*. HILITZER also listed '?*Hysterium culmigenum* var. *cerealium* Fr., Syst. mycol. 2: 591 [as 593] (1823)' as a synonym of *L. secalis*; this variety was never published by FRIES, who simply listed 'Cerealium' as one of the hosts of his species *H. culmigenum*.

The specimen on which both *Lophodermium arundinaceum* var. *tritici* and *Lophodermellina tritici* were based matches *L. culmigenum* morphologically. ROUMEGUÈRE (1892) provided no description and TEHON (1935) did not provide a Latin description nor diagnosis. GRANITI (1952) cited WEESE (1933) as making the combination *Lophodermium tritici*, but this citation is incorrect and no other record of WEESE having made the combination could be found.

Additional specimens examined: Australia: Bowral to Moss Vale, roadside 2 km N of town, on *Bromus catharticus*, 11 Oct. 1988, *M. J. Priest & J. Walker* 88/75 (**DAR** 63880b). **Canada:** BRITISH COLUMBIA: Vancouver I, Cordova Bay, on *Elymus mollis*, 30 Jun. 1957, *W.G. Ziller s. num.* (**DAOM** 57491); *idem loc.*, on *Dactylis glomerata*, 28 May 1960, *W.G. Ziller s. num.* (**DAOM** 71618, as *L. arundinaceum*); ONTARIO: York Co.: Nashville, on *Triticum vulgare*, 15 May 1949, *R.F. Cain s. num.* (**TRTC** 23075, as *L. tritici*); *idem loc.*, on *Triticum vulgare*, 15 May 1949, *R.F. Cain s. num.* (**TRTC** 23333, as *L. arundinaceum* f. *tritici*); *idem loc.*, on *Phleum pratense*, 7 Jun. 1953, *R.F. Cain s. num.* (**TRTC** 24460, as *L arundinaceum*); *idem loc.*, on *Poa*, 27 Jun. 1954, *R.F. Cain s. num.* (**DAOM** 40797, as *L. arundinaceum* f. *culmigenum*); nr Gros Morne National Park, The Arches, on sea coast, on *Elymus mollis*, 2 Aug. 1983, *J.A. Parmelee* 5977b (**GZU**, ex **DAOM** 193653). **Chile:** LA ARAUCANIA: Collipulli, on *Agrostis*, 25 Oct. 1985, *P.F. Cannon s. num.* (**IMI** 337908, as *Lophodermium* sp.); *s. loc.*, on *Holcus*, 28 Oct. 1985, *D.W. Minter s. num.* (**IMI** 337906, as *Lophodermium* sp.). **Czech Republic:** Mahren, Mahrisch-Weibkirchen, Bartelsdorf, on *Aira flexuosa*, Jul. 1937, *F. Petrak s. num.* (*Reliquiae petrakianae* no. 470; **PDD** 54459); Mahrisch-Weibkirchen, on *Aira flexuosa*, Aug. 1913, *F. Petrak s. num.* (*Reliquiae petrakianae* no. 2464; **PDD** 61377); BOHEMIA: prope vicum Klínec apud Praha, on *Secale cereale*, 5 May 1928, *A. Hilitzer s. num.* (**PRM** 705171). **England:** North Wootton, on grass, *s. dat.*, *C.B. Plowright s. num.* (**HBG**, as *L. arundinaceum*); DEVON: Dawlish Warren, on *Ammophila arenaria*, 9 Apr. 1982, *O. Petrini & J. Webster s. num.* (**PDD** 60126); DORSET: Studland, on *Typha*, 27 May 1961, *B.C. Sutton s. num.* (**IMI** 86883, as *L. arundinaceum*). **France:** Nemes, on *Ammophila arenaria*, 25 Jul. 1871, *P. Magnus s. num.* (**HBG**, as *L. arundinaceum*); Boine, on *Ammophila arenaria*, 8 Jul. 1871, *P. Magnus s. num.* (**HBG**, as *L. arundinaceum*). **Germany:** Schlesien, bei Schwarzbach, Isergebirge, on *Aira flexuosa*, 18 Jul. 1922, *H. Sydow s. num.* (Sydow, *Mycotheca germanica* no. 2147; **PDD** 55995; **GZU**); POTSDAM: Brandenburg, Sophienstadt bei Ruhlsdorf, Kreis Nieder-Barnim, on *Aira flexuosa*, 10 Aug. 1920, *H. Sydow & P. Sydow s. num.* (Sydow, *Mycotheca germanica* no. 1599; **PDD** 55992); Westend. [*sic*], Charlottenburg, on *Elymus arenarius*, Aug. 1890, *P. Sydow s. num.* (Sydow, *Mycotheca marchica* no. 3057; **HBG**, as *L. arundinaceum* f. *culmigenum*); BERLIN: on *Elymus arenarius*, Sep. 1885, *P. Sydow s. num.* (Sydow, *Mycotheca marchica* no. 855; **HBG**, as *L. arundinaceum*); Zehlendorf, on *Secale cereale*, *s. dat.*, *E. Loew & W. Zopf s. num.* (Sydow, *Mycotheca marchica* no. 25; **HBG**, as *L. arundinaceum*); Westerland, on *Elymus arenarius*, 31 Jun. 1897, *O. Franz s. num.* (**HBG**, as *L. arundinaceum* var. *culmigenum*); on *Elymus arenarius*,

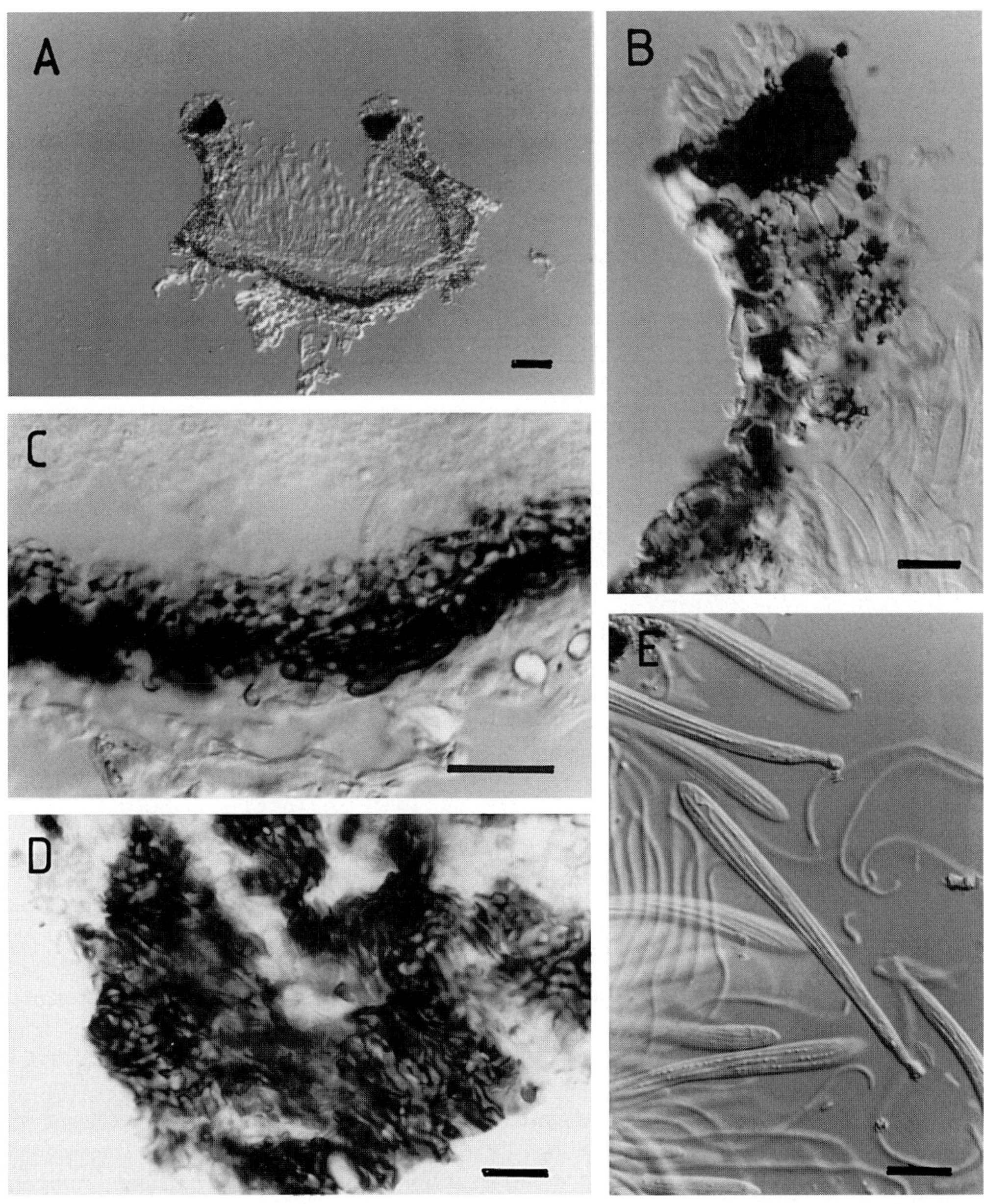

Fig. 43. *Lophodermium* cf. *culmigenum* from Greece (**IMI** 324159). **A**, ascoma in vertical section (bar = 50 µm); **B**, upper wall of ascoma in vertical section; **C**, lower wall of ascoma in vertical section; **D**, lower wall of ascoma in squash mount; **E**, asci and paraphyses (bars = 20 µm).

Aug. 1900, *Zimmerman s. num.* (**HBG**, as *L. arundinaceum* f. *culmigenum*); *s. loc.*, on *Secale cereale*, *s. dat.*, *coll. nec num.* (Rehm, *Ascomyceten* no. 2557; **S**); *s. loc.*, *dat. nec coll.* (Rehm, *Ascomyceten* no. 580; **K**). **Greece:** Agios Giannis, on grass, 21 Mar. 1988, *D.W. Minter s. num.* (**IMI** 324157, as *L. gramineum*); Levadi, on grass, 23 Mar. 1988, *D.W. Minter s. num.* (**IMI** 324158, as *L. gramineum*); Skoteina, on grass, 23 Mar. 1988, *D.W. Minter s. num.* (**IMI** 324159, as *L. gramineum*). **Italy:** Conegliano, on *Triticum*, 1877, *Spegazzini s. num.* (*Mycotheca veneta* no. 1173; **K**). **New Zealand:** CHATHAM IS: Rekohu: Nikau Bush, on *Juncus*, 18 Nov. 1992, *P.R. Johnston* C20 (**PDD** 64553); Rekohu: Hapupu, on *Ammophila arenaria*, 20 Nov. 1992, *P.R. Johnston* C71 (**PDD** 64554); Rekohu: Port Hutt road, Lake Tennant, on *Ammophila arenaria*, 23 Nov. 1992, *P.R. Johnston* C107 (**PDD** 64555); DUNEDIN: Kaka Point Reserve, on *Festuca*, 14 May 1984, *P.R. Johnston* R502 (**PDD** 49362); MID CANTERBURY: Craigieburn Forest Park, Cave Creek, on *Festuca novae-zelandiae*, 23 Feb. 1988, *P.R. Johnston s. num.* (**PDD** 48482); Cass, nr University Hut, on *Festuca novae-zelandiae*, 22 Feb. 1988, *P.R. Johnston* R771 (**PDD** 48464); SOUTHLAND: Awaroa Bog, on ?*Bromus*, 9 May 1984, *P.R. Johnston* R522 (**PDD** 49366); STEWART I: Lee Bay, on *Ammophila arenaria*, 4 May 1984, *P.R. Johnston* R578 (**PDD** 49368); TAUPO: Taupo-Napier road, nr Te Haroto, Turangakumu Reserve, on *Holcus lanatus*, 26 Mar. 1984, *P.R. Johnston* R416 (**PDD** 49360). **Pakistan:** Kagan Valley: Saifal Maluk, on *Festuca*, 30 Aug. 1959, *S. Ahmad s. num.* (**IMI** 81970, as *Lophodermium*). **Scotland:** *s. loc.*, on *Deschampsia*, 23 Feb. 1990, *D.W. Minter s. num.* (**IMI** 337927). **Sweden:** GÖTEBORGS: Strömstad, Koster-Archipel, Bohuslän, on *Ammophila arenaria*, 10 Jul. 1990, *C. Scheuer s. num.* (**GZU**); GOTLAND: Fårö Parish: Gotska Sandön, dunes W of Gamla Gården, on *Ammophila arenaria*, 29 Jun. 1984, *K. & L. Holm* 3145d (**PDD** 60124, **ZT**); *idem loc.*, on *Ammophila arenaria*, 30 Jun. 1984, *K. & L. Holm* 3159d (**PDD** 60125, **ZT**); ÖSTERGÖTLAND: Gryt Parish: nr Strömmen, on *Festuca ovina*, 22 Jul. 1958, *J.A. Nannfeldt* 15378a (**UPS**, as *L. festucae*); *idem loc.*, on *Deschampsia flexuosa*, 22 Jul. 1958, *J.A. Nannfeldt* 15377a (**UPS**, as *L. airarum*); UPPLAND: Bondkyrka Parish: Vardsatra, on *Phleum pratense*, 30 Jun. 1938, *S. Lundell & K.G. Ridelius s. num.* (**UPS**, as *L. phlei*); Dalby Parish: roadside *c.* 75 m SSW of Jerusalem, on grass, 18 Jul. 1985, *K. & L. Holm* 3635b (**PDD** 64551**ZT**); Uppsala, on *Agropyron repens*, 16 Jul. 1936, *S. Lundell* (*Fungi suecici praesertim upsaliensis* no. 3486a; **K**). **Switzerland:** AARGAU: Reusstal, Unterlunkhofen, on grass, 15 May 1980, *P. Crivelli* 21 (**PDD** 65111, **ZT**); SCHAFFHAUSEN: Thayungen, Talweiher, on grass, 4 May 1980, *L. Petrini* 16 (**PDD** 64552, **ZT**). **USA:** ALASKA: Glacier Bay National Monument, Cooper Sta., on *Elymus mollis*, 22 Aug. 1952, *R. Sprague* 879 (**TRTC** 33758, as *L. arundinaceum*); Glacier Bay National Monument, Bartlett Cove, road to lodge, on *Elymus arenarius*, 20 Jun. 1980, *W.B. & V.G. Cooke* 58131 (**NY**); INDIANA: *s. loc.*, on *Triticum aestivum*, 26 May 1920, *H.S. Jackson s. num.* (**TRTC**, as *L. arundinaceum*); NEW YORK: E side of Hemlock Lake, on *Triticum vulgare*, 21 May 1915, *Whetzel & Fitzpatrick s. num.* (**CUP** 9724, as *L. arundinaceum*); OREGON: Lincoln Co.: Yachats, on *Ammophila arenaria*, 10–12 Aug. 1978, *M.A. Sherwood s. num.* (Sherwood, *Phacidiales Exsiccati* no. 15, **CUP**, as *L. arundinaceum*); PENNSYLVANIA: State College, on *Triticum sativum*, 23 May 1908, *H.R. Fulton s. num.* (**CUP** 3594, as *L. arundinaceum*).

Lophodermium arundinaceum var. **dactyli** Roum., *Fungi gallici exsiccati* no. 1700 (1881), *nom. inval.*, *ICBN* Art. 41.3.

Lophodermium dactylis Tehon, *Illinois biol. Monogr.* **13**(4): 46 (1935), *nom. inval.*, *ICBN* Art. 36.1.

Typification: France: Toulouse, Jul. 1881, *C. Roumeguère s. num.* (Roumeguère, *Fungi gallici exsiccati* no. 1700; **K**!; **UPS**!; **NY**!, specimen on which the invalid names *L. arundinaceum* var. *dactyli* and *L. dactylis* were based).

Notes: Although structures resembling unopened ascomata of *Lophodermium* are present on the leaf of the specimen examined from **UPS**, the vertical section shows that it is not a *Lophodermium* but a small stromatic structure. The packet of this exsiccate from **K** has an annotation label attached which states 'No *Lophodermium* present, only Rust'.

TEHON (1935) cited, incorrectly, the place of publication of this taxon as '*Revue mycol.* **3**: 5 (1881)'; this is a reference (in '3e Année. No. 12. 1er Octobre 1881') to a listing of Roumeguère, *Fungi gallici exsiccati* centuries 16–18, citing no. 1700 simply as '*Lophod. arund.* **foliicol.*', without description. The exsiccate packet is labelled '*Lophodermium arundinaceum* f. *dactyli*', but with no accompanying description.

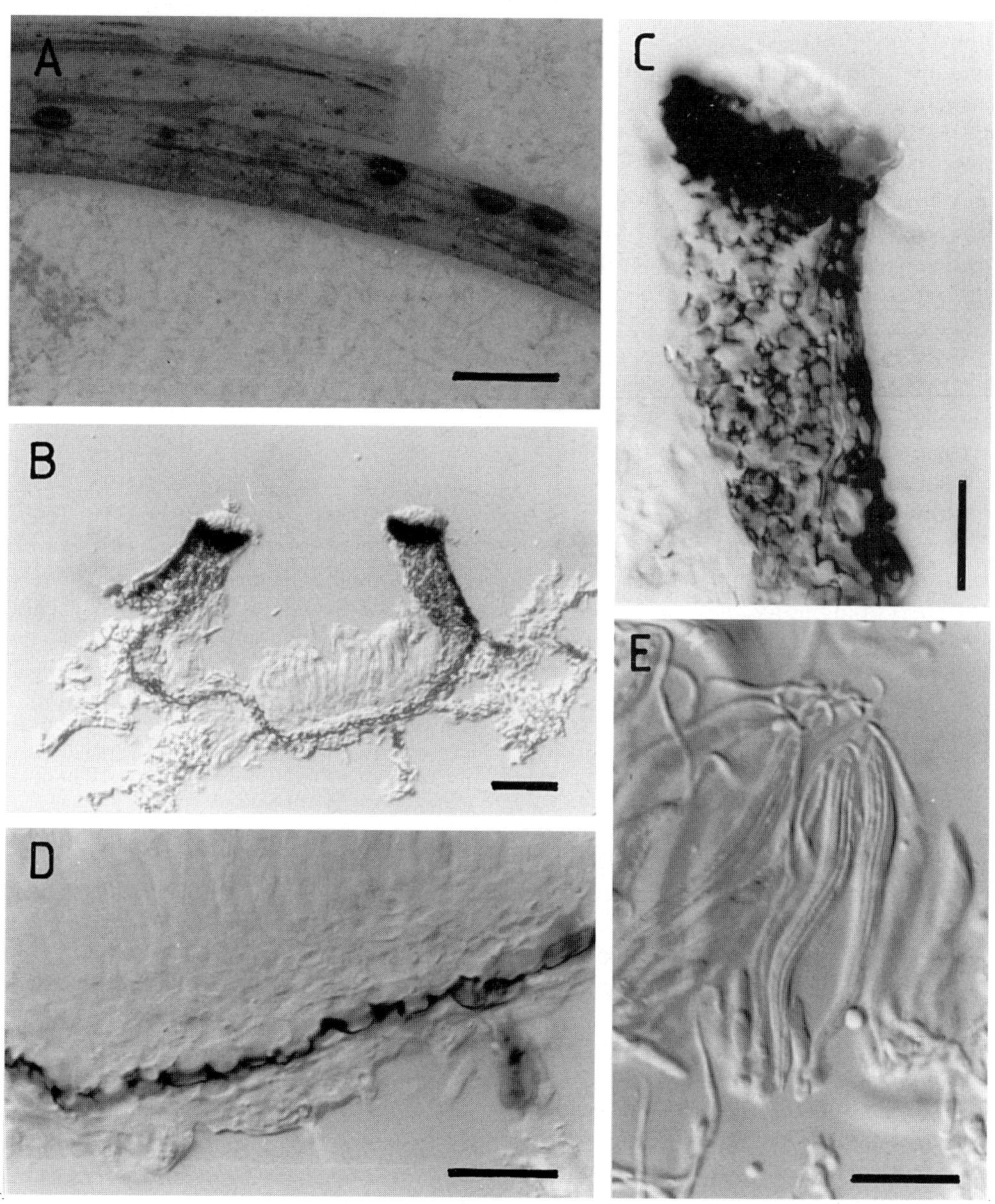

Fig. 44. *Lophodermium danthoniae* (**ILL** 25090). **A**, ascomata (bar = 1 mm); **B**, ascoma in vertical section (bar = 50 μm); **C**, upper wall of ascoma in vertical section; **D**, lower wall of ascoma in vertical section; **E**, asci and paraphyses (bars = 20 μm).

Lophodermium danthoniae Tehon, *Mycologia* **31**: 690 (1939).

Infected areas on dead, partially decomposed leaves, not paler than surrounding host tissue, not associated with zone lines, containing scattered ascomata, conidiomata not seen. *Ascomata* 0·4–0·5 × 0·2–0·3 mm, oblong-elliptical in outline, ends rounded, wall black, opening slit lined with lip cells, initially well differentiated but darkening with age. *Ascomatal insertion* subepidermal. *Covering layer* up to 45 µm thick in vertical section, comprising clypeus of dark brown hyphae within partially broken-down epidermal cells and upper wall of ascomatal stroma. *Upper wall* comprising mostly angular cells, 6–8 µm diam., with irregularly encrusted walls and group of cells with thickly-encrusted walls adjacent to lip cells. *Lower wall* 10–12 µm thick in vertical section, comprising two (or three) rows of globose to cylindrical cells, 6–10 µm diam., mostly hyaline and thin-walled, but lowermost row, or its outer wall, brown to dark brown. *Paraphyses* 1·5–2 µm diam., circinate and coiling at apex. *Asci* 60–75 × 6·5–8·5 µm, subfusoid, tapering gradually to small, rounded apex with undifferentiated wall, 8-spored, short basal stalk developing at maturity with spores extending ± to base. *Ascospores* not seen released, 25–35 × 1·5 µm, non-septate, apical gelatinous cap globose, *c.* 3 µm diam., basal gelatinous cap small, cylindrical.

Typification: USA: WISCONSIN: Brule, on *Danthonia spicata*, 11 Aug. 1934, *J.J. Davis* (**ILL** 25090!, holotype); ILLINOIS: Jackson Co.: Carbondale, on *Danthonia spicata*, 28 Apr. 1938, *G.H. Boewe* (**ILL** 27027!, paratype).

Host: *Danthonia spicata* (*Poaceae*).

Distribution: USA.

Illustrations: Figs 44, 46.

Notes: *Lophodermium danthoniae* has an ascomatal structure typical of the *actinothyrium*-group of *Lophodermium* Group A. It is similar to *L. sieglingiae*, described on *Danthonia decumbens* from Europe, and may represent the same species. However, *L. danthoniae* is known only from the small holotype collection (no *Lophodermium* was found on the paratype collection) and the spore shape characteristic of *L. sieglingiae* was not seen. *Lophodermium robergei* is also similar in hymenial dimensions. Further collections on *Danthonia* from the USA should be examined to confirm identity.

See also Notes under *L. sieglingiae*.

Terriera dracaenae (W. Phillips & Harkn.) P.R. Johnst., **comb. nov.**

Lophodermium dracaenae W. Phillips & Harkn., in COOKE & HARKNESS, *Grevillea* **12**: 84 (1883).

Hypoderma dracaenae (W. Phillips & Harkn.) Kuntze, *Revis. gen. pl.* **3**(3): 487 (1898).

Dermascia dracaenae (W. Phillips & Harkn.) Tehon, *Illinois biol. Monogr.* **13**(4): 64 (1935), *nom. inval.*, *ICBN* Art. 43.1.

Infected areas on dead leaves or stems, paler than surrounding host tissue, not associated with zone lines, containing scattered ascomata, conidiomata not seen. *Ascomata* 0·6–1 × 0·4 mm, some ± orbicular, more commonly oblong to oblong-elliptical, ends rounded, wall initially shiny black, but shelf-like zone along sides of opening slit very well developed, giving mature ascomata matt black appearance. *Ascomatal structure* typical of *Terriera* (see p. 34). *Paraphyses* 1·5–2·5 µm diam., dichotomously branched several times near apex of which some irregularly swollen, forming tight, non-gelatinous epithecium. *Asci* 130–140 (–160) × 6–7 µm, cylindrical, tapering

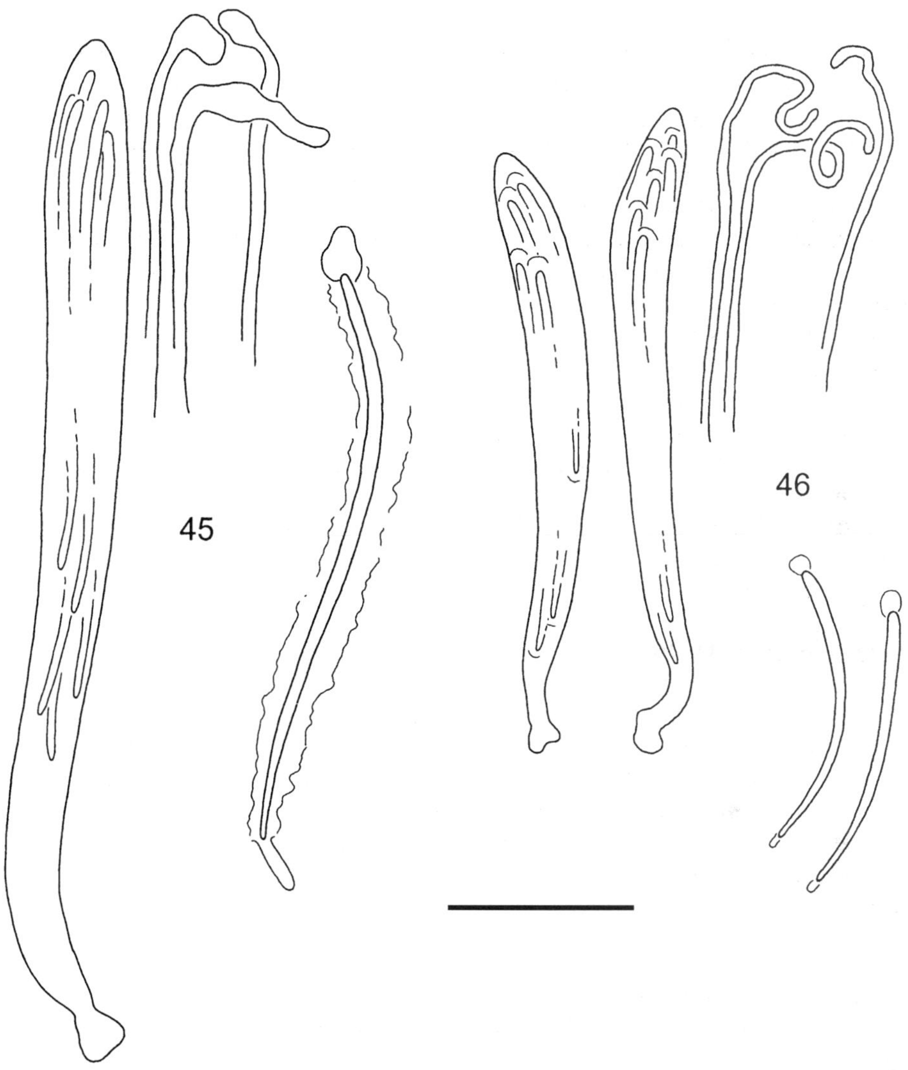

Figs 45–46. 45. *Lophodermium culmigenum.* Asci and apex of paraphyses (**PDD** 64551) and ascospores in water (**PDD** 60124). **46.** *Lophodermium danthoniae* (**ILL** 25090). Asci, apex of paraphyses and released ascospores (bar = 20 μm).

to small, rounded apex with undifferentiated wall, 8-spored, well-developed basal stalk with spores confined to upper 100–110 μm. *Ascospores* not clearly seen released, *c.* 100 × 2 μm, 1-septate, with no apparent gelatinous sheath.

Typification: USA: CALIFORNIA: *s. loc.*, on *Dracaena*, May 1881, *Harkness s. num.* (**BPI**!, Harkness Collection of Fungi no. 2514, holotype; **K**!, isotype).

Host: *Dracaena* (*Agavaceae*).

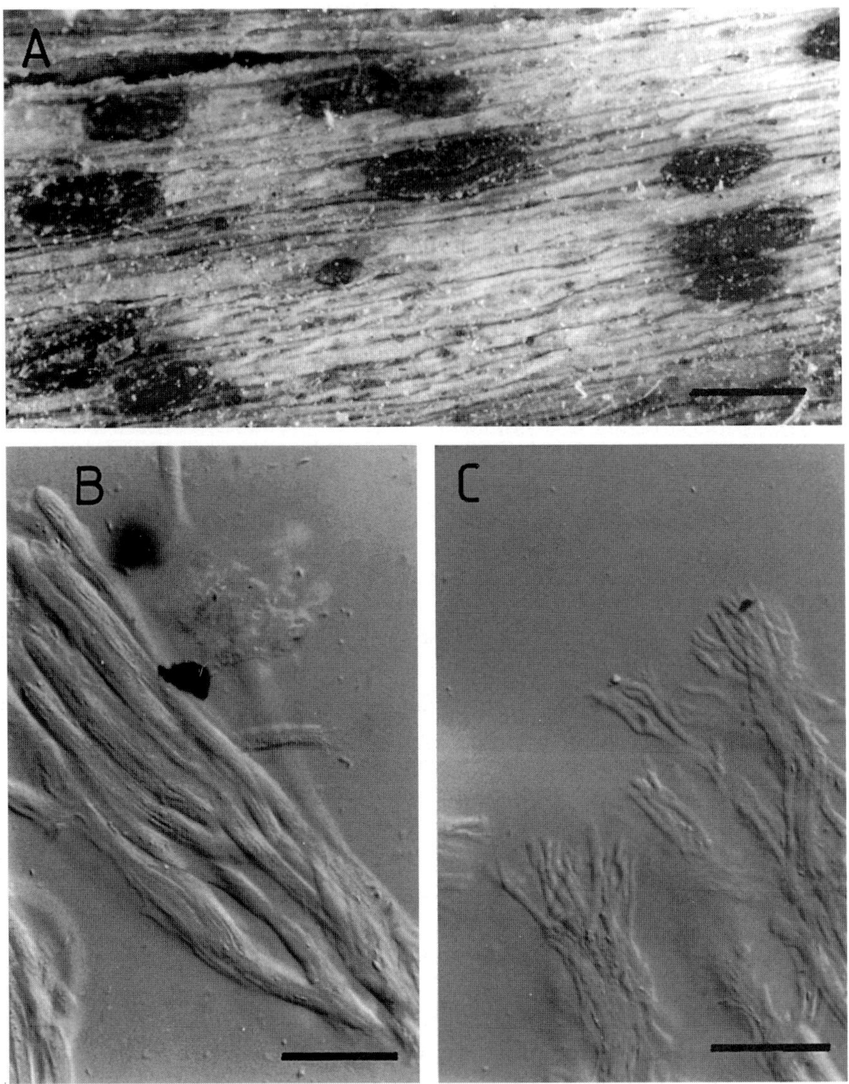

Fig. 47. *Terriera dracaenae* (**NY**, Ellis collection). **A**, ascomata (bar = 1 mm); **B**, asci; **C**, paraphyses (bars = 20 μm).

Distribution: California.

Illustration: Fig. 47.

Notes: *Terriera dracaenae* is characterized within the genus by its long asci and branching, non-gelatinous paraphyses. *Terriera arundinacea* has asci and ascospores of similar length, but the asci are wider, the ascus apex is broadly rounded and the paraphyses are unbranched. The asci of *T. breve* are similar in shape and width but a little shorter, and again the paraphyses are unbranched.

FARR *et al.* (1989) noted that this species was known only from the type collection. The collections examined from **FH** and **NY** may be isotypes.

Additional specimens examined: USA: *s. loc.*, on *Dracaena*, *s. dat.*, *H.W. Harkness s. num.* (**NY**, Ellis Collection); *s. loc.*, on dead leaves of *Dracaena*, *s. dat. nec coll.*, herb. H.W. Harkness *s. num.* (**NY**); CALIFORNIA: *s. loc. nec dat.*, *H.W. Harkness s. num.* (**FH**).

Lophium eriophori Henn., *Verh. bot. Ver. Prov. Brandenb.* **37**: 2 (1895).

Notes: NOGRASEK & MATZER (1994) noted possible similarities between what they referred to as '*Lophium*' *eriophori* Henn., *Lophodermium tumidulum* Sacc., E. Bommer & M. Rousseau and *Coccomyces insignis* P. Karst. The description of *Lophium eriophori* (HENNINGS, 1895; REHM, 1896; SACCARDO, 1899) noted conchiform ascomata, branched paraphyses, clavate asci 180–220 × 12–14 µm and ascospores 100–130 µm long. The recent collection from *Eriophorum vaginatum* cited by NOGRASEK & MATZER (1994) (**Germany:** BAYERN: Wildseefilz, Peustselau, 29 Aug. 1987, *C. Scheuer s. num.*, **GZU**!) has erumpent, subepidermal ascomata, an orange-brown hymenium, branched paraphyses, and large asci (240–270 × 14–16 µm) and ascospores (120–160 × 2–2·5 µm). It is probably conspecific with *Lophium eriophori* and, from the description provided by SHERWOOD (1980), is also very similar to *Coccomyces insignis*, a species described on *Carex pauciflora* from Finland. The type specimens of *Lophium eriophori* and *C. insignis* have not been examined.

The above collection from **GZU** is morphologically typical of *Lophodermium* Group C (see p. 40), but is distinguished from the species accepted in that group by ascus and ascospore size.

There are two collections in **UPS** labelled with the unpublished name '*Lophodermium eriophori*'. One represents *Leptostroma caricinum* (see 'Additional Specimens Examined' under *Lophodermium caricinum*). The other is a species of *Hypoderma* (**Sweden:** GÄSTRIKLAND: Gävle, Tolfforsskogen, nr Tolffors, on *Eriophorum vaginatum*, 14 Jul. 1953, *J.A. Nannfeldt s. num.* (**UPS**)) with subcuticular ascomata, well-developed lip cells, circinate paraphyses, clavate to clavate-stipitate asci, 100–115 × 10·5–12 µm and oblong-elliptical, naviculate ascospores, 25–29 × 3·5–4 µm. There was no evidence that the *Hypoderma* had an associated anamorph and the two collections represent two distinct fungi. Two collections in **NY** from North America, one each from *Eriophorum* and *Carex*, are also *Hypoderma*-like (**Canada:** NEWFOUNDLAND: Whitbourne, on *Eriophorum cyperinum*,12 Aug. 1894, *B.L. Robinson & H. Schrenk s. num.* (**NY**, as *L. caricinum*). **USA:** NEW JERSEY: Newfield, on *Carex*, 2 Sep. 1877, *s. coll.* (**NY**, as *L. caricinum*)). They differ from the European collections in having slightly longer and narrower ascospores (30–40 × 2–2·5 µm). Although these two collections are typical of *Hypoderma* in most features (JOHNSTON, 1990*d*) – lower wall of ascoma darkening before upper wall, clavate-stipitate asci, unswollen paraphyses, well-developed thick upper wall, and lip cells – they are atypical in having the epidermal cells of the host more or less intact within the upper ascomatal wall. These specimens may best be considered as representing an atypical species of *Hypoderma*.

The North American and European *Eriophorum*-inhabiting species of *Hypoderma* are distinct, representing two presumably undescribed species. Both differ from the only described '*Hypoderma*' from *Cyperaceae*, *H. scirpinum* (≡ *Hypohelion scirpinum* (DC.) P.R. Johnst.), in ascus and ascospore size, in having lip cells along the opening slit and a darkened lower ascomatal wall.

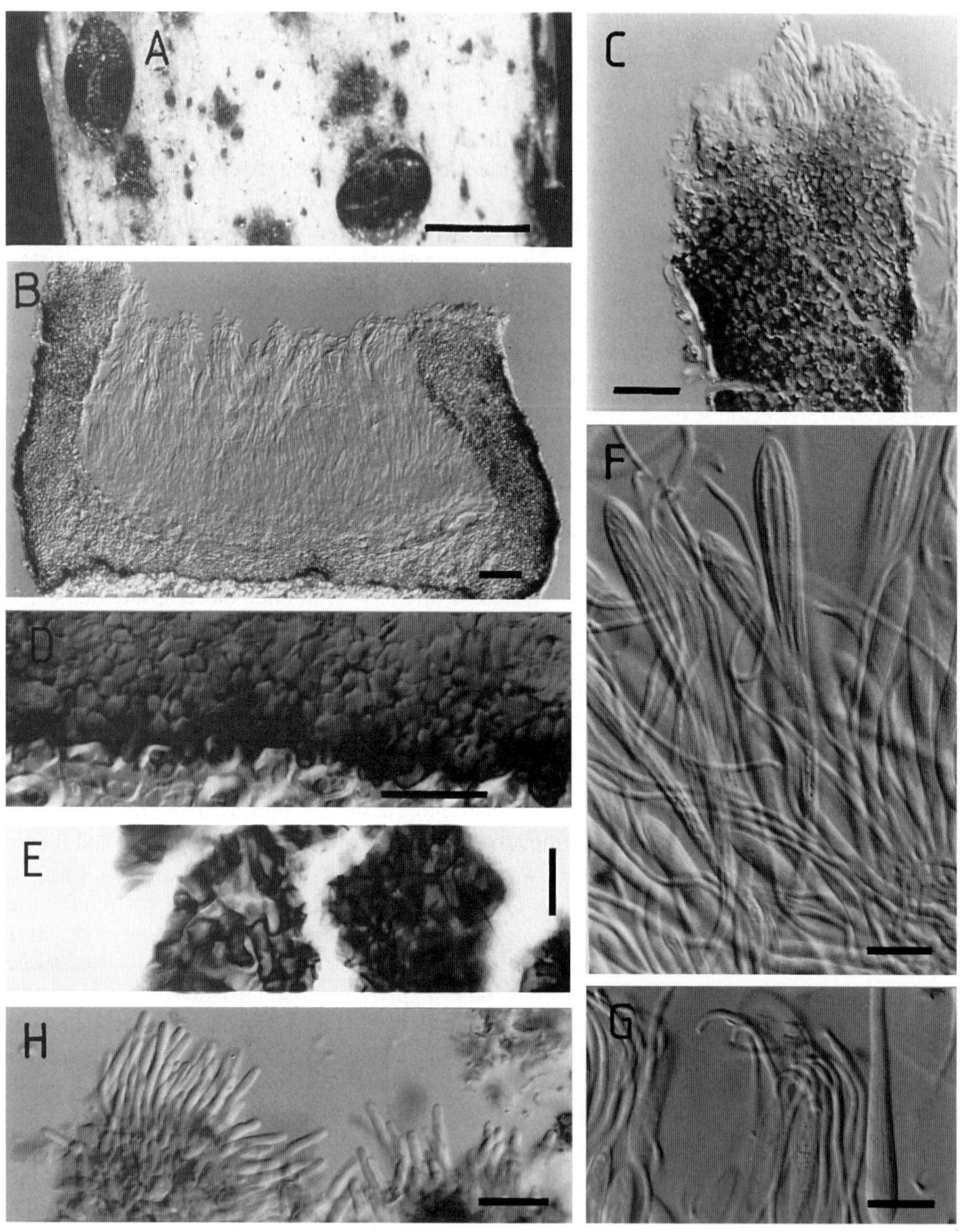

Fig. 48. *Lophodermium eucalypti* (*Sherwood et al.*, **FH**). **A**, ascomata (bar = 1 mm); **B**, ascoma in vertical section (bar = 50 μm); **C**, upper wall of ascoma in vertical section; **D**, lower wall of ascoma in vertical section; **E**, lower wall of ascoma in squash mount; **F**, asci; **G**, paraphyses; **H**, lip cells (bars = 20 μm).

Lophodermium eucalypti (Rodway) P.R. Johnst., **comb. nov.**
Colpoma eucalypti Rodway, *Pap. Proc. r. Soc. Tasm.* 1924: 92 (1925).
Lophodermium mahuianum P.R. Johnst., *N.Z. Jl Bot.* **27**: 259 (1989).

Infected areas on dead leaves, not associated with bleaching of host tissue or with zone lines, containing scattered ascomata, conidiomata not seen. *Ascomata* 0·8–1·4 × 0·5–0·7 mm, broadly elliptical to slightly irregular in outline, with margin sharply defined, wall black, single, longitudinal opening slit lined with narrow lip cells. *Ascomatal insertion* subcuticular. *Covering layer* up to 100 μm thick in vertical section, slightly thinner toward base, comprising host cuticle and upper wall of ascomatal stroma. *Upper wall* pale brown, ± uniform in structure, comprising globose cells, 5–8 μm diam., with walls slightly and irregularly thickened, one or two rows of cells adjacent to host cuticle with somewhat darker walls, unbranched lip cells lining opening slit. *Lower wall* up to 50 μm thick in vertical section, primarily comprising several rows of ± globose, hyaline- to pale brown-walled cells, 4–6 μm diam., with one or two rows of similarly-shaped cells with darkened, thickened walls on lower-side of wall, lowermost row of cells smaller, irregular in shape, globose to cylindrical, in squash mount lower wall very dark, comprising several rows of irregularly shaped, ± cylindrical cells, 4–6 μm diam., underlaid by narrow-cylindrical to hyphal cells, 3–4 μm diam., sometimes partly tangled. *Paraphyses* 2–2·5 μm diam., circinate at apex. *Asci* 150–180 × 11–12·5 μm, clavate-stipitate, tapering to subtruncate apex with undifferentiated wall, 8-spored, basal stalk with spores confined to upper 90–100 μm. *Ascospores* 50–60 × 2·5 μm, tapering gradually to base, ± straight when released, apical gelatinous cap 3–4 μm diam., basal gelatinous cap 1·5–2 μm diam., entire spore surrounded by an additional, looser gelatinous sheath 5–6 μm wide.

Typification: Australia: TASMANIA: Mt Field National Park, on *Eucalyptus*, Mar. 1922, *L. Rodway s. num.* (**HO** 32533!, holotype of *C. eucalypti*). **New Zealand:** TAUPO: Tongariro National Park, nr Mahuia Campsite, on *Nothofagus solandri* var. *cliffortioides*, 20 Nov. 1984, *P.R. Johnston* R596 (**PDD** 47022!, holotype of *L. mahuianum*).

Hosts: *Xerophyllum* (*Liliaceae*), as well as various dicotyledons and conifers, including *Agathis* (*Araucariaceae*); *Banksia, Hakea, Telopea* (*Proteaceae*); *Dracophyllum* (*Epacridaceae*); *Eucalyptus, Metrosideros* (*Myrtaceae*); *Halocarpus, Phyllocladus, Podocarpus* (*Podocarpaceae*); *Nothofagus* (*Fagaceae*).

Distribution: Tasmania, New Zealand, western North America.

Illustrations: Figs 3B, 48, 51.

Notes: The above description is based on the collections from *Xerophyllum*. This species possesses the characters traditionally diagnostic of *Lophodermium* – filiform ascospores and ascomata with a single opening slit – yet it shares all the ascomatal characters of *Hypoderma sensu* JOHNSTON (1990*d*). The combination made here is based on the traditional, widely used characters, because as yet there is no consensus that the scheme outlined by JOHNSTON (1990*a*) is an appropriate alternative. Despite this, the phylogenetic relationship of *L. eucalypti* is almost certainly with the elliptical-spored species of *Hypoderma* and the related *Lophodermium* Group 3 of JOHNSTON (1989*b*), rather than with most species of *Lophodermium s. lat.*

Lophodermium eucalypti was recorded on *Phormium* by JOHNSTON (1992, as *L. mahuianum*). Re-examination of the specimen on which this was based (**PDD** 57531) shows it to be a different species, with consistently shorter ascospores and with ascoconidia. It is probably related to the large species of *Hypoderma* found on *Dracophyllum* in New Zealand, which likewise produce ascoconidia. Although the *Phormium*-inhabiting species shares with *L. eucalypti* an unusually thick lower ascomatal wall, its structure differs, comprising several layers of hyphal cells forming

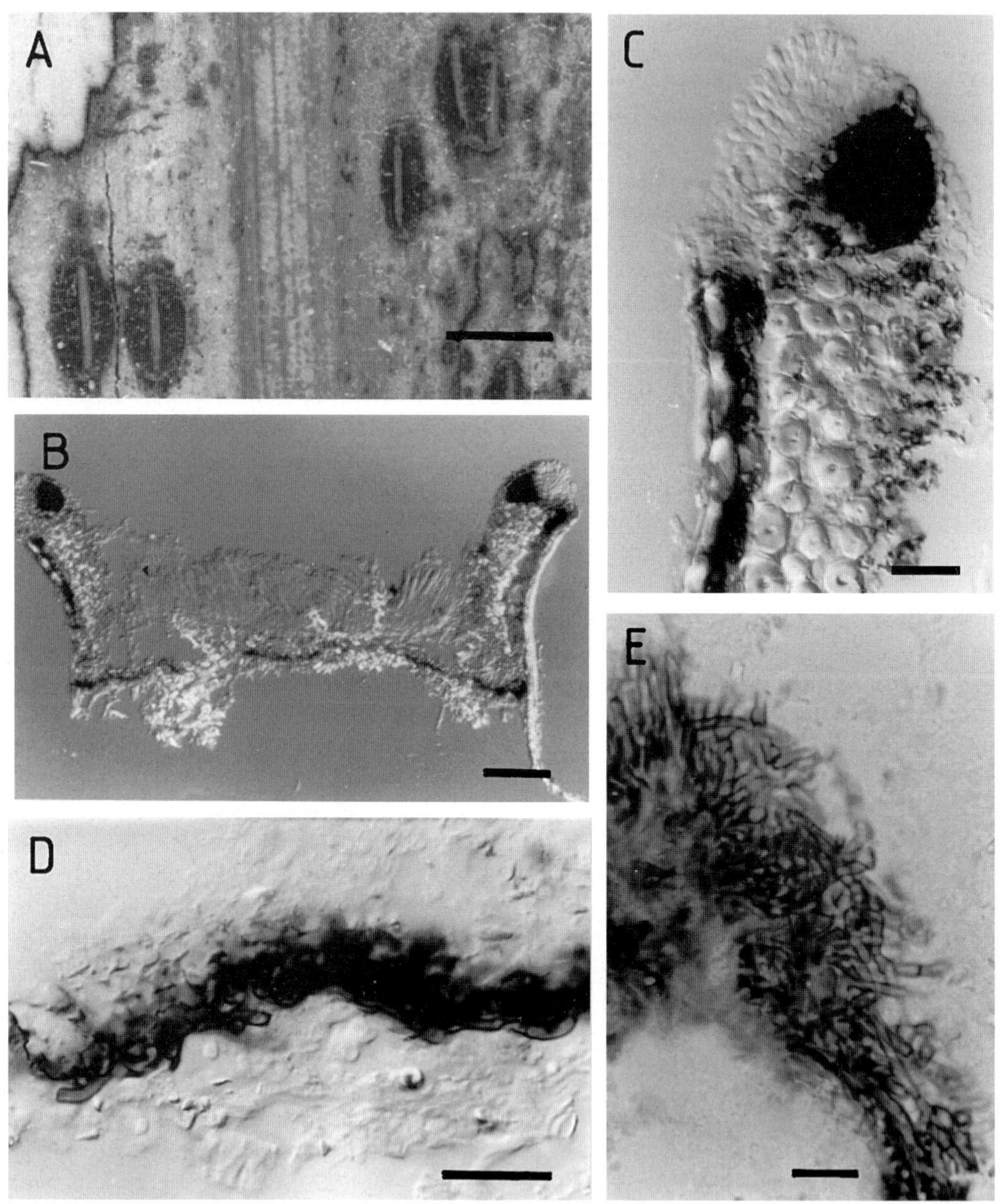

Fig. 49. *Lophodermium eximium* (Rabenhorst, *Fungi europaei* no. 2643a, **RO**). **A**, ascomata (bar = 1 mm); **B**, ascoma in vertical section (bar = 50 μm); **C**, upper wall of ascoma in vertical section; **D**, lower wall of ascoma in vertical section; **E**, lower wall of ascoma in horizontal section (bars = 20 μm).

a well-developed *textura intricata*.

Most *Rhytismataceae* have a limited host range and the diversity of monocotyledonous, dicotyledonous and coniferous hosts reported for this species is very unusual. Its broad host range could be related to its usual occurrence on well-rotted, partially decomposed leaves. Most *Rhytismataceae* form fruiting bodies on recently fallen leaves, many probably infecting the still-living leaves of their hosts initially as an endophytic phase to their life cycle. *Lophodermium eucalypti*, on the other hand, may be a true litter-inhabiter, invading the host tissue only after it has died and fallen. The *Proteaceae* reported as hosts here are new host records for the species. Several other species of *Rhytismataceae* known to be widespread in tropical and subtropical regions are found on a wide range of hosts (JOHNSTON, 1992, 1997). It is likely that with further collecting *L. eucalypti* will be found in other regions around the Pacific.

Although the herbarium packet containing the collection from *Xerophyllum* notes that the fungus is associated with dead portions of still-living leaves and may be parasitic, there are two distinct taxa of *Rhytismataceae* in the packet, the second of which does appear to be associated with more recently killed tissue. This second taxon has intra-epidermal ascomata with the upper wall uniform in thickness and structure, with no differentiated layers along the future line of opening, a poorly-differentiated lower wall comprising mostly cells with thin, pale walls, broad-cylindrical to subsaccate asci, and bifusiform to trifusiform ascospores, 1–4-septate, surrounded by a thick gelatinous sheath. This description fits none of the genera in the family but, superficially at least, most closely resembles some of the *Pinaceae*-inhabiting species with saccate asci and large ascospores with one or more constrictions along their length.

Additional specimens examined: Australia: TASMANIA: Central Plateau, nr Great Lake, Liffey Forest Reserve, on *Telopea truncata*, 28 May 1988, *P.R. Johnston* 156A (**PDD** 56311, **HO**); Central Plateau, western shore of Great Lake, Stoney Creek, on *Telopea truncata*, 27 May 1988, *P.R. Johnston* 142 (**PDD** 56317, **HO**); Central Plateau, eastern side of Great Lake, nr Mt Blackwood, on *Eucalyptus*, 28 May 1988, *P.R. Johnston* 152 (**PDD** 56310, **HO**); nr Hobart, Mt Wellington, Pipes Track, on *Hakea lissosperma*, 20 May 1988, *P.R. Johnston* 32 (**PDD** 56363, **HO**); *idem loc.*, on *Banksia marginata*, 20 May 1988, *P.R. Johnston* 33 (**PDD** 56260, **HO**); Mt Field National Park, nr Lake Fenton, on *Hakea lissosperma*, 24 May 1988, *P.R. Johnston* 96 (**PDD** 56312, **HO**); *idem loc.*, on *Eucalyptus*, 24 May 1988, *P.R. Johnston* 95 (**PDD** 56318, **HO**). **New Zealand:** MARLBOROUGH: Pelorus Bridge Scenic Reserve, on *Nothofagus solandri* var. *solandri*, 30 May 1989, *P.R. Johnston & E.M. Gibellini s. num.* (**PDD** 55275); MID CANTERBURY: Craigieburn Forest Park, nr skifield, on *Podocarpus nivalis*, 24 Feb. 1988, *P.R. Johnston* R781 (**PDD** 48451); NELSON: nr Westport, Denniston Plateau Walkway, on *Metrosideros umbellata*, 25 May 1989, *P.R. Johnston & E.M. Gibellini s. num.* (**PDD** 55273); NORTHLAND: Russell State Forest, Punaruku Rd, Hori Wehi Wehi Track, on *Agathis australis*, 11 Aug. 1988, *P.R. Johnston* R801 (**PDD** 54133); TAUPO: Kaimanawa Forest Park, Tree Trunk Gorge, on *Nothofagus solandri* var. *cliffortioides*, 19 May 1989, *P.R. Johnston s. num.* (**PDD** 55569); Tongariro National Park, Mahuia Campsite, on *Nothofagus solandri* var. *cliffortioides*, 20 Nov. 1984, *P.R. Johnston* R596 (**PDD** 47022); nr Mahuia campsite, on *Halocarpus bidwillii*, 20 Nov. 1984, *P.R. Johnston* R598 (**PDD** 49381); Tongariro National Park, Mangaturuturu–Turoa Track, on *Podocarpus nivalis*, 23 Mar. 1984, *P.R. Johnston* R456 (**PDD** 48452); Tongariro National Park, Mangaturuturu Stream, on *Dracophyllum recurvum*, 21 Nov. 1984, *P.R. Johnston* R592 (**PDD** 48615); *idem loc.*, on *Podocarpus nivalis*, 21 Nov. 1984, *P.R. Johnston* R590 (**PDD** 48619); Tongariro National Park, nr HQ, on *Nothofagus solandri* var. *cliffortioides*, 19 Mar. 1984, *P.R. Johnston* R449 (**PDD** 49251); Tongariro National Park, Silica Springs Walk, nr bridge, on *Nothofagus solandri* var. *cliffortioides*, 24 Mar. 1984, *P.R. Johnston* R493 (**PDD** 49252); WELLINGTON: Tararua Range, nr Dundas Hut, on *Phyllocladus alpinus*, Feb. 1985, *P.R. Johnston* T46 (**PDD** 47067); Dundas Hut, Ridge Track, on *Dracophyllum filifolium*, 10 Feb. 1985, *P.R. Johnston* R695c (**PDD** 49053); Ridge Track, nr clearing, on *Podocarpus totara*, 7 Feb. 1985, *P.R. Johnston* R702 (**PDD** 49382); WESTLAND: Fox Glacier, Gillespies Beach Rd, on *Phyllocladus*, 7 Apr. 1983, *P.R. Johnston* R222 (**PDD** 47384). **USA:** OREGON: Linn Co.: Cascade Mtns: Wildcat Mtn, on *Xerophyllum tenax*, *M.A. Sherwood et al. s. num.* (**FH**, as *Lophodermium* sp., *p.p.*).

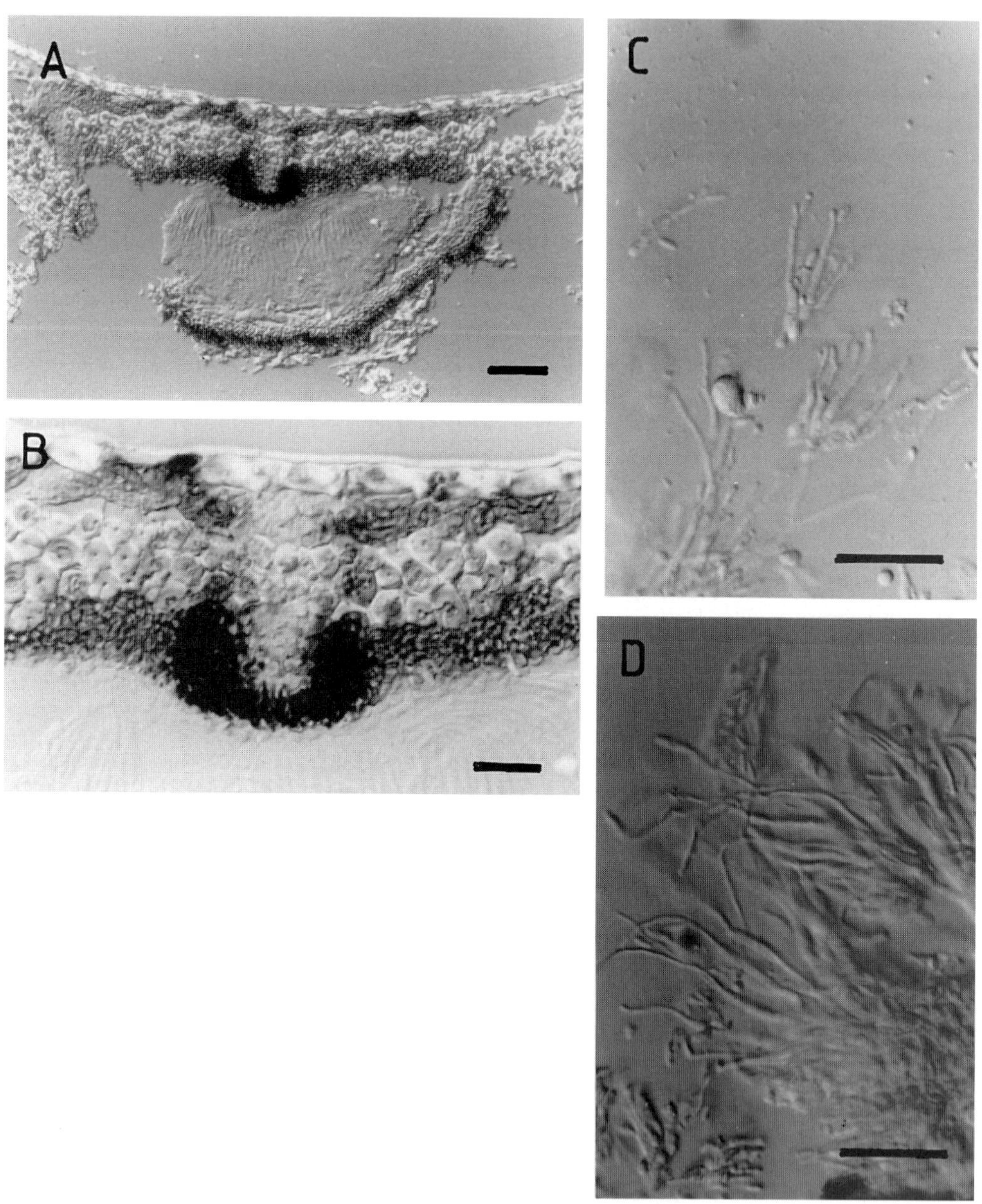

Fig. 50. *Lophodermium eximium* (Rabenhorst, *Fungi europaei* no. 2643a, **RO**). **A**, unopened ascoma in vertical section (bar = 50 µm); **B**, upper wall of unopened ascoma in vertical section; **C**, conidiogenous cells and conidia; **D**, sterile filaments near centre of conidioma (bars = 20 µm).

Lophodermium eximium Ces. ex G. Winter, *Hedwigia* **21**: 7 (1882).
Hypoderma eximium (Ces. ex G. Winter) Kuntze, *Revis. gen. pl.* **3**(3): 487 (1898).

Infected areas on dead leaves and sheaths, paler than surrounding host tissue, often associated with narrow, black zone lines, containing scattered ascomata and conidiomata. *Ascomata* initially elliptical, dark grey, with well-differentiated, narrow paler zone along future line of opening, mature ascomata 0·6–1·2 × 0·4–0·5 mm, arching above level of surrounding host leaf, broadly elliptical in outline with rounded ends, wall black, single, longitudinal opening slit, with well-developed, brownish-orange lip cells, in some ascomata lips lost soon after opening. *Ascomatal insertion* subepidermal and beneath extensive host fibre bundles adjacent to epidermis. *Covering layer* up to 85 µm thick in vertical section, comprising clypeus of dark brown, thick-walled hyphae within epidermal cells, and upper wall of ascomatal stroma. *Upper wall* poorly developed, mostly of cells with pale, slightly and irregularly encrusted walls, but group of cells with very thick, dark walls adjacent to well-developed lip cells. *Lower wall* up to 25 µm thick in vertical section, but in places ± lacking, or intermixed with partially broken-down host cells, cells appearing to be irregular in shape, small and globose or narrow-cylindrical, with thick, dark walls, in squash mount darkened cells of lower wall mostly long-cylindrical, 3 µm diam., forming layer variable in thickness, in places comprising single row of cells, in others comprising several rows forming *textura intricata*. *Paraphyses* 2–2·5 µm diam., slightly swollen and hooked to loosely circinate at apex. *Asci* 100–125 × 7·5–9·5 µm, subclavate to subfusoid, tapering gradually to small rounded apex with undifferentiated wall, 8-spored, long basal stalk at maturity with spores confined to upper 70–80 µm. *Ascospores* 60–70 × 1·5 µm, straight when released, tapering to narrow base, non-septate, apical gelatinous cap small, globose, basal gelatinous cap short-cylindrical, gelatinous sheath not observed. *Conidiomata* associated with immature ascomata, 0·2–0·3 mm diam., ± round in outline, wall pale brown to ± concolorous with surrounding host tissue near centre, grey to dark grey at edge, ostiole not obviously differentiated, over-mature conidiomata appearing darker, near centre are grouped filiform elements up to 65 µm long, *c.* 1 µm diam. over most of length, wider near base. *Conidiogenous cells* solitary, forming palisade across base of conidioma, ± cylindrical, 10–20 × 1·5–2·5 µm, proliferation sympodial with two developing conidia held at apex of conidiogenous cells. *Conidia* 3·5–5 × 1–1·5 µm, oblong-elliptical, straight, ends rounded, non-septate, hyaline.

Typification: Italy: *s. loc.*, on *Ampelodesmos tenax*, *s. coll.* (Rabenhorst, *Fungi europaei* no. 2643a; **RO**!, *lectotypus* of *L. eximium*, selected here; **K**!; **NY**!, *isolectotypi*, selected here).

Host: *Ampelodesmos tenax* (*Poaceae*).

Distribution: Italy, Algeria.

Illustrations: Figs 49, 50, 52.

Notes: *Lophodermium eximium* has an ascomatal structure and development typical of the *arundinaceum*-group of *Lophodermium* Group A. The only species known from *Ampelodesmos*, it is characterized by zone lines on the host leaf, the presence of very pale, lenticular conidiomata, the long basal stalk in mature asci, and ascospores tapering to a very narrow base. The other species in this group associated with conidiomata, *L. arundinaceum*, has conidiomata barely visible, globose and deeply immersed within the host leaf and conidiogenous cells mostly proliferating percurrently rather then sympodially.

An anamorph has not been reported previously for *L. eximium*, although conidiomata were common on all duplicates of the type collection examined.

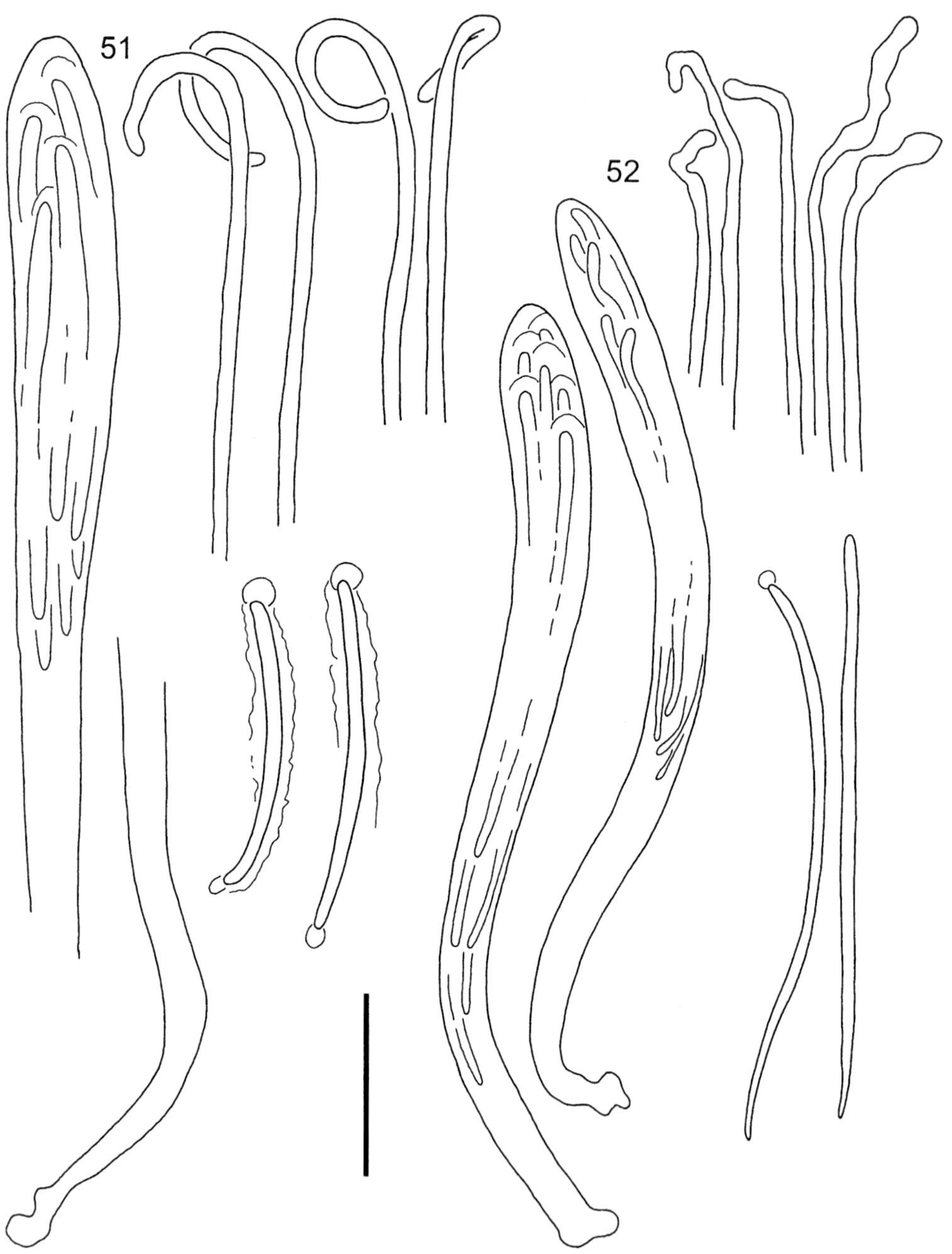

Figs 51–52. 51. *Lophodermium eucalypti* (*Sherwood et al.*, **FH**). Asci, apex of paraphyses and ascospores in water. **52.** *Lophodermium eximium.* Ascus, apex of paraphyses (Rabenhorst, *Fungi europaei* no. 2643a, **RO**) and released ascospores (**IMI** 45937) (bar = 20 μm).

Additional specimens examined: Algeria: Oued-Djer, on *Ampelodesmos mauritanica*, 15 Feb. 1915, *R. Maire s. num.* (Maire, *Mycotheca boreali-africana* no. 274; **NY**). **Italy:** *s. loc.*, on *Ampelodesmos tenax*, 30 Jan. 1951, *Cicarone s. num.* (**IMI** 45937, as *L. exiguum*).

Terriera fourcroyae (Berk. & Broome) P.R. Johnst., **comb. nov.**
Hysterium fourcroyae Berk. & Broome, *J. Linn. Soc. Bot.* **14**: 132 (1873).
Gloniella fourcroyae (Berk. & Broome) Sacc., *Syll. Fung.* **2**: 796 (1883).
Lophodermium fourcroyae (Berk. & Broome) Massee, *Grevillea* **22**: 34 (1893).
Gloniella fourcroyae var. *palmicola* Sacc., *Syll. Fung.* **2**: 769 (1883).

Infected areas on dead leaves, slightly paler than surrounding host tissue, not associated with zone lines, containing gregarious ascomata, these sometimes confluent and forming dark patches on leaf, conidiomata not seen. *Ascomata* 0·5–1·5 × 0·25–0·3 mm, oblong-elliptical in outline, irregularly oriented on leaf and often curved, ends rounded, wall black but often with pale patch near ends of ascomata and faint paler line sometimes visible along future line of opening of unopened ascomata, single longitudinal opening slit, with poorly-developed black, shelf-like area to either side, ascomata often coalescing, then sometimes appearing branched. *Ascomatal insertion* subepidermal. *Ascomatal structure* matching that of *Terriera* (see p. 37), but darkening of cells in periphysoid layer and in lower wall less well developed than usual, clypeus also poorly developed, comprising scattered dark-walled hyphae within ± intact epidermal cells. *Paraphyses* 1·5 µm diam., increasing to 2·5–4 µm at irregularly swollen apex, often dichotomously branched once or twice, forming epithecium 20–25 µm thick. *Asci* 95–110 × 6–6·5 µm, cylindrical, tapering to small, rounded apex with undifferentiated wall, 8-spored, short basal stalk with spores confined to upper 75–90 µm. *Ascospores* not seen released, 60–70 × 1·5–2 µm, 1-septate, partly released spores appearing to taper slightly at ends and to be gently coiled or sigmoid on release.

Typification: Sri Lanka: *s. loc.*, on leaves of *Furcraea* (as '*Fourcroya*'), *Thwaites* 361 (**K** 32866!, holotype of *H. fourcroyae*); *s. loc.*, *hosp. nec dat.*, *Thwaites* 307 (**K** 32867!, holotype of *Gloniella fourcroyae* var. *palmicola*).

Host: *Furcraea* (*Agavaceae*).

Distribution: Sri Lanka.

Illustration: Fig. 53.

Notes: This species clearly belongs in the genus *Terriera*, but is unusual in lacking a well-developed pale zone along the future line of opening, in having a poorly-developed clypeus and relatively poorly-developed shelf-like covering over the hymenium.

The hymenial elements of *T. fourcroyae* are similar in dimensions to those of the tropical American species *T. pandani* (Tehon) P.R. Johnst. However, as discussed on p. 36, species limits among this group of fungi remain uncertain. The phylogenetic significance of this apparent similarity, and of the macroscopic differences between the two species, is at present unknown.

The type specimen of *Gloniella fourcroyae* var. *palmicola* matches *T. fourcroyae* morphologically and the host substratum appears to be the same in both collections.

Cannon & Minter (1986) reported multi-septate ascospores for *T. fourcroyae*. Although ascospore septation was unclear in the material examined, only one septum appeared to be present.

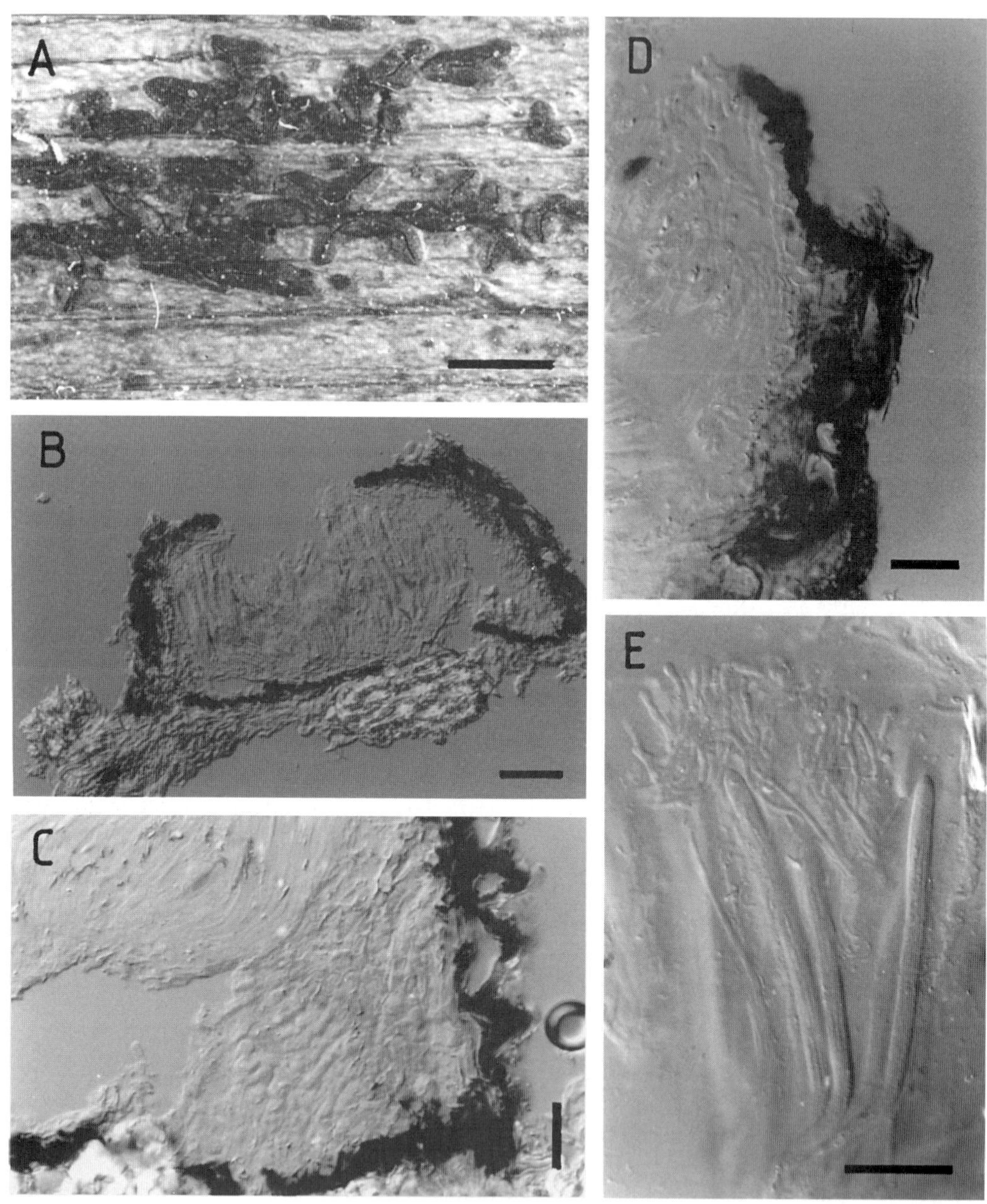

Fig. 53. *Terriera fourcroyae* (holotype of *H. fourcroyae*, **K**). **A**, ascomata (bar = 1 mm); **B**, ascoma in vertical section (bar = 50 µm); **C**, detail of lower corner of ascoma in vertical section; **D**, upper wall of ascoma in vertical section; **E**, asci and paraphyses (bars = 20 µm).

Terriera fuegiana (Speg.) P.R. Johnst., **comb. nov.**
Lophodermium fuegianum Speg., *Boln Acad. nac. Cienc. Córdoba* **11**: 250 (1888).
Hypoderma fuegianum (Speg.) Kuntze, *Revis. gen. pl.* **3**(3): 487 (1898).

Infected areas on dead leaves, not associated with bleaching of host tissue or zone lines, containing groups of ascomata, conidiomata not seen. *Ascomata* 0·5–0·7 × 0·3–0·4 mm, oblong-elliptical to broad-elliptical in outline, ends rounded, wall black, with narrow pale zone along future line of opening, single, longitudinal opening slit, with narrow, black, flattened area on either side. *Ascomatal insertion* subepidermal. *Ascomatal structure* typical of *Terriera* (see p. 37). *Paraphyses* 1·5 µm diam., undifferentiated or slightly swollen at apex, embedded in gel. *Asci* 75–95 × 7–10 µm, cylindrical to subfusoid, apex broadly rounded, 8-spored, spores extending almost to base. *Ascospores* not seen released, 60–65 × 1·5–2·5 µm, tapering slightly to ends, 1-septate, with no gelatinous sheath.

Typification: Argentina: TIERRA DEL FUEGO: Isla de los Estados, Agaia, on *Rostkovia grandiflora* (as *Marsippospermum grandiflorum*), Jun. 1882, *C. Spegazzini s. num.* (**LPS** 1007!).

Host: *Rostkovia grandiflora* (*Juncaceae*).

Distribution: Tierra del Fuego.

Illustration: Fig. 54.

Notes: In comparison with most species of *Terriera*, the shelf-like extension of the upper wall across the top of the hymenium is poorly developed in *T. fuegiana*. This species is distinguished by its short, wide asci; those of *T. asteliae* are of a similar size, but differ in having a broadly truncate rather than rounded apex.

The type collection is small and in poor condition. A specimen collected by Dusén (**BPI**) is from the same host and general locality, and the description and notes here are based on data from both collections. Two distinct species of *Rhytismataceae* are present on the Dusén collection, the other being *Duplicaria antarctica*. They can be distinguished macroscopically, *D. antarctica* having larger ascomata and being associated with conidiomata.

SPEGAZZINI (1887) provided a detailed description and made illustrations on the herbarium packet. He gave dimensions of asci (60–65 × 5 µm) and ascospores (40 × 1·5 µm) and illustrated cylindrical asci having a broadly rounded, slightly thickened apex with a small central pore. Results from examination of the type material largely agree with SPEGAZZINI's data in respect of the macroscopic appearance of the ascomata and ascus, ascospore and paraphysis shape, although the dimensions of asci and ascospores cited are smaller than those observed here.

Additional specimen examined: Argentina: TIERRA DEL FUEGO: Isla Desolarion, Puerto Angusto, on *Rostkovia* (as *Marsippospermum*), 1896, *P. Dusén s. num., p.p.* (**BPI**, as *Lophodermium antarcticum*).

Lophodermium fusiforme P.R. Johnst., **sp. nov.**

Etymology: Referring to the shape of the ascomata.

A *L. arundinaceo* ascomatibus fusiformibus, cellulis labiorum debiliter evolutis, cellulis atrobrunneis parietis inferioris ascomatum late cylindricalis vel subglobosis et in strato unico dispositis, apicibus paraphysium subclavatis differt.

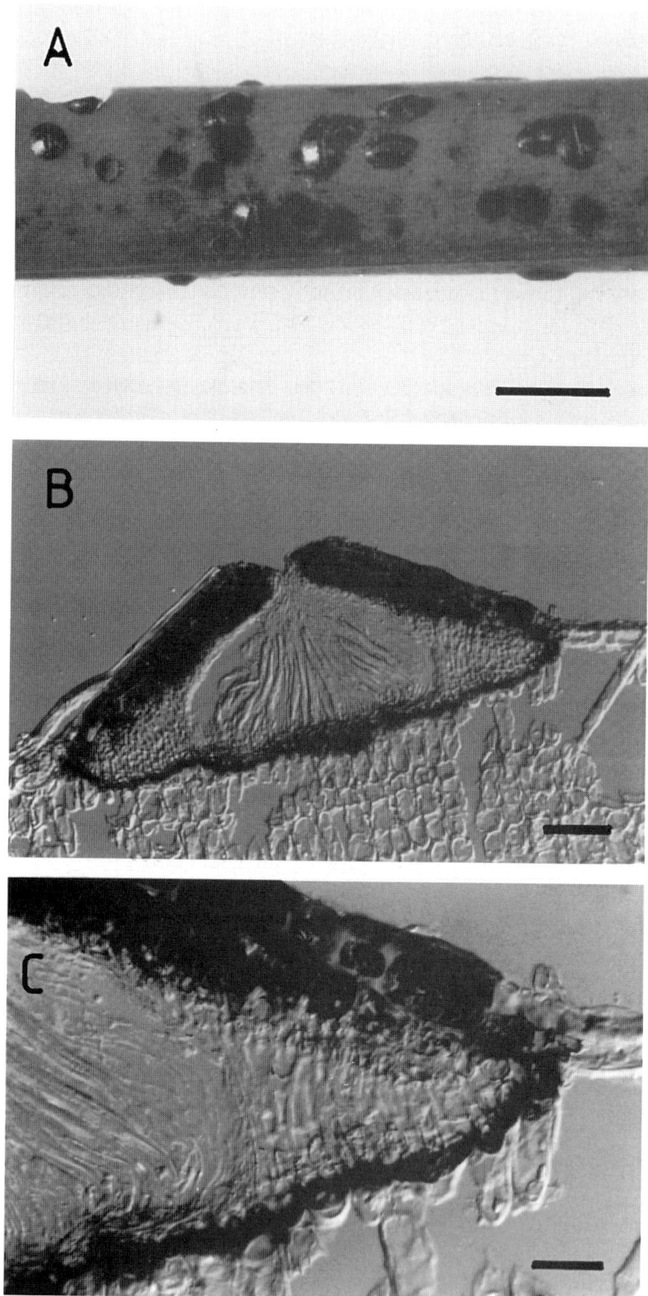

Fig. 54. *Terriera fuegiana*. **A**, ascomata (**LPS** 1007) (bar = 1 mm); **B**, ascoma in vertical section (*Dusén s. num.*, **BPI**) (bar = 50 µm); **C**, detail of lower corner of ascoma in vertical section (*Dusén s. num.*, **BPI**) (bar = 20 µm).

Infected areas on dead leaves, not associated with bleached areas or zone lines, containing scattered ascomata and some collections with structures resembling conidiomata of *Rhytismataceae*. *Ascomata* 0·6–1·0 × 0·3–0·4 mm, elliptical in outline, ends ± acute, margin not sharply differentiated, wall black, with no pale zone along future line of opening, single longitudinal opening slit lined with poorly-differentiated dark lip cells. *Ascomatal insertion* subepidermal. *Covering layer* of ascomata 50–60 µm thick in vertical section, comprising clypeus of dark-walled hyphae within partially broken-down epidermal cells, and upper wall of ascomatal stroma. *Upper wall* up to 50 µm thick in vertical section, comprising mostly globose cells 8–10 µm diam., with walls slightly and irregularly encrusted with dark brown material, but with group of very dark-walled cells adjacent to poorly-developed lips. *Lower wall* up to 10–15 µm thick in vertical section, comprising three or four rows of globose cells, lowermost row with all walls, or only the lower wall, darkened, in squash mount darkened cells of lower wall broad-cylindrical to globose, 6–10 (–12) µm diam., arranged in single layer across base of ascoma. *Paraphyses* 2 µm diam., undifferentiated or swollen to 3–4 µm at knob-like apex, forming regular palisade above developing asci. *Asci* 105–120 (–130) × (8·5–) 9–10·5 (–11) µm, subclavate, tapering to small rounded apex with undifferentiated wall, 8-spored, well-developed basal stalk at maturity with spores confined to upper 90–100 µm. *Ascospores* (45–) 60–75 (–90) × (1·5–) 2 µm, straight when released, tapering gradually to base, non-septate, apical gelatinous cap 5–6 × 3 µm, broad-cylindrical to pyriform, bent in side view, basal gelatinous cap narrow-cylindrical, 5–6 × 1·5–2 µm, gelatinous sheath 5–7 µm diam. *Conidioma*-like structures round in outline, 0·2 mm diam., pale orange-brown, lenticular in vertical section. *Conidia* and *conidiogenous cells* not seen.

Typification: Switzerland: GRAUBÜNDEN: Bergün, Val Tuors, on *Dactylis glomerata*, 14 Jul. 1980, *L. Petrini* 55 (**ZT**!, *holotypus*; **PDD** 65727!, *isotypus*).

Hosts: *Agrostis*, *Anthoxanthum*, *Brachypodium*, *Calamagrostis*, *Dactylis*, *Festuca*, *Helictotrichon*, *Koeleria*, *Poa*, *Sesleria* (*Poaceae*); *Carex* (*Cyperaceae*).

Distribution: Switzerland.

Illustrations: Figs 5B, 55.

Notes: *Lophodermium fusiforme* has an ascomatal structure typical of the *actinothyrium*-group of *Lophodermium* Group A, although the lip cells are poorly developed relative to all other species in this group. It is characterized by ascomata acute to the ends, poorly-developed lips and paraphyses undifferentiated to swollen and clavate. Asci and ascospores are similar in size to those of *L. culmigenum* and *L. nitidum*, which are, however, distinguished by paraphysis shape and the structure of the lower wall of the ascomata.

This species has been found only in the Swiss Alps, although known from a wide range of hosts.

Additional specimens examined: Switzerland: GLARUS: Braunwald, Gumen, on *Carex*, 26 Jul. 1986, *O. Petrini* 156.2 (**PDD** 65740, **ZT**); GRAUBÜNDEN: Oberengadin, Zuoz, San Batrumieu, on *Sesleria coerulea*, 15 Jul. 1980, *L. Petrini* 53 (**PDD** 65722, **ZT**); *idem loc.*, on grass, 15 Jul. 1980, *L. Petrini* 108 (**PDD** 65723, **ZT**); *idem loc.*, on *Koeleria pyramidata*, 15 Jul. 1980, *L. Petrini* 56 (**PDD** 65725, **ZT**); *idem loc.*, on *Agrostis*, 15 Jul. 1980, *L. Petrini* 109 (**PDD** 65726, **ZT**); Zuoz, Crasta, on *Koeleria pyramidata*, 15 Jul. 1980, *L. Petrini* 113 (**PDD** 65724, **ZT**); Bergün, Val Tuors, on ?*Calamagrostis*, 14 Jul. 1980, *L. Petrini* 120 (**PDD** 65728, **ZT**); *idem loc.*, on *Sesleria coerulea*, 14 Jul. 1980, *L. Petrini* 121 (**PDD** 65729, **ZT**); *idem loc.*, on *Helictotrichon pubescens*, 14 Jul. 1980, *L. Petrini* 122 (**PDD** 65730, **ZT**); Unterengadin, Prui-Laret, on *Calamagrostis varia*, 27 Aug. 1984, *L. Petrini* 134 (**PDD** 65731, **ZT**); Unterengadin, Itan, Alp Clünas, on *Sesleria coerulea*, 31 Aug. 1984, *L. Petrini* 139 (**PDD** 65732, **ZT**); Sufers, Davons, on *Calamagrostis villosa*, 20 Aug. 1981, *L. Petrini* 146 (**PDD** 65733, **ZT**); TICINO: Airolo, Nante, on *Festuca*, 22 Jul. 1980, *L. Petrini* 177 (**PDD** 65734, **ZT**); *idem loc.*, on *Dactylis glomerata*, 22 Jul. 1980, *L. Petrini* 175 (**PDD** 65735, **ZT**); San Gottardo, on grass, 21 Jul. 1980, *L. Petrini* 182 (**PDD** 67193); *idem loc.*, 21 Jul 1980, *L. Petrini* 187 (**PDD** 65736, **ZT**); San Gottardo, Motta Bartola, on *Brachypodium prunatum*, 21 Jul. 1980, *L. Petrini* 183 (**PDD** 65737, **ZT**); Alpe di Piora, on *Anthoxanthum odoratum*, 27 Jul. 1980, *L.*

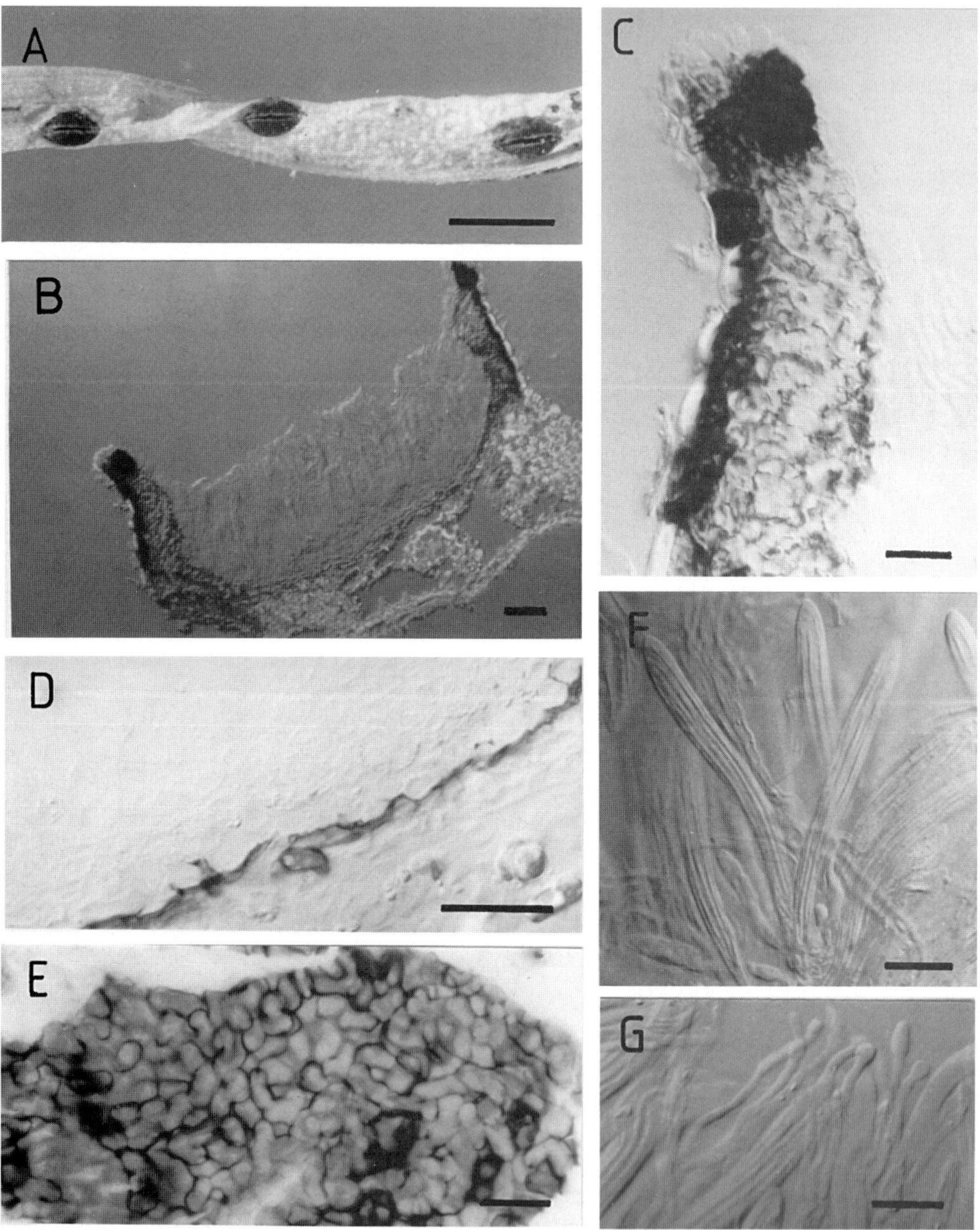

Fig. 55. *Lophodermium fusiforme* (**A–D**, **F**, **PDD** 65722; **E**, **G**, **PDD** 65736). **A**, ascomata (bar = 1 mm); **B**, ascoma in vertical section (bar = 50 µm); **C**, upper wall of ascoma in vertical section; **D**, lower wall of ascoma in vertical section; **E**, lower wall in squash mount; **F**, asci; **G**, paraphyses (bars = 20 µm).

Petrini 198.3 (**PDD** 65738, **ZT**); *idem loc.*, on *Dactylis glomerata*, 26 Jul. 1980, *L. Petrini* 212.1 (**PDD** 65739, **ZT**); WALLIS: Lötschental, Ried-Weissenried, on *Poa*, 9 Jul. 1979, *L. Petrini* 4 (**PDD** 65284, **ZT**).

Lophodermium gramineum (Fr.) Chevall., *Fl. gén. env. Paris* **1**: 435 (1826).
Hysterium gramineum Pers., in MOUGEOT & NESTLER, *Stirp. crypt.* no. 368 (1813), *nom. inval.*, *ICBN* Art. 41.3.
Hysterium culmigenum var. *gramineum* Fr., *Syst. mycol.* **2**: 591 (1823).
Lophodermium arundinaceum var. *gramineum* (Fr.) Duby, *Mém. Soc. Phys. Hist. nat. Genève* **16**: 59 (1862).
Lophodermium arundinaceum var. *gramineum* Fuckel, *Symb. mycol.*: 257 (1870), *nom. inval.*, *ICBN* Art. 32.1.
Lophodermellina graminea (Fr.) Höhn., *Hedwigia* **62**: 79 (1920).
Lophodermium velatum Berk., *Grevillea* **22**: 15 (1893).
Hypoderma velatum (Berk.) Kuntze, *Revis. gen. pl.* **3**(3): 487 (1898).
Lophodermium culmigenum var. *festucae* Roum., *Revue mycol.* **19**: 150 (1897).
Lophodermium festucae (Roum.) Terrier, *Beitr. Krypt.-Fl. Schweiz* **9**(2): 24 (1942).
Dermascia festucae (Roum.) Tehon, *Illinois biol. Monogr.* **13**(4): 65 (1935), *nom. inval.*, *ICBN* Art. 43.1.
Lophodermium proximellum Mouton, *Bull. Soc. r. Bot. Belg.* **39**: 49 (1900).
Lophodermium latisporum Terrier, *Beitr. Krypt.-Fl. Schweiz* **9**(2): 32 (1942), *nom. inval.*, *ICBN* Art. 36.1.
Dermascia latispora Tehon, *Illinois biol. Monogr.* **13**(4): 60 (1935), *nom. inval.*, *ICBN* Art. 43.1.
Lophodermium salisburgense Petr., *Sydowia* **16**: 179 (1963).

Infected areas on dead leaves, not associated with bleached areas or zone lines, containing scattered ascomata, conidiomata not seen. *Ascomata* up to 0·6 mm long, broadly oblong-elliptical, ± ovate or rhomboid in outline, wall varying in colour from pale grey to black, unopened ascomata with poorly-developed pale zone along future line of opening, opened ascomata with no lip cells along single, longitudinal opening slit. *Ascomatal insertion* subepidermal. *Covering layer* up to 65 µm thick in vertical section, typically with clypeus of dark-walled hyphae within epidermal cells (lacking in some collections, where epidermal cells remain empty) and upper wall of ascomatal stroma. *Upper wall* up to 50 µm thick (sometimes very poorly developed), comprising mostly pale, angular to globose cells, 6–8 µm diam., with walls slightly and irregularly encrusted, lined with well-developed, loose layer of hyaline, ± monilioid periphysoids (sometimes lost in ascomata which have been open for some time), but with groups of dark cells with thickly encrusted walls along edge of opening slit. *Lower wall* up to 15 µm thick in vertical section, comprising two or three rows of large, thin-walled, globose cells, with only outer wall of lowermost row of cells darkened, in squash mount darkened cells of lower wall globose, 8–12 µm diam., forming single layer across base of ascoma. *Paraphyses* 1·5–2 µm diam., undifferentiated or slightly and irregularly swollen near apex. *Asci* 90–120 × (10–) 11–13 µm, broad-cylindrical to subsaccate, tapering suddenly to small, truncate apex with undifferentiated wall, 8-spored, spores extending to base. *Ascospores* (45–) 65–90 (–110) × 2–2·5 µm, tapering slightly to base, apical gelatinous cap globose, 3 µm diam., basal gelatinous cap tiny, 1–2 µm diam., gelatinous sheath *c.* 10 µm diam.

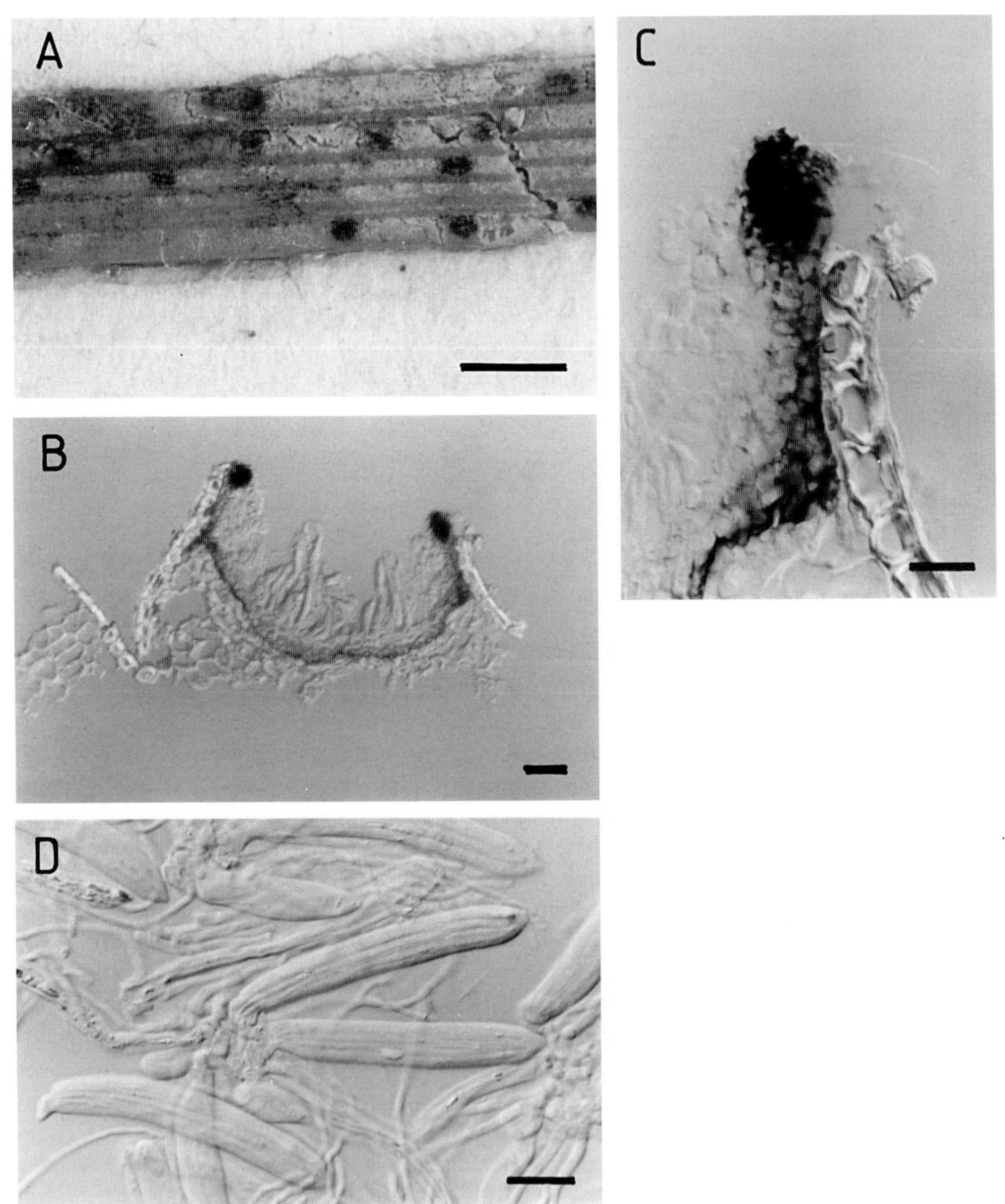

Fig. 56. *Lophodermium gramineum* (lectotype, Mougeot & Nestler, *Stirp. crypt.* no. 368, **K**). **A**, ascomata (bar = 1 mm); **B**, ascoma in vertical section (bar = 50 μm); **C**, upper wall of ascoma in vertical section; **D**, asci (bars = 20 μm).

Typification: Germany: *s. loc.*, on *Poa compressa*, *s. coll.* (Mougeot & Nestler, *Stirpes cryptogamae vogeso-rhenanae* no. 368, **K**!, *lectotypus* of *H. culmigenum* var. *gramineum*, selected here; **IMI**!; **FH**!, *isolectotypi*, selected here). **France:** *s. loc.*, on *Festuca rubra*, May 1897, *F. Fautrey s. num.* (Roumeguère, *Fungi gallici exsiccati* no. 7143; **UPS**!, *lectotypus* of *L. festucae*, selected here). **Portugal:** MADEIRA: *s. loc.*, on *Festuca*, *s. dat. nec coll.* (**K**!, holotype of *L. velatum*, as *Hysterium velatum*). **Belgium:** nr Liège, Beaufays, on *Poa compressa*, Jun. 1898, *V. Mouton s. num.* (Rehm, *Ascomyceten* no. 1324; **K**!, *lectotypus* of *L. proximellum*, selected here; **BR**!, *isolectotypus* of *L. proximellum*, selected here). **Austria:** SALZBURG: Kolm-Saigurn, on *Agrostis alpina*, 1 Jul. 1936, *K. Ronniger s. num.* (**W**!, *lectotypus* of *L. salisburgense*, selected here); TIROL: Sulden, Ortler, on grass, Apr. 1884, *Rehm s. num.* (Rehm, *Ascomyceten* no. 775; **FH**!; **K**!, specimen on which the invalid name *L. latisporum* was based).

Hosts: *Agrostis*, *Anthoxanthum*, *Calamagrostis*, *Elymus*, *Festuca*, *Nardus*, *Poa*, *Sesleria* (*Poaceae*); *Carex* (*Cyperaceae*).

Distribution: Europe, Russia, Madeira, North America, India.

Illustrations: Figs 4A, 56, 57, 60.

Notes: *Lophodermium gramineum* is typical of the *alpinum*-group of *Lophodermium* Group A in ascomatal structure. Although difficult to distinguish macroscopically from *L. alpinum*, it is easily distinguished microscopically from all other species in the group by its strictly cylindrical asci which taper suddenly to the foot-like base, and its undifferentiated to slightly swollen paraphyses.

A wide range of ascus and ascospore lengths has been accepted in this species. Although the range in individual collections was much smaller – some having ascospores restricted to the lower end of the range and others restricted to the higher end – the various collections were otherwise indistinguishable.

Several authors (TEHON, 1935; CANNON & MINTER, 1986; FARR *et al.*, 1989) have suggested that *L. gramineum* may be confined to species of *Poa*; however, host substratum cannot be used to help identify this species.

FUCKEL (1870) appeared to propose a new variety, *L. arundinaceum* var. *gramineum*, in that no basionym was cited. However, he included the type of *L. gramineum* among the specimens cited under this name, thus rendering it superfluous. He also cited *Fungi rhenani* no. 740 under this name. The duplicates of this specimen examined contain a few immature *Lophodermium* ascomata which, in vertical section, were typical of *L. culmigenum*.

The lectotype collection of *L. salisburgense* is typical of *L. gramineum*. However, PETRAK (1963) cited two of Ronniger's specimens, only one of which has been examined here. PETRAK gave ascus (46–65 × 13–17 µm) and ascospore (23–45 × 2·5–3 µm) dimensions shorter than those determined here on the type material, closely matching those typical of *L. alpinum*. Macroscopically, *L. gramineum* and *L. alpinum* are indistinguishable and often present together in single collections. It is possible that one (or both) of Ronniger's collections contains ascomata of *L. alpinum* as well as *L. gramineum*.

JOHNSTON (1989*b*, 1992) and WALKER (1980) incorrectly reported *L. gramineum* from New Zealand and Australia, respectively.

Additional specimens examined: Belgium: Chaumes, on *Poa compressa*, *V. Mouton* 535 (**BR**). **Canada:** QUÉBEC: Portes de l'Infer, Parc des Laurentides, on *Agrostis*, 7 Jul. 1980, *O. Petrini s. num.* (**PDD** 60250, **ZT**). **Czech Republic:** BOHEMIA: Ounuz, Sedlec, on *Calamagrostis epigeios*, 13 Jul. 1948, *M. Svrček s. num.* (**PRM** 756745, as *L. apiculatum*). **Georgia:** YUGO-OSTINSKAYA: Belotskii distr.: Nishnü Pereral, on *Festuca valesiaca* subsp. *sexatilis*, 16 Jul. 1930, *Bush s. num.* (**K**). **Germany:** RHEINLAND-PFALZ: Nassau, Hattenheim, on '*Cynodontis Dactylon*', *Fuckel s. num.* (Fuckel, *Fungi rhenani* no. 740; **UPS**; **DAOM**). **Portugal:** MADEIRA: *R.T. Low s. num.* (**K**, as *Hysterium velatum*). **Russia:** URALSKOI OBL.: *s.*

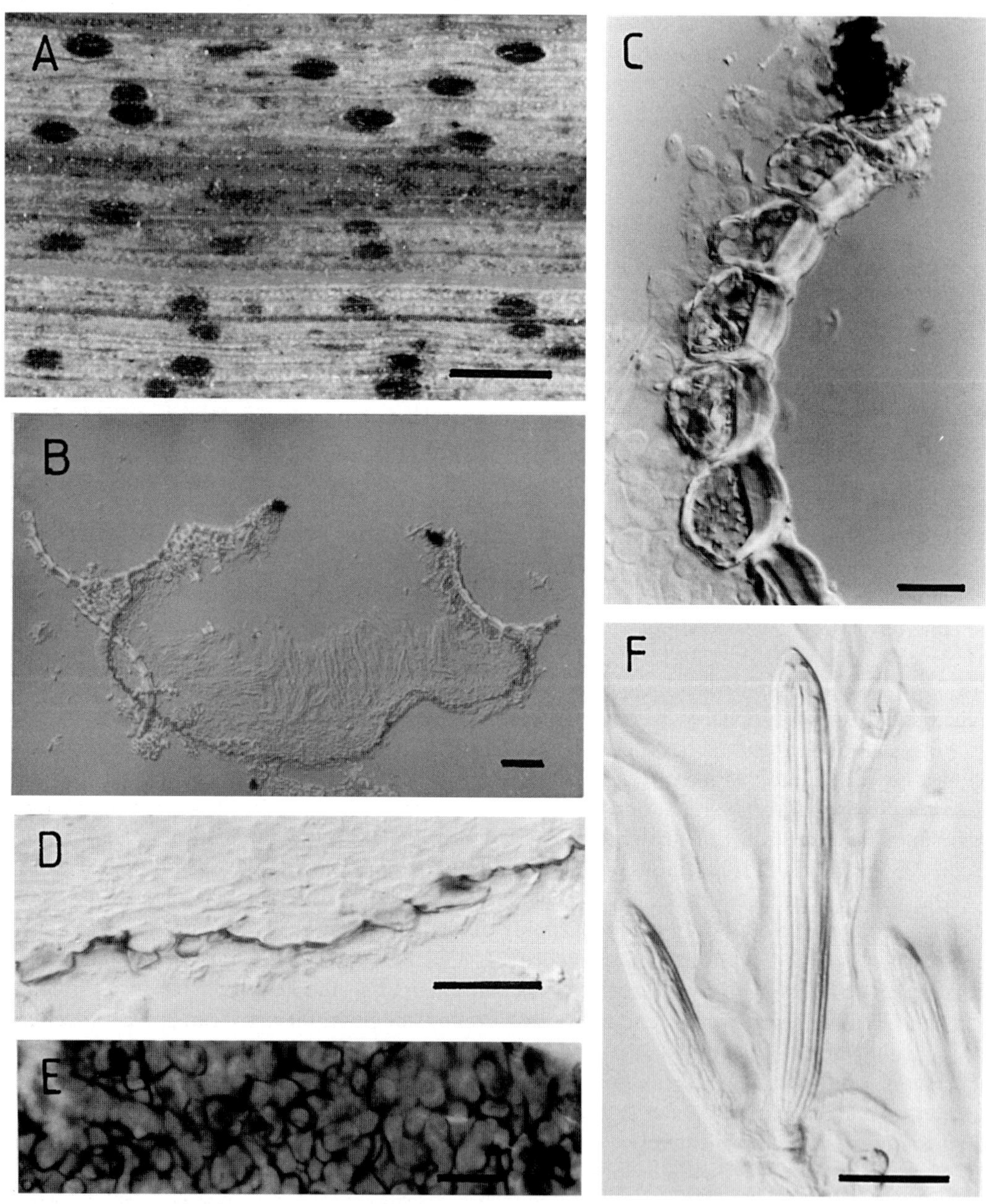

Fig. 57. *Lophodermium gramineum.* **A**, ascomata (**PRM** 756745) (bar = 1 mm); **B**, ascoma in vertical section (**PDD** 59581) (bar = 50 µm); **C**, upper wall of ascoma in vertical section (**PDD** 59581); **D**, lower wall of ascoma in vertical section (**PDD** 59581); **E**, lower wall of ascoma in squash mount (**PDD** 59585); **F**, ascus (**PDD** 59576) (bars = 20 µm).

loc., on *Calamagrostis epigeios*, 25 Jul. 1929, *s. coll.* (**K**). **Slovakia:** Malá Fatra, Fatranský Kirván, Meškalka, on *Calamagrostis varia*, 28 Jun.–2 Jul. 1947, *M. Svrček s. num.* (**PRM** 756739, as *L. apiculatum*); Muráňská vysočina, on *Calamagrostis varia*, 28 Jul. 1947, *M. Svrček s. num.* (**PRM** 756738, as *L. apiculatum*); nr Glac., on *Calamagrostis montana*, 24 Jul. 1947, *M. Svrček s. num.* (**PRM** 756737, as *L. apiculatum*). **Sweden:** JÄMTLAND: Åre Parish: Brännan, on *Calamagrostis purpurea*, 2 Aug. 1951, *J.A. Nannfeldt* 11663a (*Fungi suecici upsaliensis*, as *L. apiculatum*; **K**); Torne Lappmark, Jukkasjärvi, on *Sesleria varia*, 2 Jul. 1986, *A. Nograsek s. num.* (**GZU**, as *L. festucae*). **Switzerland:** GRAUBÜNDEN: Alvaneu, on *Sesleria coerulea*, 11 Jun. 1980, *L. Petrini* 37 (**PDD** 59584, **ZT**); Davos Dorf, 1600 m, on *Festuca ?rubra*, 11 Jun. 1980, *L. Petrini* 38 (**PDD** 59573, **ZT**); Davos Wolfgang, 1630 m, on *Agrostis gigantea*, 12 Jun. 1980, *L. Petrini* 44 (**PDD** 60251, **ZT**); Oberengadin, Zuoz, Castlatsch, 1780 m, on *Festuca*, 15 Jul. 1980, *L. Petrini* 57 (**PDD** 59578, **ZT**); Zuoz, San Batrumieu, 1680 m, on *Festuca*, 15 Jul. 1980, *L. Petrini* 103.2 (**PDD** 59575, **ZT**); *idem loc.*, on *Poa*, 15 Jul. 1980, *L. Petrini* 105.1 (**PDD** 59579, **ZT**); *idem loc.*, on *Poa*, 15 Jul. 1980, *L. Petrini* 106 (**PDD** 59572, **ZT**); Bergin, Val Tuors, 1850 m, on *Nardus stricta*, 14 Jul. 1980, *L. Petrini* 52.1 (**PDD** 59574, **ZT**); Unterengaden, Samnaun, Alp Trida, on *Poa*, 28 Aug. 1984, *O. Petrini* 141 (**PDD** 59580, **ZT**); Unterengaden, Alp Clunas, on *Sesleria coerulea*, 31 Aug. 1984, *L. Petrini* 138 (**PDD** 59583, **ZT**); Ofenpass, Murtarol, on *Nardus stricta*, 4 Oct. 1986, *L. Petrini* 132.2 (**PDD** 59581, **ZT**); TICINO: Val Bedretto, All'Aqua, on *Carex fusca*, 21 Jul. 1980, *L. Petrini* 162 (**PDD** 59589, **ZT**); Airolo, Alpe di Ruvina, 1775 m, on *Nardus stricta*, 22 Jul. 1980, *L. Petrini* 167.2 (**PDD** 59576, **ZT**); Airolo, Nante, on *Sesleria coerulea*, 22 Jul. 1980, *L. Petrini* 173 (**PDD** 59585, **ZT**); Alpe di Piora, on *Festuca rubra*, 26 Jul. 1980, *L. Petrini* 213 (**PDD** 59577, **ZT**); Alpe di Piora, on *Anthoxanthum odoratum*, 27 Jul. 1980, *L. Petrini s. num.* (**PDD** 60253, **ZT**). **USA:** COLORADO: Cabin Canyon, 2700 m, on *Calamagrostis canadensis*, 12 Jul. 1905, *F.E. & E.S. Clements s. num.* (Clements, *Cryptogamae formationum coloradensium* no. 48; **CUP**); Larkspur Dell, 2600 m, on *Elymus ambiguus*, 20 Jul. 1905, *F.E. & E.S. Clements s. num.* (Clements, *Cryptogamae formationum coloradensium* no. 47; **CUP**); NEW MEXICO: Colfax Co.: nr Ute Park, on *Calamagrostis hyperborea*, 23 Aug. 1923, *P.C. Standley* 13624 (**CUP** 15-118); WASHINGTON: Yakima Co.: Wodanthal, Mt Adams, on *Calamagrostis*, 3 Oct. 1902, *W.N. Suksdorf* 606 (Shaw & Cooke, *Reliquiae suksdorfiana* no. 30; **CUP**; **TRTC**, as *L. arundinaceum*).

Lophodermium grandialpinum P.R. Johnst., **sp. nov.**

Etymology: Referring to the large asci and ascospores of this *L. alpinum*-like species.

A *L. alpino* apicibus paraphysium clavatis non circinatis, ascis subclavatis vel clavatis, 120–140 × 13–16 μm, ascosporis 100–130 × 2·5–3 μm differt.

Infected areas on dead leaves, not associated with bleaching of host tissue or zone lines, containing groups of ascomata, conidiomata not seen. *Ascomata* 0·4–0·7 × 0·2–0·4 mm, oblong-elliptical in outline, ends rounded, margin not sharply defined, wall dark grey to black, with narrow paler zone along future line of opening, no lips along single, longitudinal opening slit. *Ascomatal insertion* subepidermal. *Covering layer* up to 65 μm thick in vertical section, comprising clypeus of dark-walled hyphae within epidermal cells, and upper wall of ascomatal stroma. *Upper wall* up to 50 μm thick in vertical section, comprising two layers, outer layer 30 μm thick, of angular to globose cells, mostly pale with irregularly encrusted walls, but with patch of very dark cells adjacent to opening slit, inner layer of ± moniliform periphysoids, lost after ascomata have been open for some time. *Lower wall* up to 15 μm thick in vertical section, comprising two or three rows of globose cells, 5–10 μm diam., lowermost row with darkened walls, in squash mount darkened cells of lower wall variable in shape, broad-cylindrical to subglobose, 5–11 μm diam., arranged in single row across base of ascoma. *Paraphyses* 1·5 μm diam., irregularly swollen to 2–3 μm diam. at ± clavate apex, initially forming regular palisade, becoming irregularly and tightly tangled with age, then forming poorly-developed excipulum. *Asci* 120–140 × 13–16 μm, subclavate to clavate, tapering to ± rostrate apex with undifferentiated wall, 8-spored, short basal stalk at maturity with spores extending ± to base. *Ascospores* 100–130 × 2·5–3 μm, tapering to very narrow, tail-like base, apical gelatinous cap broad-cylindrical, *c.* 7 × 5 μm,

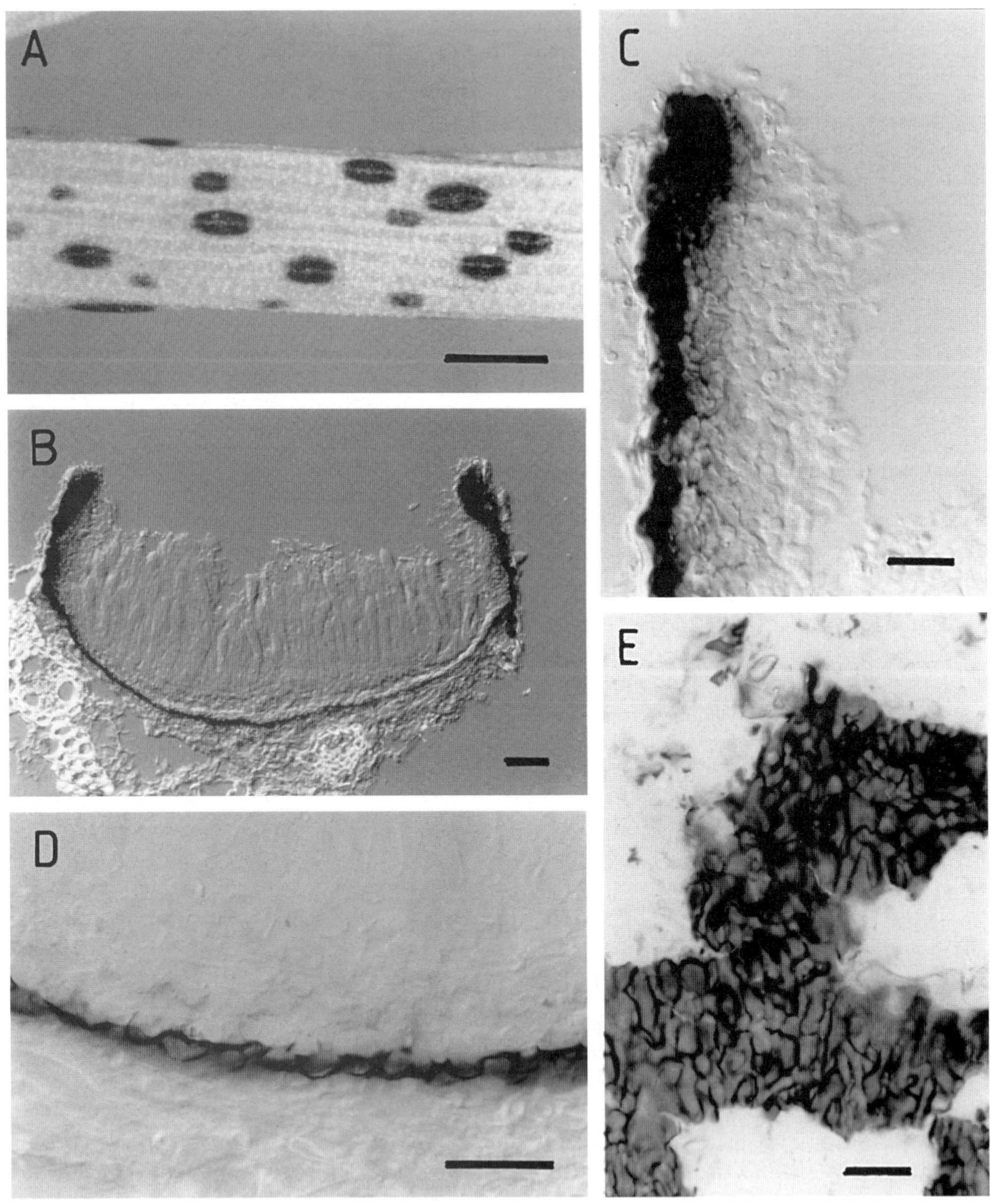

Fig. 58. *Lophodermium grandialpinum* (**PDD** 65057). **A**, ascomata (bar = 1 mm); **B**, ascoma in vertical section (bar = 50 μm); **C**, detail of upper wall of ascoma in vertical section; **D**, detail of lower wall of ascoma in vertical section; **E**, lower wall of ascoma in squash mount (bars = 20 μm).

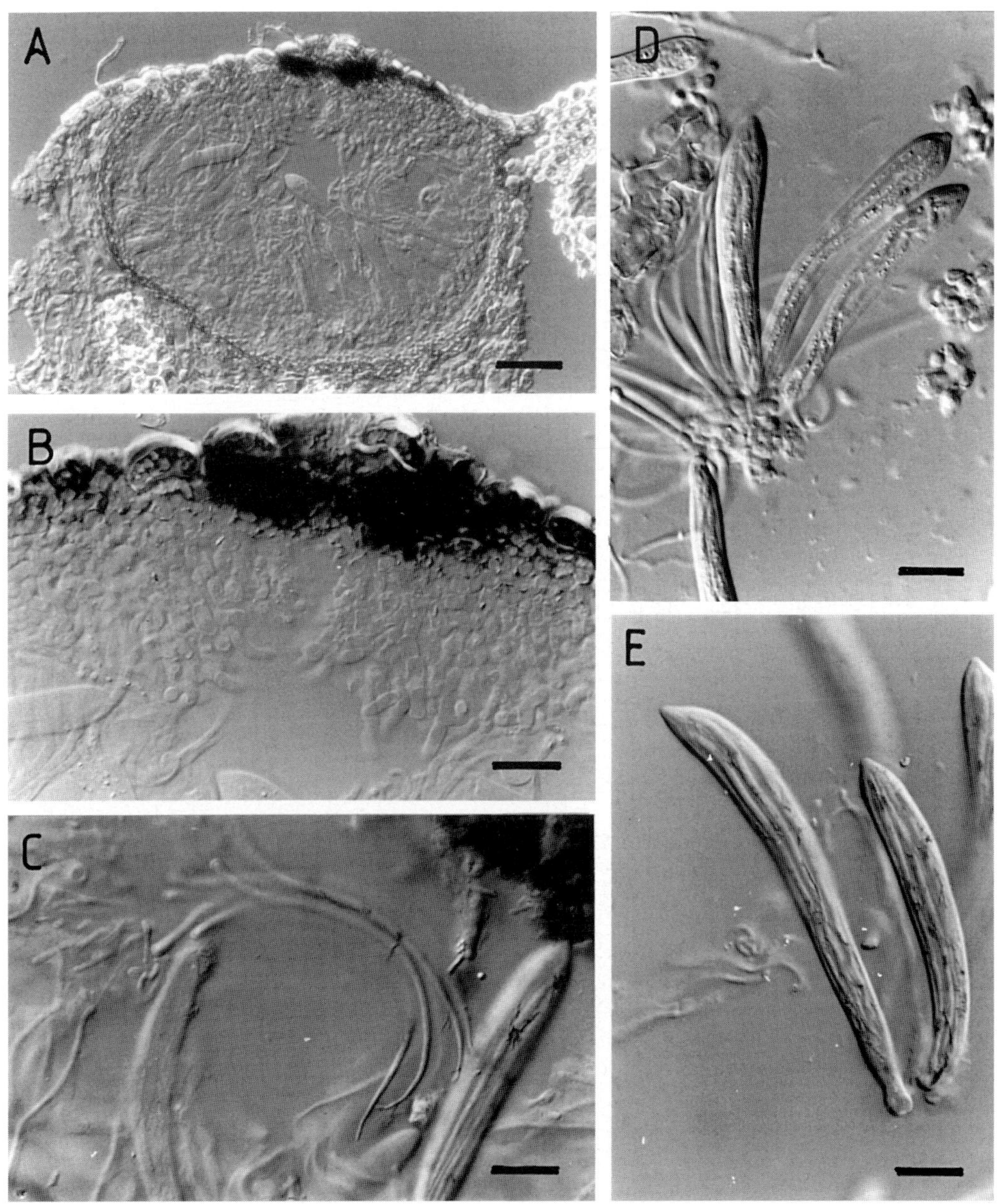

Fig. 59. *Lophodermium grandialpinum* (**A**, **B**, **D**, **PDD** 65057; **C**, **E**, *Lind*, 17 Mar. 1903, **HBG**). **A**, unopened ascoma in vertical section (bar = 50 µm); **B**, detail of upper wall of unopened ascoma in vertical section, with broad layer of periphysoids; **C**, released ascospores, tapering to more or less acute base; **D**, asci (slightly immature, with no basal stalk); **E**, asci with basal stalk starting to develop (bars = 20 µm).

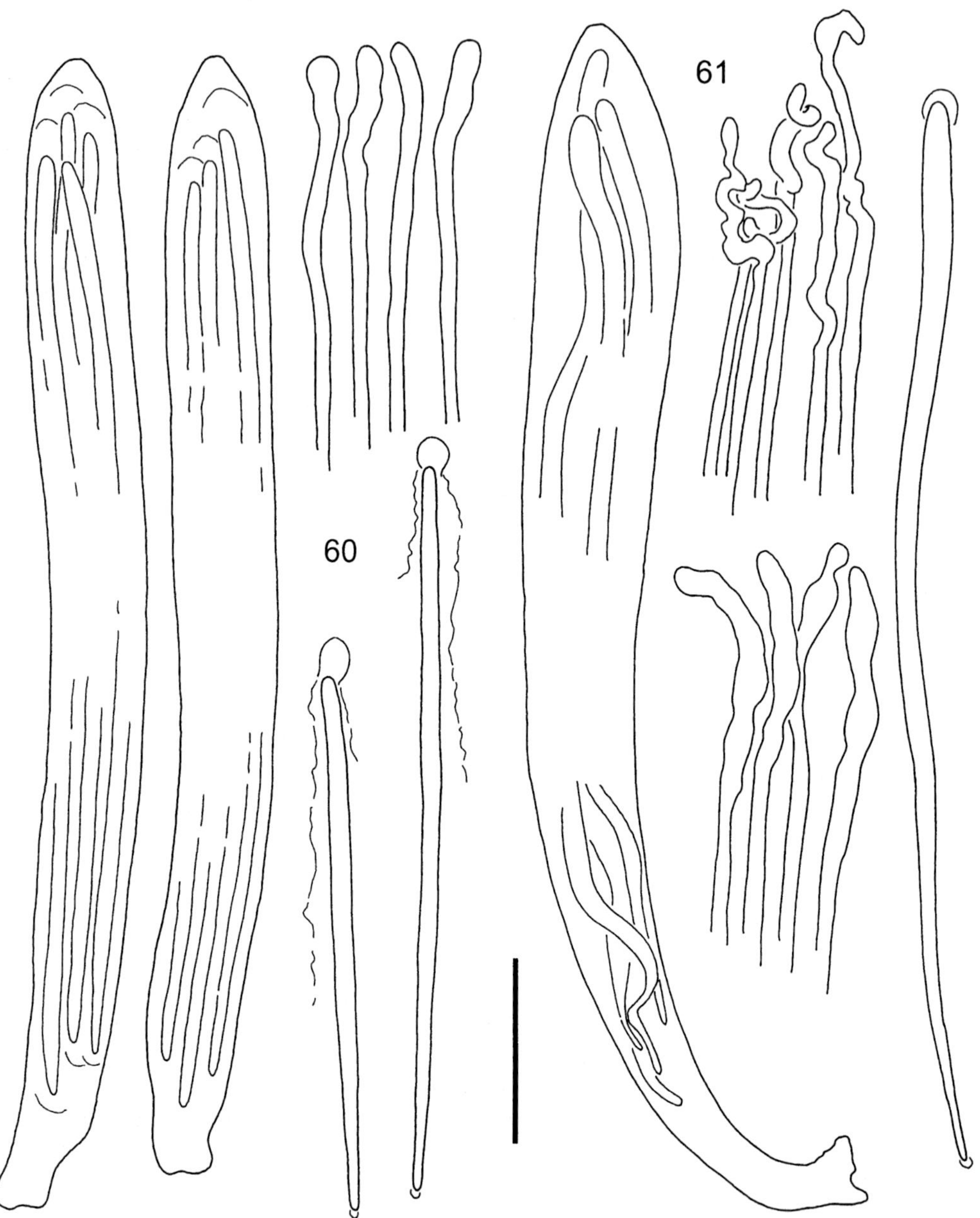

Figs 60–61. 60. *Lophodermium gramineum.* Asci and apex of paraphyses (**PDD** 59578) and ascospores in water (**PDD** 60253). **61.** *Lophodermium grandialpinum.* Ascus, ascospore and apex of paraphyses (at top) (*Lind*, 17 Mar. 1903, **HBG**), apex of paraphyses (at bottom) (**PDD** 65057) (bar = 20 μm).

basal gelatinous cap small, globose, 2–3 μm diam., gelatinous sheath 8–10 μm thick.

Typification: Sweden: UPPLAND: Dalby Parish: nr Viggeby, on ?*Elytrigia repens*, 25 May 1985, *K. & L. Holm* 3540a (**ZT**, *holotypus*; **PDD** 65057, *isotypus*).

Hosts: *Elytrigia*, *Festuca*, *Sesleria* (*Poaceae*).

Distribution: Europe.

Illustrations: Figs 58, 59, 61.

Notes: *Lophodermium grandialpinum* is typical of the *alpinum*-group of *Lophodermium* Group A, within which it is distinguished by its large asci and ascospores, the latter tapering to a very narrow base. Spore size and shape are similar to *L. sesleriae*, but the two species are distinguished by ascus size and shape.

Additional specimens examined: Austria: NIEDERÖSTERREICH: Hainburg, Braunsberg, on *Sesleria varia*, May 1940, *F. Petrak s. num.* (*Reliquiae petrakianae* no. 1225; **GZU**). **Denmark:** JYLLAND: Viborg, on *Festuca duriuscula*, 17 Mar. 1903, *J. Lind s. num.* (**HBG**, as *L. arundinaceum*). **Spain:** BALEARES: Mallorca: Ca'n Picafort, on grass, 19 May 1985, *J. Poelt s. num.* (**GZU**).

Lophodermium griseum (Schwein.) Petr., *Reliquiae petrakianae* no. 471 (1979), *nom. inval.*, *ICBN* Art. 33.2.

Hysterium griseum Schwein., *Trans. Am. phil. Soc.* Ser. 2, **4**(2): 245 (1832).

Sporomega grisea (Schwein.) Cooke, *Bull. Buffalo Soc. nat. Sci.* **3**: 36 (1875).

Typification: 'vulgatissimum sub subepidermide ramorum juniorum Smilacinum frigore enecatorum, Bethl.' (SCHWEINITZ, 1832).

Notes: SCHWEINITZ (1832) cited no type specimen, and no Schweinitz material of *H. griseum* was located. *Reliquiae petrakianae* no. 471, distributed under the name *Lophodermium griseum*, contains both immature and over-mature ascomata. The ascomata appear rhytismataceous, are elliptical in outline, acute to the ends, 1–1·5 × 0·3–0·4 mm, with walls pale brown and an indistinct paler zone along the future line of opening. Mature asci and ascospores were not seen. Vertical sections show the ascomatal primordium to comprise *textura intricata*, the unopened ascomata to be subepidermal, the upper wall darkening before the lower wall and lined with a broad layer of periphysoids arranged in compact columns across the central part of the ascoma (**Fig. 62A–D**).

The appearance of the immature ascomata in vertical section provides no clear indication of a relationship with any other monocotyledon-inhabiting species of *Lophodermium*. The well-organized columnar arrangement of the periphysoids is most suggestive of those seen in a group of *Pinaceae*-inhabiting species: *L. nanakii* P.F. Cannon & Minter (see CANNON & MINTER, 1986, fig. 19; OSORIO & STEPHAN, 1991, fig. 2, as *L. piceae*), *L. consociatum* Darker (**Fig. 62E**) and *L. lacerum* Darker (**Fig. 62F**), or those *Ericaceae*-inhabiting species discussed and illustrated by JOHNSTON (1988*b*). This apparent morphological similarity is, however, unlikely to reflect a phylogenetic relationship between these ecologically diverse groups. Indeed, DICOSMO *et al.* (1984) illustrated a similar arrangement within the upper wall of several species of *Phacidium*.

The combination *L. griseum* (Schwein.) Petr. does not appear to have been published; it was not listed in SAMUELS (1985).

Another specimen on *Smilax*, annotated as '*L. smilacinum* Petrak, n.sp.' in W (**USA:** HAWAII: Hawaii I, Olinda Pipe Line, on *Smilax*, 29 Dec. 1928, *Shear & Stevens* 618, **W**), represents a different species to that in *Reliquiae petrakianae* no. 471. The herbarium packet was labelled as a type, yet the name was apparently never published and was not mentioned in SAMUELS (1985). The Hawaiian specimen is typical of species of *Terriera* in ascomatal structure. It has longer asci than other species in this genus from Hawaii, but is not described as a new species here because

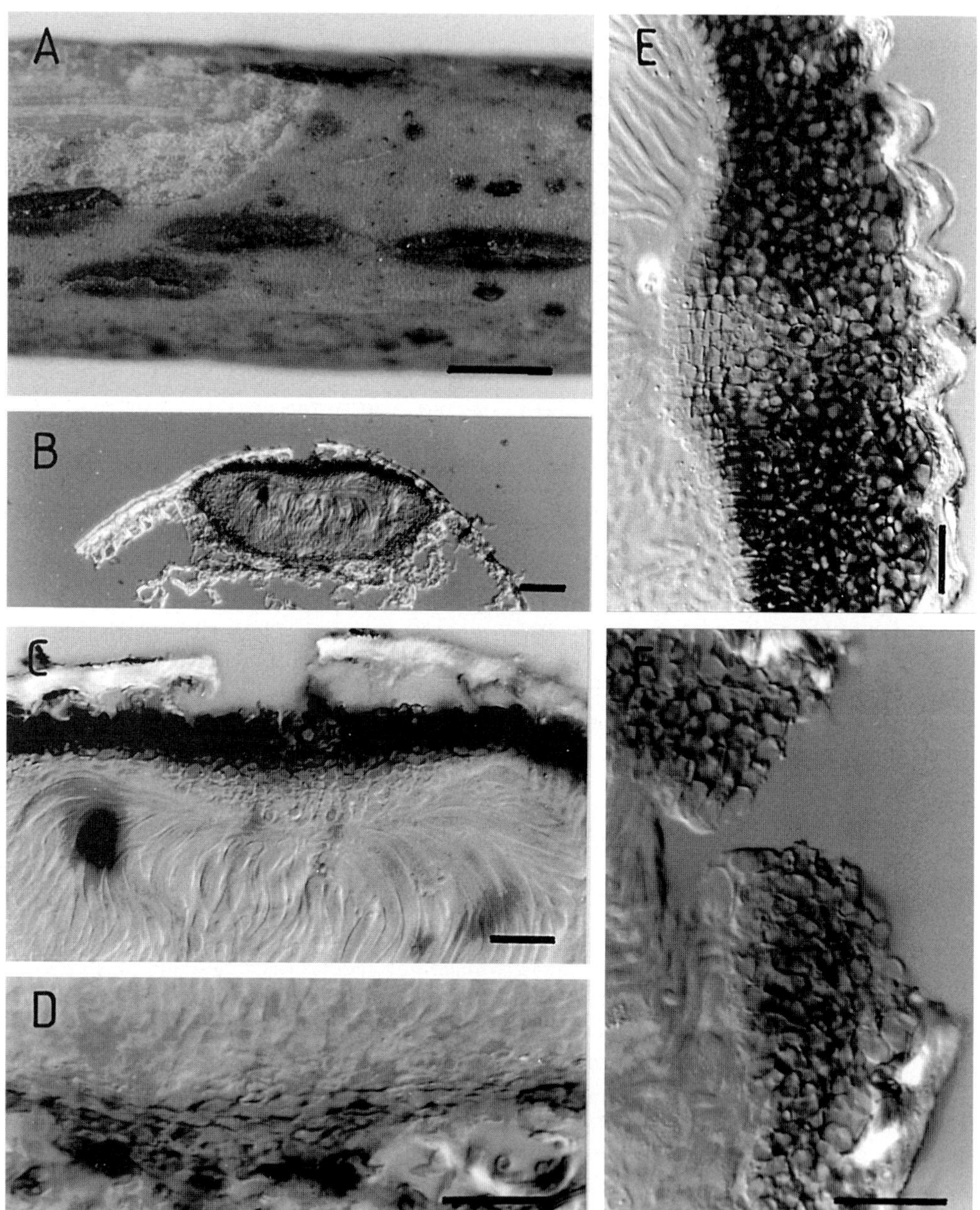

Fig. 62. A–D. *Lophodermium griseum* (*Reliquiae Petrakianae* no. 471, **CUP**). **A**, ascomata (bar = 1 mm); **B**, unopened ascoma in vertical section (bar = 50 μm); **C**, upper wall of unopened ascoma in vertical section; **D**, lower wall of unopened ascoma in vertical section. **E.** *Lophodermium consociatum* (**IMI** 250594). Upper wall of unopened ascoma in vertical section. **F.** *Lophodermium lacerum* (**PDD** 25637). Upper wall of unopened ascoma (broken during sectioning) in vertical section (bars = 20 μm).

of the uncertainty still surrounding species limits in *Terriera*. It is very similar in size and shape of asci and ascospores to *T. dracaenae*, which differs in having branched paraphyses.

Specimens examined: ***Lophodermium griseum*: USA:** MARYLAND: Beltsville, on *Smilax*, 28 May 1950, *F. Petrak s. num.* (*Reliquiae petrakianae* no. 471; **CUP**; **PDD**).

***Lophodermium consociatum*: USA:** OREGON: Linn Co.: Santiam Pass, old lava flow 2 miles SW of Sahalie Falls, on *Abies amabilis*, 15 Aug. 1978, *M.A. Sherwood & L.H. Pike s. num.* (**TNS** F-194591, as *L. uncinatum*). **Canada:** BRITISH COLUMBIA: Vancouver I, Bamfield, Frederick Lake, on *Abies amabilis*, 23 May 1961, *D.G. Collins* (**IMI** 250594).

***Lophodermium lacerum*: Canada:** ONTARIO: Lake Timagami, Bear I., on *Abies balsamea*, 20 Jun. 1924, *G.D. Darker s. num.* (**PDD** 25637 ex **DAOM** 109286).

Lophodermium hauturuanum P.R. Johnst., *N.Z. Jl Bot.* **27**: 255 (1989).

Infected areas on dead leaves, slightly paler than surrounding host tissue, usually partially surrounded by broad, dark zone lines, containing scattered ascomata and conidiomata. *Ascomata* 0·5–2·5 × 0·2–0·3 mm, oblong to sublinear in outline, ends rounded, unopened ascomata with wall grey to dark grey, paler in small patch near either end and in broad zone along future line of opening, open ascomata with wall grey, dark line around outside edge and often along margin of opening slit, longitudinal opening slit extending along most of ascoma, often with short side slits near ends. *Ascomatal insertion* subepidermal. *Covering layer* 20–25 μm thick in vertical section, comprising mostly host cuticle and clypeus of dark-walled hyphae within partially broken-down epidermal cells, with poorly-developed upper wall of ascomatal stroma. *Upper wall* comprising layer, 5–10 μm wide, of angular to globose, pale brown, thin-walled cells, 4–6 μm diam., near opening slit inner part of wall often lined with scattered, elongate periphysoids, lip cells poorly developed, tops of lip cells and periphysoids becoming encrusted with dark brown material. *Lower wall* variable in thickness, constricted by host fibre bundles, between fibre bundles comprising one or two layers of angular to cylindrical cells, 4–5 μm diam., with several rows of dark-walled hyphae, 2–3 μm diam., forming layer of *textura intricata* beneath them but not across top of fibre bundles. *Paraphyses* 2–2·5 μm diam., increasing gradually in width to the ± clavate apex, embedded in common gelatinous matrix containing numerous small, irregularly shaped, crystal-like inclusions, gel and crystals soluble in KOH. *Asci* 100–125 × 4·5–6·5 μm, cylindrical to subclavate, tapering to small rounded apex with thickened wall and broad apical pore, 8-spored, short basal stalk. *Ascospores* 70–85 × 1·5 (–2) μm, tapering gradually to base, 12–15 μm from base bent at *c.* 120° with small, barb-like gelatinous appendage on outside of bend. *Conidiomata* 0·5 mm diam., round in outline, initially concolorous with surrounding host tissue, becoming pale orange-brown, subepidermal, in vertical section lenticular in outline, upper wall comprising one or two layers of pale brown to hyaline cells, lower wall 10 μm thick, comprising cells with dark, thickened walls. *Conidiogenous cells* lining lower wall, 7–10 × 2–3 μm, solitary, cylindrical to flask-shaped, proliferation percurrent, with wall thickened and flaring at apical conidiogenous locus. *Conidia* 3–4 × 1 μm, cylindrical, hyaline, non-septate.

Typification: New Zealand: COROMANDEL: Little Barrier I, Shag Track, on *Gahnia*, 15 Jun. 1984, *P.R. Johnston* LB11 (**PDD** 46189!).

Host: *Gahnia* (*Cyperaceae*).

Distribution: New Zealand.

Illustrations: Figs 14C, 63, 65.

Notes: In ascomatal structure *L. hauturuanum* is typical of *Lophodermium* Group E (see p. 43). It is very similar to *L. unciniae* which, however, has smaller ascomata and conidiomata and

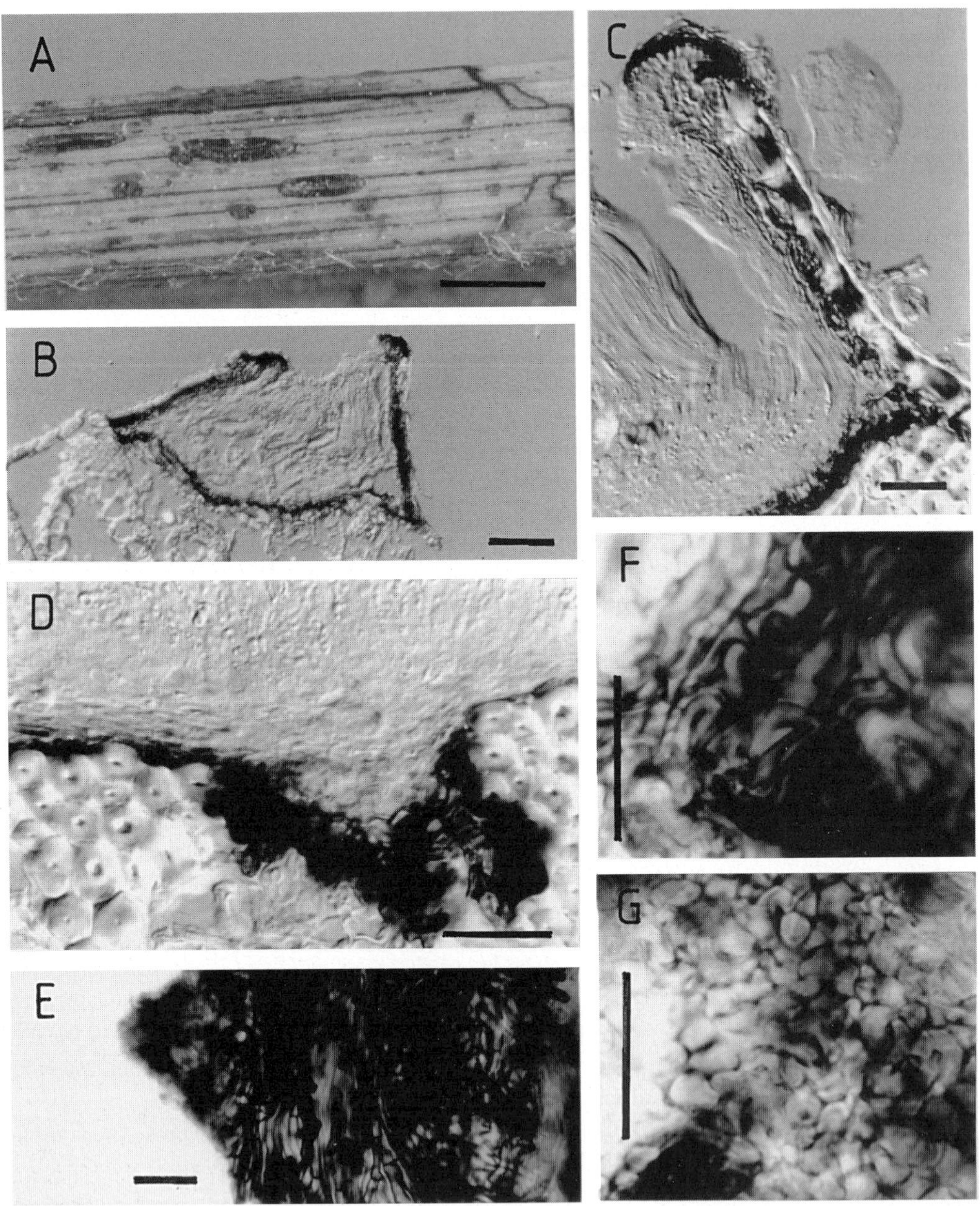

Fig. 63. *Lophodermium hauturuanum* (**A–C**, **E–G**, **PDD** 49353; **D**, **PDD** 62292). **A**, ascomata (bar = 1 mm); **B**, ascoma in vertical section (bar = 50 μm); **C**, upper wall of ascoma in vertical section; **D**, lower wall of ascoma, squash mount; **E**, lower wall of ascoma in vertical section; **F**, clypeus-like layer within epidermal cells, squash mount; **G**, cells within upper wall of ascoma, squash mount (bars = 20 μm).

shorter asci and ascospores. JOHNSTON (1994) also noted a difference in the position of the distinctive bend seen in the released ascospores of both species. It is distinguished from *L. inclusum*, also found on *Gahnia* in New Zealand, by its narrower ascomata which are cylindrical in outline and its distinctive ascomatal structure.

Additional specimens examined: New Zealand: AUCKLAND: Waitakere Ranges, Scenic Drive, on *Gahnia*, 8 Oct. 1984, *P.R. Johnston* R582A (**PDD** 46165); Waitakere Ranges, Sharps Bush, on *Gahnia*, 19 Jul. 1983, *P.R. Johnston* R331 (**PDD** 49353); Waitakere Ranges, on *Gahnia*, 2 Jun. 1993, *P.R. Johnston s. num.* (**PDD** 62292); BULLER: Greymouth, Pt Elizabeth Walk, on *Gahnia*, 1 May 1985, *P.R. Johnston* R643 (**PDD** 46938); COROMANDEL: Little Barrier I, Awaroa Stream, on *Gahnia*, 6 Apr. 1988, *P.R. Johnston s. num.* (**PDD** 54800); NORTHLAND: Omahuta Forest, Kauri Reserve, on *Gahnia*, 22 Oct. 1987, *P.R. Johnston* R753 (**PDD** 53857); TAUPO: Tongariro National Park, Ohakune Mtn Rd, Blyth Track, on *Gahnia*, 20 May 1989, *P.R. Johnston s. num.* (**PDD** 55566); WAIKATO: nr Waharoa, Kahikatea Reserve, on *Gahnia*, 28 Jul. 1989, *P.R. Johnston s. num.* (**PDD** 56344).

Lophodermium herbarum Chevall., *Fl. gén. env. Paris* **1**: 437 (1826).

Typification: No type was indicated by CHEVALLIER (1826), but he noted the species as being common on leaves of plants such as *Aconitum napellus*, *Epilobium spicatum* and *Umbelliferae*.

Hosts: *Aconitum* (*Ranunculaceae*); *Epilobium* (*Onagraceae*); *Apiaceae* (*fide* CHEVALLIER, 1826).

Distribution: Europe.

Notes: *Lophodermium herbarum* Chevall. has been regarded by most authors as a combination based on *Hysterium herbarum* Fr. but, as noted by FRIES (1828: 145), CHEVALLIER was describing a different fungus. *Hysterium herbarum* (≡ *L. herbarum* (Fr.) Fuckel) is found on *Convallaria* (*Liliaceae*), while *L. herbarum* Chevall. is reported from dicotyledonous hosts. Few species of *Lophodermium* are found on both dicotyledons and monocotyledons and, although no type specimen exists for *L. herbarum* Chevall., it is likely that two distinct species are involved.

Lophodermium herbarum (Fr.) Fuckel is a later homonym of *L. herbarum* Chevall. and hence is illegitimate. However, in the spirit of the changes proposed for the *Code* at the Yokohama Botanical Congress (GREUTER & NICOLSON, 1993; GREUTER *et al.*, 1994), changes of name purely for nomenclatural reasons should be avoided if possible. To preserve current usage, *L. herbarum* (Fr.) Fuckel should be retained for the species occurring on *Convallaria*, while those occurring on *Aconitum* and *Epilobium* can be referred to *L. tumidum* var. *napelli* and *L. ciliatum*, respectively (see below).

FRIES (1828) considered Chevallier's fungus to be *Hysterium commune* Fr. (= *Hypoderma rubi*), and CHEVALLIER noted that his fungus had a similar appearance to *Hypoderma* (as *Lophodermium*) *rubi*.

Two other species of *Lophodermium* have been described from the same hosts as CHEVALLIER reported for *L. herbarum*: *L. tumidum* var. *napelli* Rehm on *Aconitum*, and *L. ciliatum* Speg. & Roum. on *Epilobium*. They are macroscopically very similar to *Hypoderma rubi*, although both differ from the latter microscopically. Both *Lophodermium tumidum* var. *napelli* and *L. ciliatum* have subcuticular black ascomata, elliptical in outline with broadly rounded ends, well-developed lip cells, circinate paraphyses, large clavate asci and filiform ascospores, but differ from one another in a number of features. In the former the ascomata are associated with dark brown to black conidiomata, the upper wall contains a restricted group of very dark-walled cells near the well-developed lips, the lower wall comprises one or two rows of large, short-cylindrical to

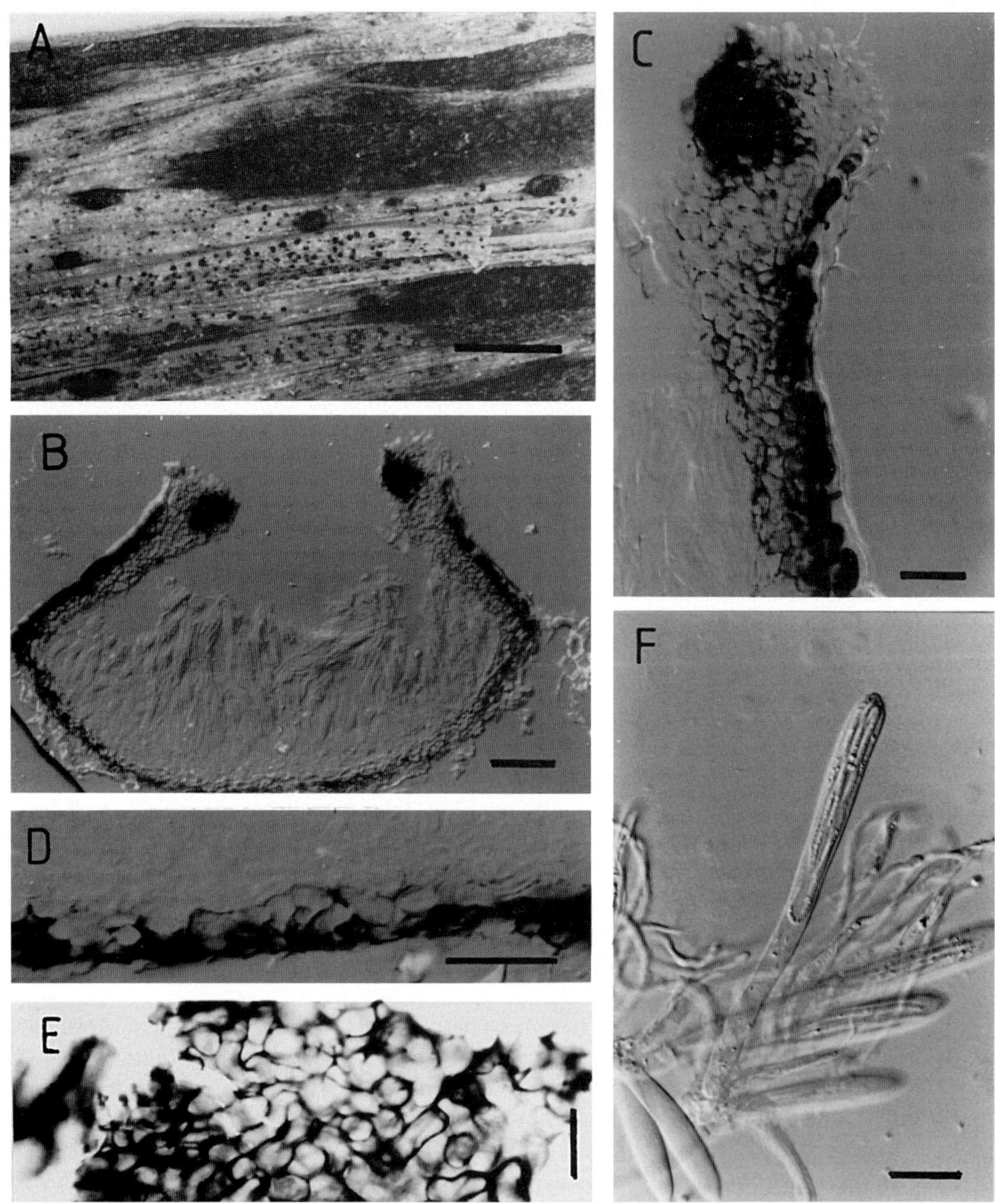

Fig. 64. *Lophodermium herbarum* (**A**, Fries, *Scleromyceti sueciae* no. 96, **K**; **B–D**, **PDD** 59592; **E**, **F**, **PDD** 59591). **A**, ascomata (bar = 1 mm); **B**, ascoma in vertical section (bar = 50 μm); **C**, upper wall of ascoma in vertical section; **D**, lower wall of ascoma in vertical section; **E**, lower wall of ascoma in squash mount; **F**, asci (bars = 20 μm).

angular/globose cells and the ascospores are 90–100 µm long. In *L. ciliatum* the ascomata are not associated with conidiomata, the upper wall is more or less uniformly dark, while the lower wall comprises several layers of long-cylindrical, partly tangled cells and ascospores are 60–70 µm long. It is likely that *L. herbarum* Chevall. is conspecific with one of these two later-described species, neither of which matches *L. herbarum* (Fr.) Fuckel from *Convallaria*.

Type material of *L. aconiti* Losa, also described from *Aconitum*, has not been located, but the description by LOSA (1947) is of a fungus with smaller asci (65–75 × 7–7·5 µm) and ascospores (25–35 × 1·5–2 µm) than *L. tumidum* var. *napelli* (asci 140–160 × 11–12 µm).

Two other species of *Lophodermium* have been reported from *Apiaceae*, but neither is likely to represent *L. herbarum* Chevall. *Lophodermium ginzbergeri* Petr. was described from *Eryngium* from tropical South America and, from the description given by SHERWOOD (1980), appears to be conspecific with *Coccomyces pampeanus* Speg., originally described from the same host. *Lophodermium mangatepopense* P.R. Johnst. is a Southern Hemisphere species known from several different host families.

Specimens examined: ***Lophodermium ciliatum*: Belgium:** nr Malmédy, on *Epilobium angustifolium, s. dat. nec coll.* (Roumeguère, *Fungi gallici exsiccati* no. 662; **K**, isotype); *s. loc.*, on *Epilobium angustifolium, s. dat.*, *?Libert s. num.* ('*Hysterium ciliatum* Lib. n.sp. in herb.'; **K**). **Norway:** OPPLAND: Lake Bygdin, on *Epilobium* (as *Chamaenerion*) *angustifolium*, 16 Aug. 1956, *R.W.G. Dennis s. num.* (**K**).

***Lophodermium tumidum* var. *napelli*: Austria:** TIROL: Mittelberg, on *Aconitum*, Aug. 1856, *Rehm s. num.* (**S**, holotype).

***Lophodermium ginzbergeri*: Brazil:** RIO DE JANEIRO: *s. loc.*, on *Eryngium paniculatum*, 27 Oct. 1927, *A. Ginzberger* 77 (**W**, holotype).

Lophodermium herbarum (Fr.) Fuckel, *Symb. mycol. Nachtr.* **2**: 50 (1873), *nom. illegit., ICBN* Art. 53.1.

Hysterium herbarum Fr., *Syst. mycol.* **2**: 593 (1823).

Aporia herbarum (Fr.) Duby, *Mém. Soc. Phys. Hist. nat. Genève* **16**: 64 (1862).

Hypoderma herbarum (Fr.) Kuntze, *Revis. gen. pl.* **3**(3): 487 (1898).

Infected areas on dead leaves, sometimes associated with bleaching of host tissue, but no zone lines, containing groups of ascomata and what appear to be conidiomata. *Ascomata* 0·5–0·7 × 0·25–0·4 mm, broadly elliptical in outline, edge sharply defined and raising abruptly at margin above level of surrounding leaf, wall black, with no pale zone along future line of opening, single longitudinal opening slit conspicuously shorter than the ascoma, lined with yellowish lips. *Ascomatal insertion* subepidermal. *Covering layer* up to 60 µm thick in vertical section, comprising clypeus of dark brown, thick-walled hyphae, 3 µm diam., within host epidermal cells, and upper wall of ascomatal stroma. *Upper wall* up to 45 µm thick, comprising mostly globose, pale brown cells, 6–8 µm diam., with irregularly encrusted walls and restricted group of very dark cells adjacent to well-developed lips. Poorly-developed periphysoids present in unopened ascomata, lost after ascomata open. *Lower wall* 15–20 µm thick in vertical section, comprising two or three rows of ± globose cells, the lowermost row with thickened and darkened walls, in squash mount darkened cells of lower wall globose to short-cylindrical, 8–12 µm diam., arranged in one or two rows across base of ascoma. *Paraphyses* 1·5–2 µm diam., undifferentiated or slightly and irregularly swollen near apex. *Asci* 115–130 × 10·0–11·5 µm, subclavate, tapering to subtruncate apex with undifferentiated wall, 8-spored, long basal stalk at maturity with spores confined to

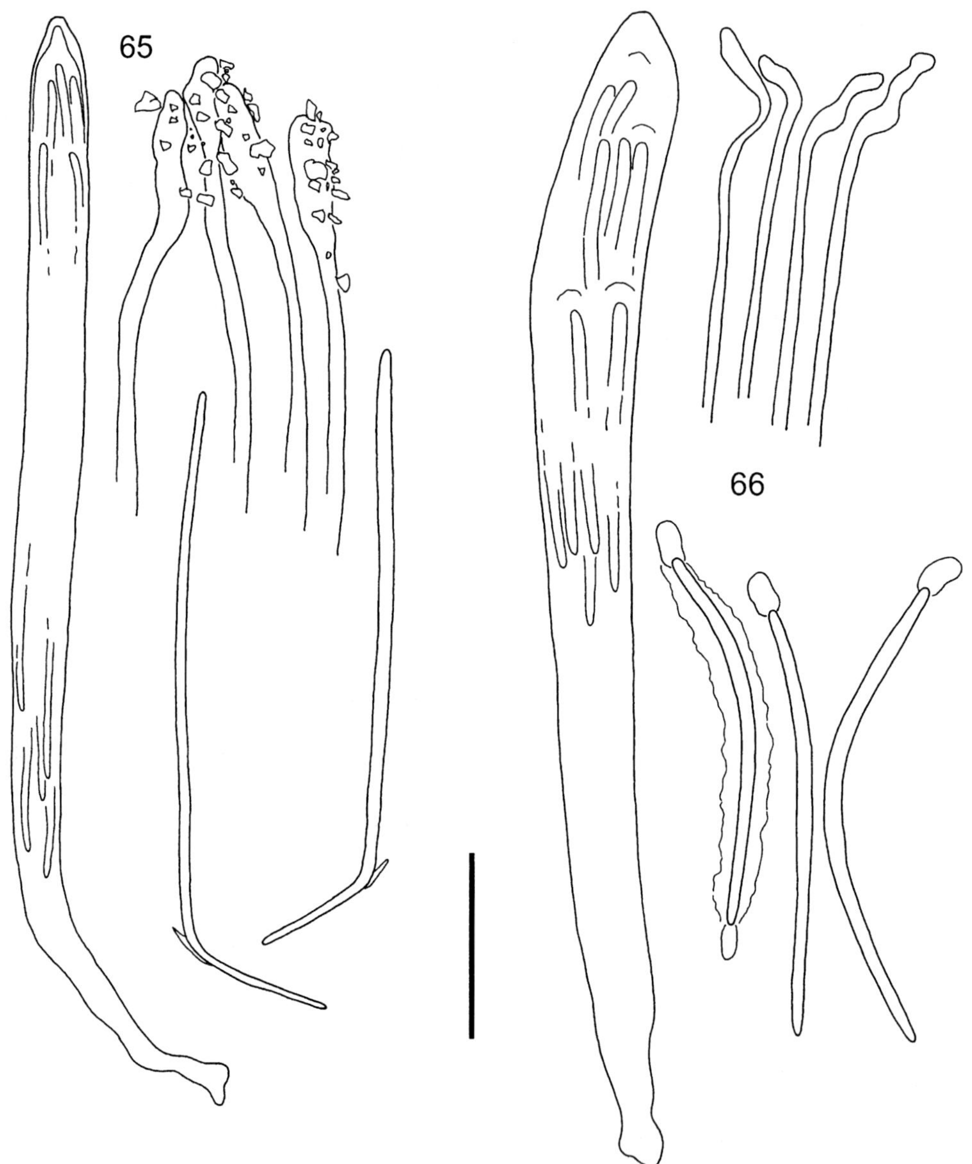

Figs 65–66. 65. *Lophodermium hauturuanum* (**PDD** 55566). Asci, apex of paraphyses and released ascospores in water. **66.** *Lophodermium herbarum* (**PDD** 59592). Ascus, apex of paraphyses and released ascospores (some with gel sheaths lost) in water (bar = 20 µm).

upper 60–80 µm. *Ascospores* 35–50 × 2 µm, tapering slightly to base, apical gelatinous cap subglobose, 5 µm diam., basal gelatinous cap short-cylindrical, gelatinous sheath 5–6 µm diam.

Typification: *S. loc., hosp. nec coll.* (Fries, *Scleromyceti sueciae* no. 96; **UPS**!, *lectotypus*, selected here; **K**!, *isolectotypus*, selected here).

Host: *Convallaria* (*Liliaceae*).

Distribution: Europe.

Illustrations: Figs 64, 66.

Notes: *Lophodermium herbarum* has an ascomatal structure typical of the *actinothyrium*-group of *Lophodermium* Group A (see p. 33), within which it is characterized by having very long asci relative to the ascospores (hence a long basal stalk) and large and more or less globose darkened cells of the lower wall. It is the only species of *Lophodermium* known from *Convallaria*. The type specimen is immature, but macroscopically and in vertical section it is indistinguishable from the recent collections examined. It is very similar to *L. iridicolum* Petr.; see Notes under that species. For notes on the nomenclature of this species see *L. herbarum* Chevall.

HÖHNEL (1906) and TEHON (1935) recorded *Allium* as a host for *L. herbarum*; this was based on the opinion that *L. herbarum* and *L. alliaceum* (described from *Allium*) were synonymous. However, isotype material of *L. alliaceum* differs in ascus and ascospore size and in the structure of the lower wall of the ascoma.

Additional specimens examined: Sweden: UPPLAND: Dalby Parish: Jeriko, on *Convallaria majalis*, 7 May 1985, *K. & L. Holm* 3511 (**PDD** 59591, **ZT**); Dalby Parish: Jerusalem, on *Convallaria majalis*, 12 Jun. 1977, *K. & L. Holm* 1107 (**PDD** 59592, **ZT**); *idem loc.*, on *Convallaria majalis*, 25 May 1985, *K. & L. Holm* 3541 (**PDD** 59593, **ZT**).

Lophodermium inclusum P.R. Johnst., *N.Z. Jl Bot.* **27**: 256 (1989).

Infected areas on dead leaves, paler than surrounding host tissue, often associated with incomplete, black zone lines, containing ascomata and conidiomata. *Ascomata* 1·3–2 × 0·4–0·6 mm, elliptical in outline, ends ± acute, unopened ascomata with wall dark grey with irregular paler patches, opened ascomata with wall mostly pale grey, but with dark grey to black line along outside edge and along single, longitudinal opening slit. *Ascomatal insertion* subepidermal, usually also beneath well-developed fibre bundles immediately below epidermis. *Covering layer* of opened ascomata up to 40 µm thick in vertical section, comprising clypeus within epidermal cells and upper wall of ascomatal stroma with embedded fibre bundles. *Upper wall* of unopened ascomata (when paraphyses first becoming differentiated, at up to 15 µm long) 15–25 µm thick in vertical section, comprising angular to globose, dark- and slightly thick-walled cells, 3–6 µm diam., lined with globose to short-cylindrical, hyaline periphysoids, in opened ascomata upper wall 15–25 µm thick, comprising dense, black tissue with no obvious cellular structure, periphysoids lost, no differentiated cells along edge of opening slit. *Lower wall* differentiating after upper wall, in vertical section comprising several layers of irregular-sized, angular to globose cells with thick, dark walls, in squash mount lower wall in immature ascomata (paraphyses *c.* 25 µm long) comprising two (or three) rows of angular to globose cells with irregularly darkened walls, at maturity becoming very dark, with two or three layers of angular to cylindrical, dark, thick-walled cells, 6–10 µm diam., and with irregular layer of *textura intricata* beneath, comprising one to several layers of hyphae, 3–5 µm diam., with thick, very dark walls. *Reduced excipulum* surrounding hymenium, comprising paraphysis-like elements, differentiated by being more closely septate than the paraphyses and by being embedded in common, thick gelatinous matrix. *Paraphyses* 1·5 µm diam., swelling to 2·5–3 µm near apex, forming regular palisade above asci. *Asci* 95–130 × 5·5–7·5 µm, cylindrical to subclavate, tapering to small, rounded apex with slightly thickened wall, 8-spored, empty basal stalk at maturity with spores confined to upper 80–90 µm, ascus development sequential. *Ascospores* 55–80 × 1·5 µm, non-septate, straight when released, with tiny gelatinous frill at base (when mounted in water). *Conidiomata* round in outline,

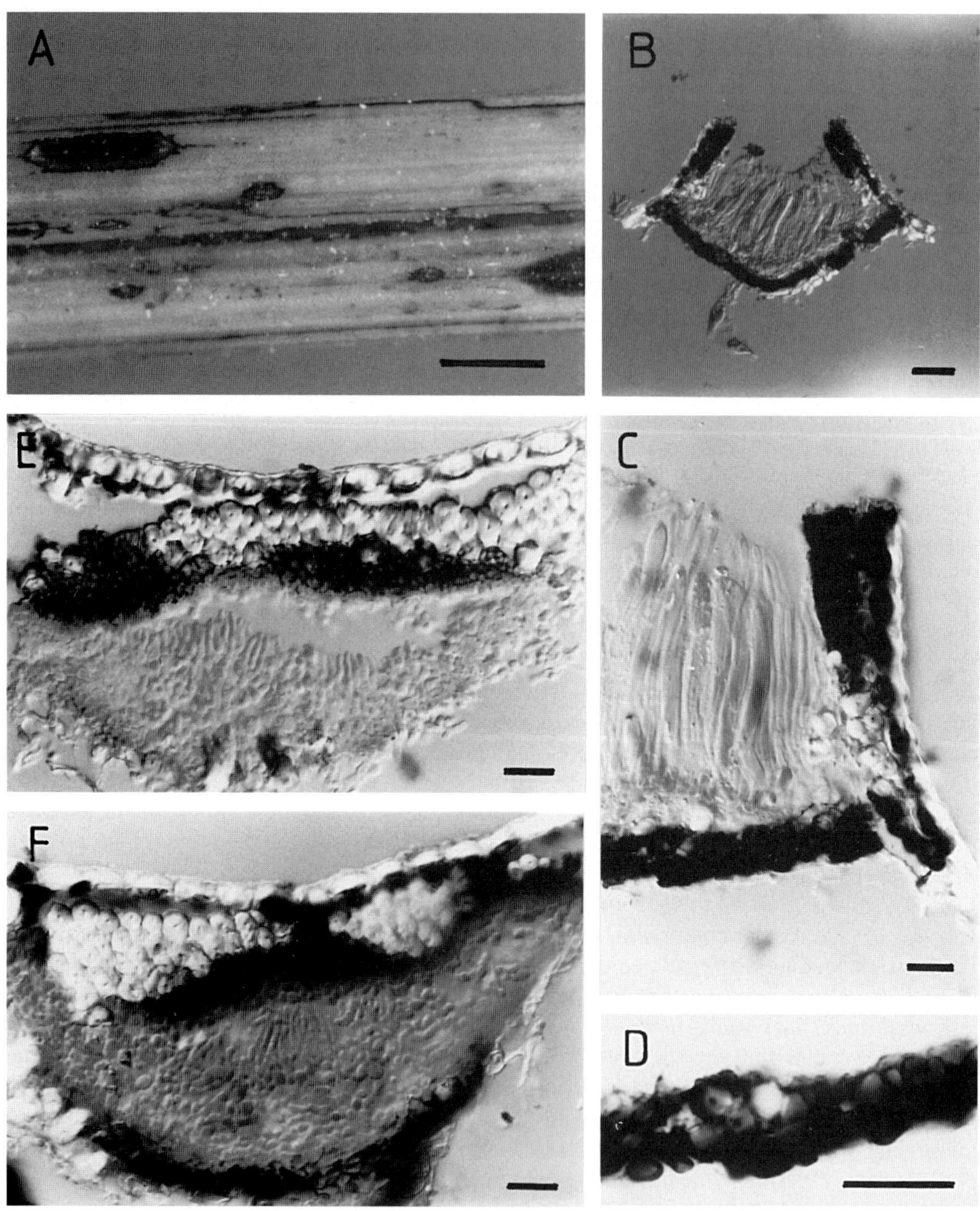

Fig. 67. *Lophodermium inclusum* (**A**, **PDD** 46168; **B–F**, **PDD** 46167). **A**, ascomata and conidiomata (bar = 1 mm); **B**, ascoma in vertical section (bar = 50 µm); **C**, upper wall of ascoma in vertical section; **D**, lower wall of ascoma in vertical section (bars = 20 µm); **E**, very immature ascoma in vertical section, lower wall not yet darkened; **F**, immature ascoma in vertical section, with darkened lower wall, no differentiated asci (bars = 50 µm).

0·3–0·5 mm diam., wall pale brown, dark brown line at edge, upper wall poorly developed, lower wall lined with palisade of conidiogenous cells. *Conidiogenous cells* 9–13 × 2–2·5 µm, cylindrical to flask-shaped, proliferation sympodial, often with two conidia held at apex. *Conidia* 4–4·5 × 1–1·5 µm, hyaline, straight, non-septate.

Typification: New Zealand: BULLER: nr Greymouth, Pt. Elizabeth Walk, on *Gahnia*, 1 May 1985, *P.R. Johnston et al. s. num.* (**PDD** 46937!, holotype).

Host: *Gahnia* (*Cyperaceae*).

Distribution: New Zealand.

Illustrations: Figs 67, 69.

Notes: The relationship of *L. inclusum* with other monocotyledon-inhabiting species is uncertain. Features such as the very dark upper wall, lack of differentiated cells around the ascomatal opening, upper wall darkening before lower wall and small gelatinous cap on ascospores, provide a combination of characters not seen in any other species examined.

The other species of *Lophodermium* occurring on *Gahnia* in New Zealand, *L. hauturuanum*, is distinguished by its narrower, cylindrical ascomata developing above rather than below the fibre bundles in the host leaf, and different structure to the upper wall of the ascoma. Paraphyses, asci and ascospores are of similar size in the two species. Ascospores differ in appearance when released, those of *L. hauturuanum* having a characteristic bend near the base and a small gelatinous appendage at the point of the bend rather than at the end of the spore, as in *L. inclusum*.

Additional specimens examined: New Zealand: AUCKLAND: Waitakere Ranges, Scenic Drive, on *Gahnia*, 8 Oct. 1984, *P.R. Johnston* R582c (**PDD** 46167); Waitakere Ranges, Sharps Bush, on *Gahnia*, 3 Feb. 1984, *P.R. Johnston* R381 (**PDD** 46166); BULLER: nr Blackball, Croesus Track, on *Gahnia*, 4 May 1985, *P.R. Johnston* R651 (**PDD** 46164); COROMANDEL: Little Barrier I, Summit Track, on *Gahnia*, 7 Apr. 1988, *P.R. Johnston s. num.* (**PDD** 54859); SOUTHLAND: Catlins, Lake Wilkie, on *Gahnia*, 9 May 1995, *P.R. Johnston* R942.1 (**PDD** 64752); TAUPO: Tongariro National Park, Ohakune Mtn Rd, Blyth Track, on *Gahnia*, 20 May 1989, *P.R. Johnston s. num.* (**PDD** 55567); WAIKATO: nr Waharoa, Kahikatea Reserve, on *Gahnia*, 28 Jul. 1989, *P.R. Johnston s. num.* (**PDD** 56345); WESTLAND: Fox Glacier, Lake Gault, on *Gahnia*, 10 Apr. 1983, *P.R. Johnston s. num.* (**PDD** 43978, 46168, 49351, 49352).

Lophodermium iridicolum Petr., *Annls mycol.* **20**: 10 (1922).

Infected areas on dead leaves, slightly paler than surrounding host tissue, not associated with zone lines, containing discrete groups of ascomata and conidiomata. *Ascomata* 0·5–0·9 × 0·3–0·4 mm, elliptical in outline, wall pale to dark grey, with no paler line along future line of opening, single, longitudinal opening slit, with well-developed lip cells, lips initially pale, becoming reddish with age. *Ascomatal insertion* subepidermal. *Covering layer* up to 60–70 µm thick in vertical section, comprising clypeus of dark-walled hyphae forming within epidermal cells, and upper wall of ascomatal stroma. *Upper wall* up to 60 µm thick, comprising ± globose, mostly pale brown, slightly and irregularly thick-walled cells, 8–10 µm diam., with small patch of very dark, very thick-walled cells in inner half of wall adjacent to lip cells. *Lower wall* 15–25 µm thick in vertical section, comprising two or three rows of globose cells, 8–12 µm diam., mostly hyaline and thin-walled except for lowermost wall of lowest row, which is darkened and slightly thickened. *Paraphyses* 1·5 µm diam., irregularly swollen up to 4 µm diam. near apex, extending 15–20 µm beyond asci. *Asci* 90–110 × 9–10·5 µm, broad-cylindrical to subclavate, tapering gradually to rounded apex with undifferentiated wall, 8-spored, short basal stalk at maturity with spores

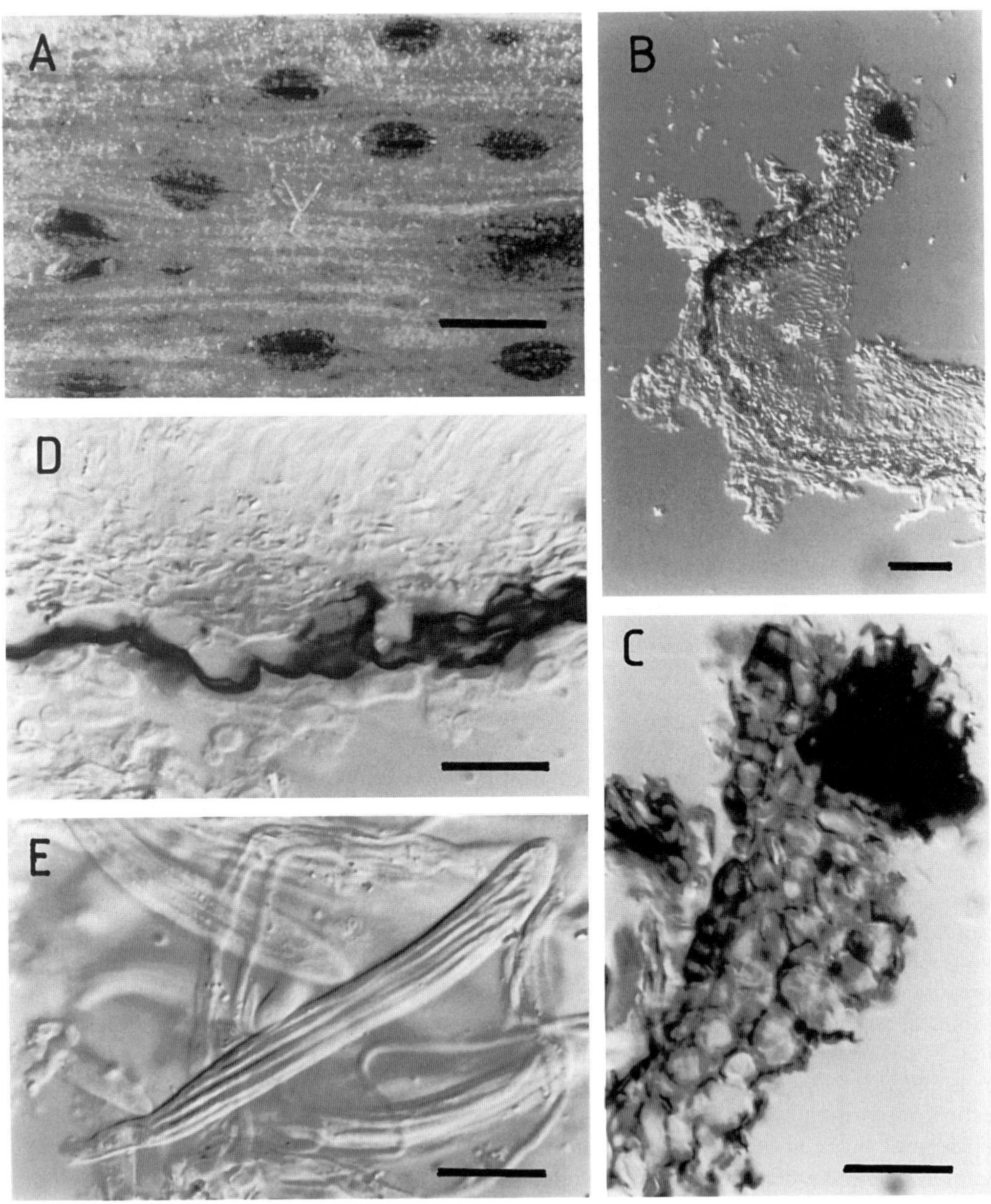

Fig. 68. *Lophodermium iridicolum* (holotype, **W**). **A**, ascomata (bar = 1 mm); **B**, one side of ascoma in vertical section (bar = 50 μm); **C**, upper wall of ascoma in vertical section; **D**, lower wall of ascoma in vertical section; **E**, ascus (bars = 20 μm).

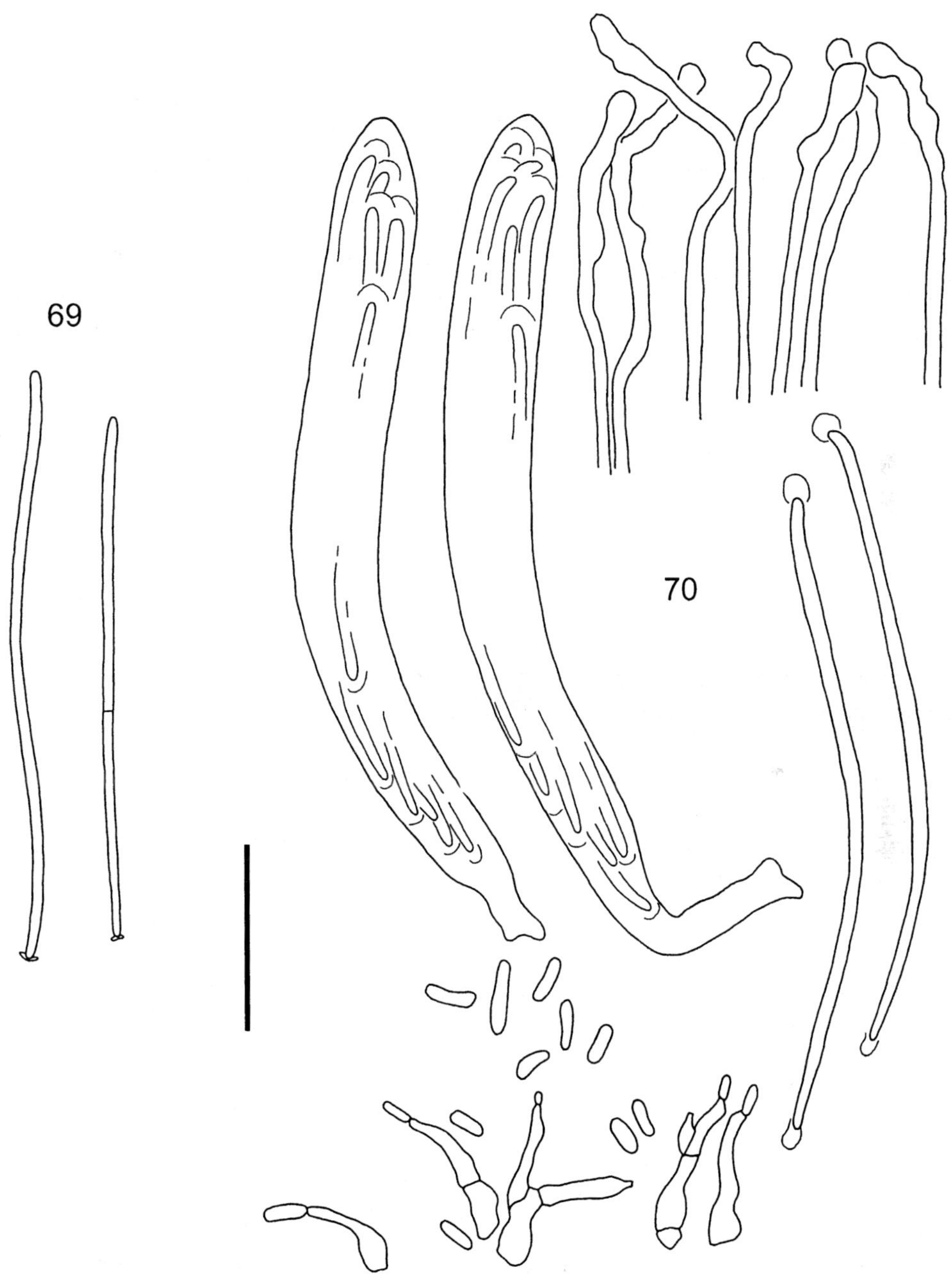

Figs 69–70. 69. *Lophodermium inclusum* (**PDD** 64752). Ascospores in water with small basal gelatinous cap. **70.** *Lophodermium iridicolum* (holotype, **W**). Asci, apex of paraphyses, ascospores and conidiogenous cells and conidia (bar = 20 µm).

confined to upper 80–95 µm. *Ascospores* 60–70 × 2 µm, tapering slightly to base, apical gelatinous cap globose, 4–5 µm diam., basal gelatinous cap cylindrical, *c.* 3 × 1·5 µm, gelatinous sheath *c.* 5 µm diam. *Conidiomata* scattered, 0·2–0·5 × 0·1–0·2 mm, oblong in outline, sometimes confluent, pale orange-brown, in vertical section upper wall lacking, lower wall 10–15 µm thick, comprising several rows of hyphal cells, 2–3 µm diam., with gelatinized walls, and above it single layer of angular, hyaline, thin-walled cells on which conidiogenous cells are held, at margin of conidioma is small group of pale brown, thin-walled, angular cells. *Conidiogenous cells* 10·5–17 × 2–3·5 µm, swollen at base, with long, cylindrical upper part, method of proliferation not clearly seen, although only one developing conidium held at apex of conidiogenous cells. *Conidia* 4·5–8·5 × 1·5–2 µm, oblong-elliptical to oblong, ends rounded, non-septate, hyaline.

Typification: Albania: Skutari, at foot of mountain, on leaves of *Iris*, 12 Oct. 1918, *F. Petrak s. num.* (**W**!, as *L. iridis*).

Host: *Iris* (*Iridaceae*).

Distribution: Albania.

Illustrations: Figs 68, 70.

Notes: *Lophodermium iridicolum* has an ascomatal structure typical of the *actinothyrium*-group of *Lophodermium* Group A. PETRAK (1922) compared his species with *Lophodermium herbarum* (Fr.) Fuckel: the two species are morphologically similar, but *L. herbarum* differs in having slightly wider, longer asci and shorter, narrower ascospores. Although some collections of *L. herbarum* examined had structures resembling the conidiomata of *L. iridicolum*, conidiogenous cells and conidia were never seen.

Terriera javanica (Penz. & Sacc.) P.R. Johnst., **comb. nov.**
Lophodermium javanicum Penz. & Sacc., *Malpighia* **11**: 529 (1897).

Infected areas on dead leaves, paler than surrounding host tissue, associated with ± complete narrow, black zone lines, containing ascomata and conidiomata. *Ascomata* 0·4–1·5 × 0·2–0·3 mm, oblong-elliptical to sublinear in outline, ends ± acute, margin not sharply defined, stippled, unopened ascomata with narrow pale zone along future line of opening, open ascomata with wall shiny black, single, longitudinal opening slit with narrow, black, shelf-like zone along either side. *Ascomatal insertion* subepidermal. *Ascomatal structure* typical of *Terriera* (see p. 37). *Paraphyses* 1·5 µm diam., swelling slightly to 2·5–3·5 µm at subclavate apex, embedded in thick gel, tangled and often branched near apex, extending 10–15 µm beyond asci as dense epithecium. *Asci* 85–95 × 5·5–7 µm, cylindrical, tapering to rounded apex with undifferentiated wall, 8-spored, basal stalk developing at maturity with spores restricted to upper 65–75 µm. *Ascospores* not clearly seen, 50–60 × 1·5 µm. *Conidiomata* 0·2 mm diam., round in outline, concolorous with surrounding leaf surface except for small black spot near centre of conidioma. *Conidiogenous cells* cylindrical to flask-shaped, proliferation sympodial, often with two conidia held at apex. *Conidia* 3·5–4·5 × 1 µm, oblong-elliptical with rounded ends, non-septate, hyaline.

Typification: Indonesia: JAVA: Tjibodas, on *Elettaria*, 27 Feb. 1897, *O. Penzig s. num.* (**W**!, *lectotypus*, selected here; **BO** 3749!, *isolectotypus*, selected here).

Illustrations: Figs 71, 73.

Host: *Elettaria* (*Zingiberaceae*).

Distribution: Java.

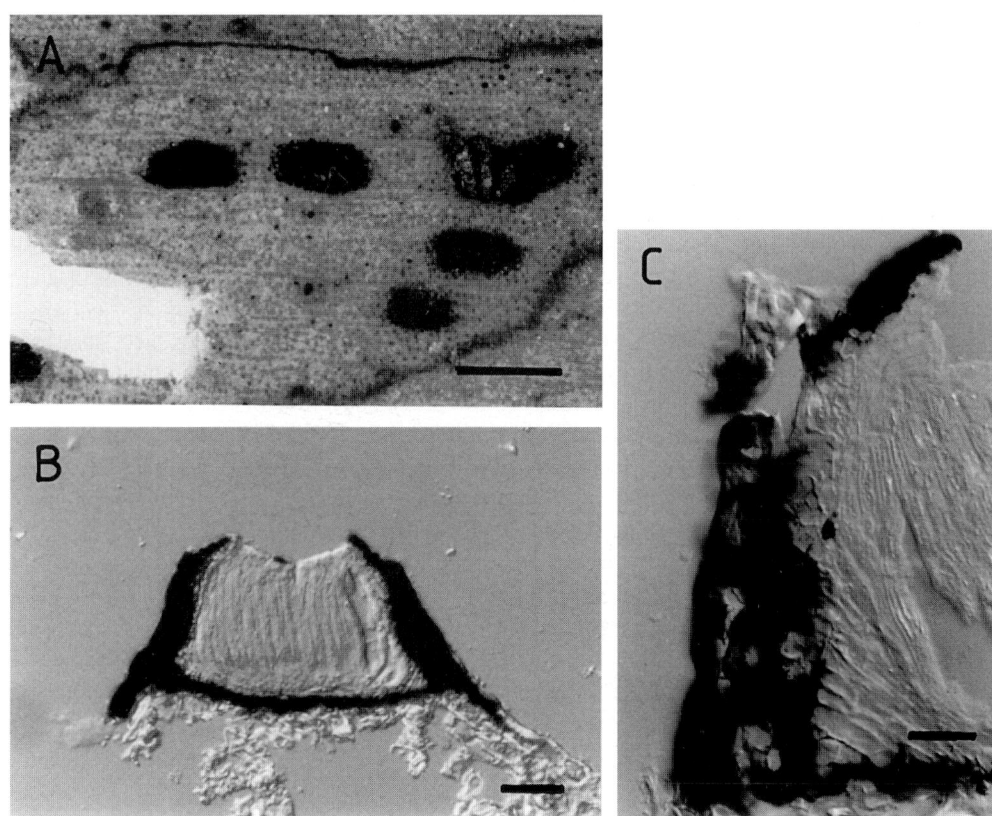

Fig. 71. *Terriera javanica* (lectotype, **W**). **A**, ascomata (bar = 1 mm); **B**, ascoma in vertical section (bar = 50 μm); **C**, detail of margin of ascoma in vertical section (bar = 20 μm).

Notes: *Terriera javanica* is characterized by the more or less complete zone lines surrounding that part of the leaf invaded by the fungus. Of the two other species with zone lines, *T. clithris* has pale ascomata and longer asci with a truncate apex, while *L. andropogonis* lacks conidiomata and has a broadly rounded ascus apex.

Terriera javanica is very similar to two dicotyledon-inhabiting species of *Lophodermium*, the names of which have yet to be combined into *Terriera*: *L. planchoniae* Rehm (described from tropical Asia) and *L. acacicolum* Tehon (described from Hawaii). Both have ascomata associated with zone lines and similar-sized asci and ascospores. If these species were to be placed in synonymy, the epithet *javanica* would have priority.

Two separate species of *Terriera* are present on the lectotype specimen, distinguished in part by a difference in macroscopic appearance of the ascomata. Most are as described above, but there are some groups of ascomata which are oblong in outline with rounded ends, have dark brown to dull black walls and a sharply-defined margin. In the first group with blacker, diffuse-margined ascomata, the clypeus-like layer above the ascomatal stroma extends beyond the margins of the ascoma, while in the second group the clypeus is restricted to those epidermal cells immediately above the ascomatal stroma. The macroscopic difference appears to correlate with ascus size, those with the paler ascomata having asci 100–120 µm long, as compared to 80–95 µm in *T. javanica*. This second species appears to match *T. pandani*, described from central America.

Rhytisma juncicola Rehm, *Hedwigia* **21**: 116 (1882).

Infected areas on dead leaves, not associated with bleached areas or zone lines, containing scattered ascomata, conidiomata not seen. *Ascomata* irregular in shape, wall black, opening slit single or branched depending on ascomatal shape, without lip cells. *Ascomatal insertion* subcuticular. *Covering layer* up to 65 µm thick in vertical section, comprising host cuticle and upper wall of ascomatal stroma. *Upper wall* comprising angular cells with thickened, dark brown walls in outer half, inner half with hyaline, cylindrical, tangled periphysoids embedded in gel. *Lower wall* up to 30 µm thick in vertical section, comprising several rows of cells, the inner rows pale, the outer rows dark-walled, irregular in shape, in squash mount darkened cells of lower wall long-cylindrical to hyphal, forming *textura intricata*-like layer across base of ascoma, subhymenial layer between hymenium and lower wall very well developed. *Paraphyses* 2–3 µm diam., irregularly swollen and tangled at apex, forming dense epithecium above asci. *Asci* 115–125 × 11·5–13·5 µm, clavate to subclavate, tapering to small, truncate apex with undifferentiated wall, 8-spored, empty basal stalk with spores confined to upper 90–100 µm. *Ascospores* not seen released, 50–60 × 2–2·5 µm, tapering gradually to base, non-septate, apical gelatinous cap visible within ascus.

Typification: Austria: TIROL: Kühtai, on *Juncus hostii*, Aug. 1874, *Rehm s. num.* (**S**!, holotype).

Illustrations: Figs 72, 74.

Host: *Juncus* (*Juncaceae*).

Distribution: Europe.

Notes: *Rhytisma juncicola* is characteristic of *Lophodermium* Group D as described on p. 42.

Although *Rhytisma* is clearly not the appropriate genus for this species, no new combination is proposed here because of questions about correct generic and species limits among these fungi. This species may be conspecific with *Coccomyces coronatus*, the only apparent difference being that the paraphyses in the former are irregularly swollen and tangled at the apex, while in the latter they form a regular palisade.

See also Notes under *L.* cf. *luzulae*.

Additional specimens examined: ***Coccomyces coronatus*: England:** YORKSHIRE: Hackfall, on *Quercus*, Sep. 1948, *S.J. Hughes & J. Webster s. num.* (**PDD** 14713). **USA:** NEW HAMPSHIRE: Carroll Co.: Redstone, on *Quercus*, 5 Sep. 1963, *H.E. & M.E. Bigelow s. num.* (**NY**).

Lophodermium *cf.* **juncinum** (Jaap) Terrier ex E. Müll., *Beitr. Krypt.-Fl. Schweiz* **15**: 15 (1977).
Lophodermium arundinaceum var. *juncinum* Jaap, *Annls mycol.* **15**: 103 (1917).

Infected areas on dead leaves, not associated with bleaching of host tissue or zone lines, containing scattered ascomata, conidiomata not seen. *Ascomata* 0·6–0·8 × 0·3 mm, oblong-elliptical in outline, ends broadly rounded, becoming erumpent, at maturity with most of upper

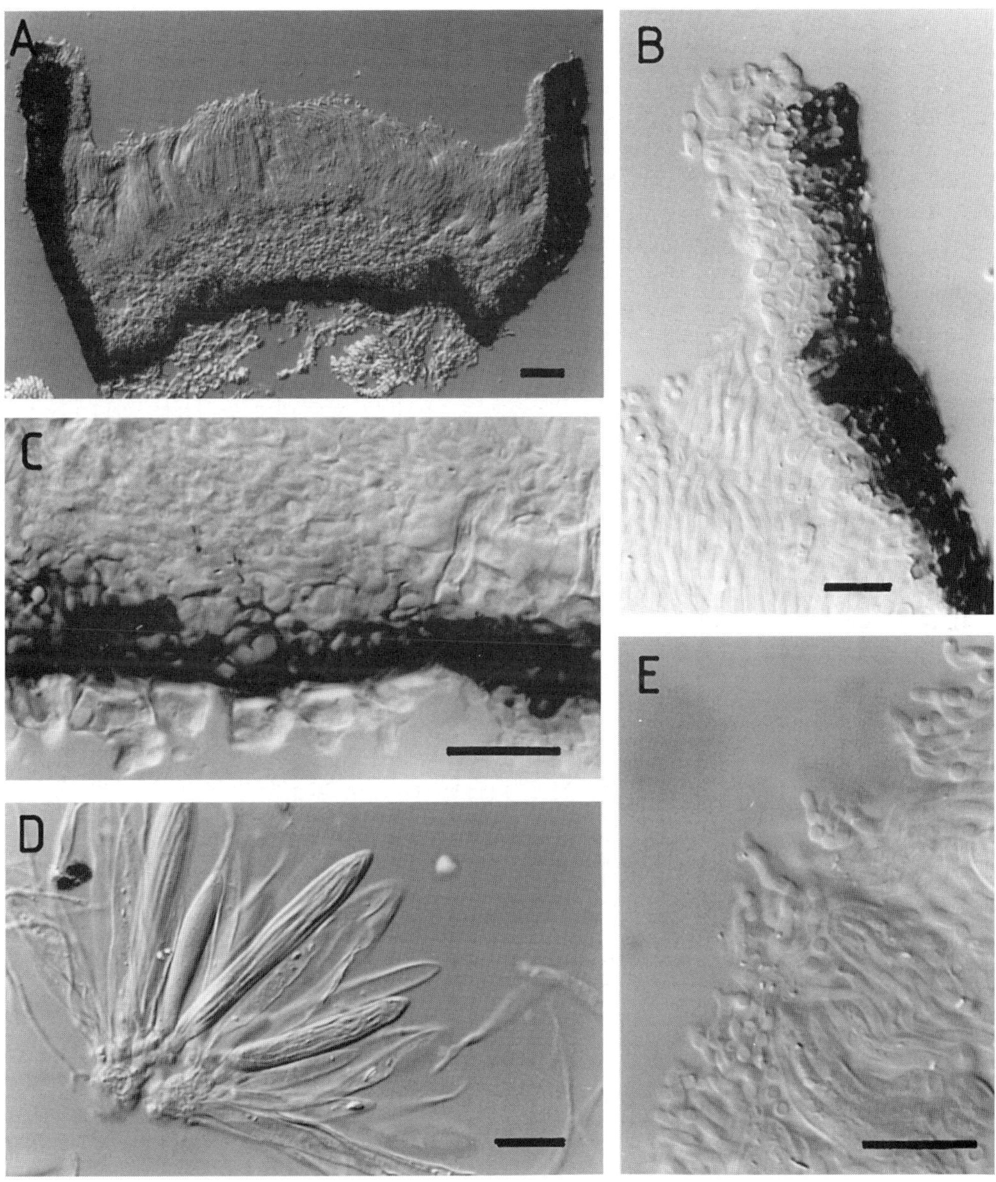

Fig. 72. *Rhytisma juncicola* (holotype, **S**). **A**, ascoma in vertical section (bar = 50 µm); **B**, upper wall of ascoma in vertical section, showing periphysoids; **C**, lower wall of ascoma in vertical section; **D**, asci; **E**, tangled paraphysis tips (bars = 20 µm).

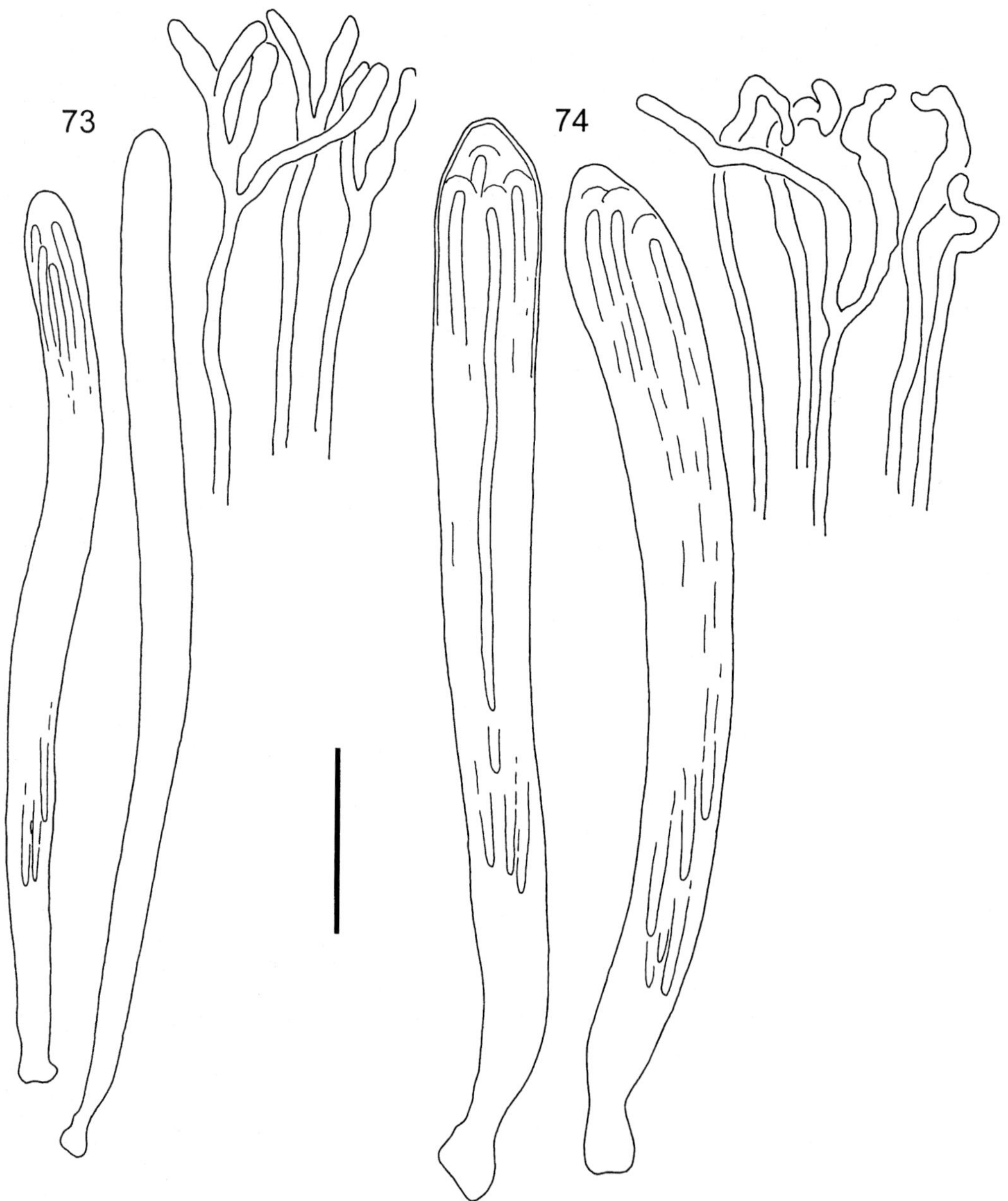

Figs 73–74. 73. *Terriera javanica* (lectotype, **W**). Asci and apex of paraphyses. **74.** *Rhytisma juncicola* (holotype, **S**). Asci and apex of paraphyses (bar = 20 μm).

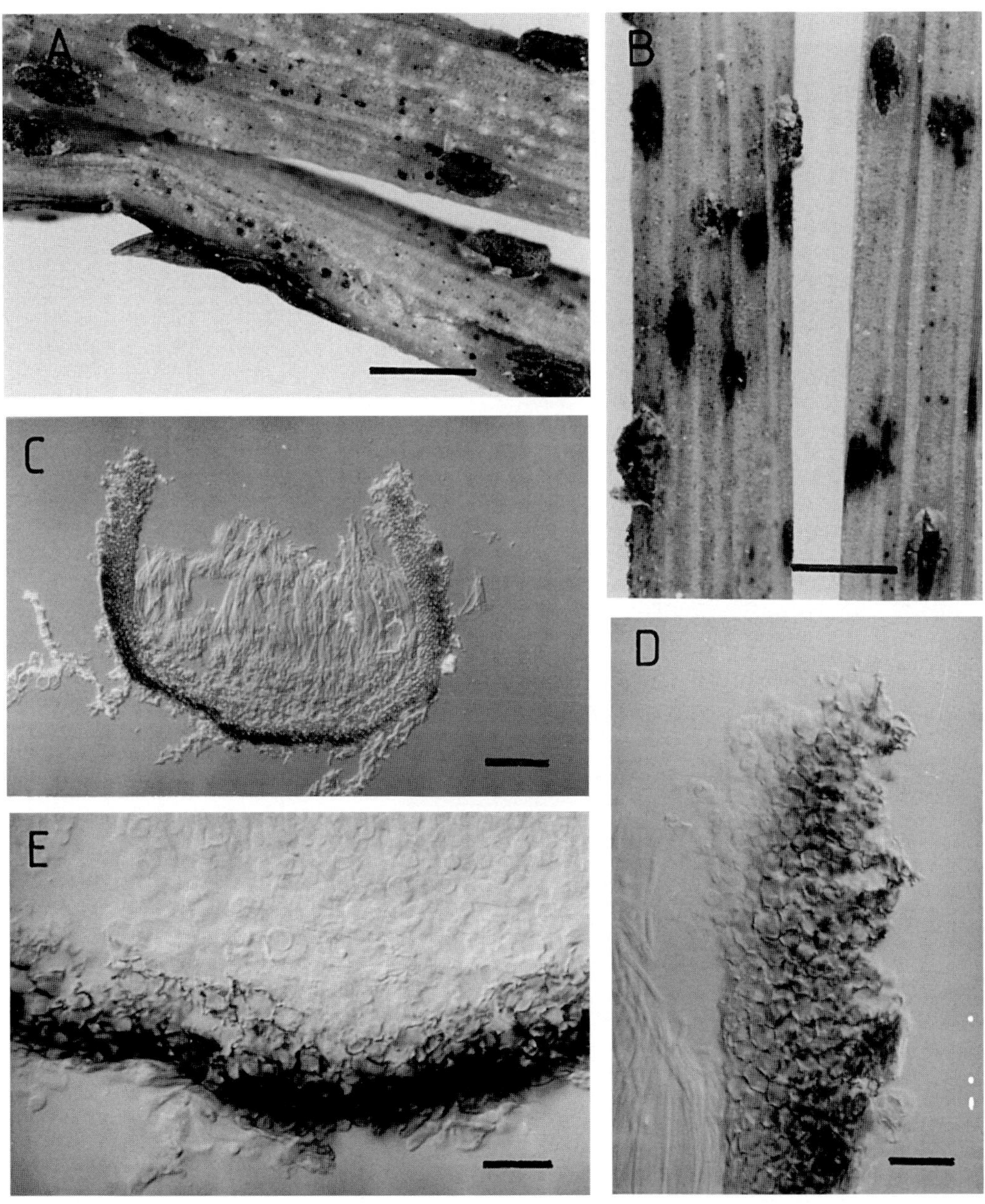

Fig. 75. *Lophodermium* cf. *juncinum* (**A**, **C–E**, *Müller*, 9 Sep. 1962, **ZT**; **B**, *Müller*, 25 Aug. 1980, **ZT**). **A**, ascomata (bar = 1 mm); **B**, ascomata (bar = 1 mm); **C**, ascoma in vertical section (bar = 50 μm); **D**, upper wall of ascoma in vertical section; **E**, lower wall of ascoma in vertical section (bars = 20 μm).

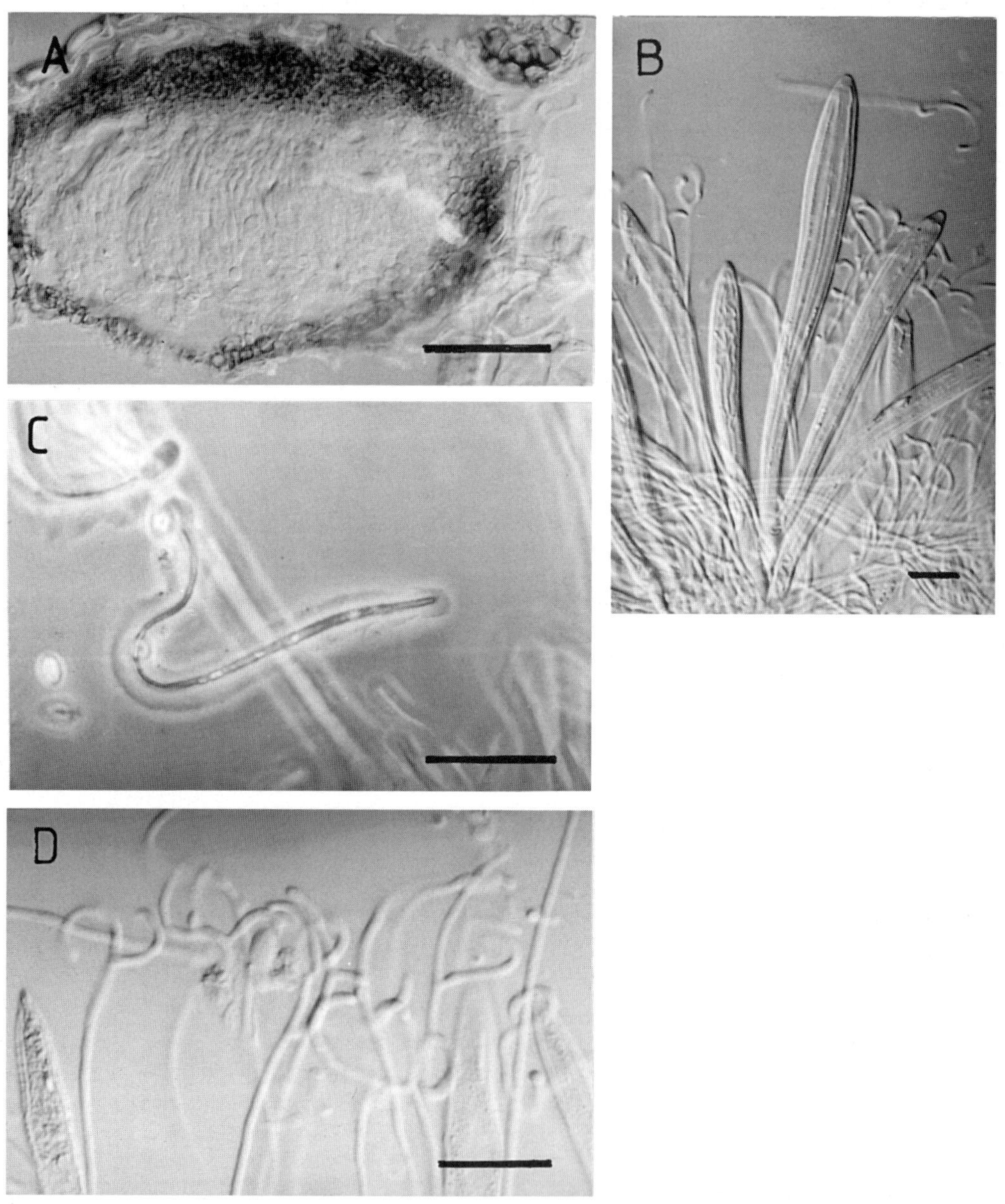

Fig. 76. *Lophodermium* cf. *juncinum* (**A**, *Müller*, 27 Jul. 1956, **ZT**; **B–D**, *Müller*, 25 Aug. 1980, **ZT**). **A**, unopened ascoma in vertical section (bar = 50 µm); **B**, asci; **C**, released ascospores with gelatinous caps and sheath; **D**, paraphyses (bars = 20 µm).

part not covered by host tissue, wall dark grey, hymenium dull orange (pigmentation apparently associated with aborted asci), opening by single longitudinal slit with no lip cells. *Ascomatal insertion* initially subepidermal, covering host tissue eroding as ascoma develops, with upper wall of ascoma becoming exposed. *Upper wall* of unopened ascomata (where paraphyses starting to elongate but asci not yet visible) *c.* 30 μm thick, in vertical section comprising several rows of ± globose, thin-walled, pale brown cells, 2·5–4 μm diam., in opened ascomata upper and lower walls apparently continuous, with ascoma appearing cupulate in vertical section, part of wall above hymenium 30–60 μm thick, comprising five to seven rows of angular to globose cells, outer two or three rows of cells 4–7 μm diam., with dark brown, thick walls, inner rows of cells 7–12 μm diam., with pale brown, thin walls. *Lower wall* 20 μm thick, comprising three or four rows of globose cells, 4–6 μm diam., with slightly thickened, dark brown walls. *Paraphyses* 1·5 μm diam., branched several times and circinate at apex. *Asci* 160–220 × 12·5–17 μm, clavate, tapering gradually to small, subtruncate apex with slightly thinner wall, 8-spored, short basal stalk at maturity with spores confined to upper 110–130 μm. *Ascospores* 95–120 × 1·5–2·5 μm, tapering slightly to ends, apical gelatinous cap globose, 4·5–5 μm diam., basal gelatinous cap very small, knob-like to cylindrical, gelatinous sheath 5–10 μm diam.

Typification: 'auf Halmen von *Juncus jaquinii* L., Furkapass, 3.8.1905, wiedergeunden, 30.7.1910' (MÜLLER, 1977).

Host: *Juncus* (*Juncaceae*).

Distribution: Europe.

Illustrations: Figs 75, 76.

Notes: JAAP's type specimens have not been seen and his very brief description of the species provides few clues to its appearance, either macroscopic or microscopic. The above description is based on material in **ZT** collected by MÜLLER from the European Alps, the type locality of *L. juncinum*. The collections cited by MÜLLER (1977) under *L. juncinum* contain three species, discussed further under *L.* cf. *luzulae*. The collections cited below are typical of the species chosen here to represent *L. juncinum sensu* Müller. This species is known only from *Juncus*. The macroscopic features described by MÜLLER match these specimens, although the microscopic features more closely resemble those of another species found on *Juncus* and other hosts, discussed below under *L.* cf. *luzulae*.

Lophodermium cf. *juncinum* is typical of *Lophodermium* Group C (see p. 39). The two species in this group differ from each other primarily in the depth at which they develop within the host tissue: *L.* cf. *juncinum* is subepidermal, *L. tumidulum* subcuticular. The apparent difference in the arrangement of the upper and lower walls of the two species (mature ascomata of *L. juncinum* appearing to have a single wall layer surrounding the entire ascoma) may possibly be explained by the difference in ascomatal insertion and the associated influence on ascomatal shape of the difference in pressure exerted by the host tissue. Ascomata of *L. juncinum* develop within the cavity beneath the host epidermis, whereas *L. tumidulum* is constricted between the cuticle and epidermis, causing the lower part of the ascoma to become 'flattened' and hence appearing to remain as a distinct lower wall. There is no evidence that the difference in insertion relates to differences in host, collections typical of both species having been reported from a single host, *Juncus trifidus*.

For the present, the difference in ascomatal insertion and the small differences in ascus and ascospore size are taken to justify retaining *L.* cf. *juncinum* and *L. tumidulum*.

Specimens examined: Austria: KÄRNTEN: Kreuzeck-Gruppe, on *Juncus trifidus*, 18 Aug. 1981, *C. Scheuer s. num.* (**GZU**, as '*Rhytisma*' *juncicolum*); SALZBURG: Radstädter, Tauern, on *Juncus trifidus*, 24 Jul. 1982, *C. Scheuer s. num.* (**GZU**, as

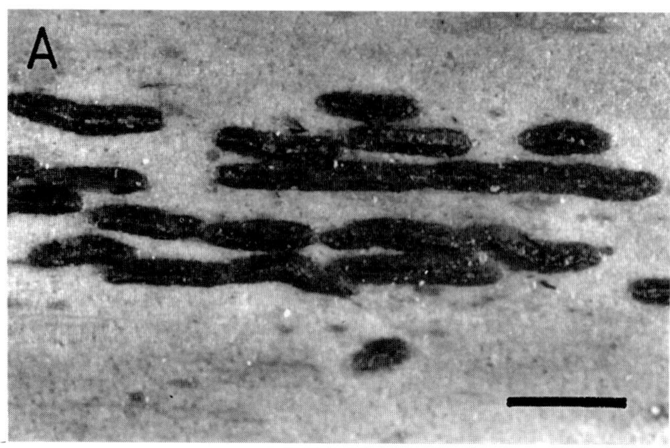

Fig 77. *Terriera latiascus* (**PDD** 58076). Ascomata (bar = 1 mm).

'Rhytisma' juncicolum). **Switzerland:** GRAUBÜNDEN: Albulapass, Murtel, on *Juncus jaquini*, 25 Aug. 1980, *E. Müller s. num.* (**ZT**); Bergui Val Tuors, on *Juncus jaquini*, 27 Jul. 1956, *E. Müller s. num.* (**ZT**); WALLIS: Aletschwald, Moräne unterhalb Silbersand, on *Juncus jaquini*, 9 Sep. 1962, *E. Müller s. num.* (**ZT**); Aletschreservat bei Brig, Moräne unterhalb Silbersand, on *Juncus jaquini*, 10 Sep. 1972, *E. Müller s. num.* (**ZT**); Aletsch glacier, on *Juncus jaquini*, 10 Sep. 1970, *R.W.G. Dennis & E. Müller s. num.* (**K**).

Terriera latiascus P.R. Johnst., **sp. nov.**

Etymology: Referring to the broad asci.

A *T. cladophila* ascis 80–95 × 7–8·5 μm, ascosporis 40–50 × 2–2·5 μm differt.

Infected areas on dead leaves, paler than surrounding host tissue, not associated with zone lines, containing numerous, gregarious ascomata, conidiomata not seen. *Ascomata* 0·6–1 × 0·25 mm, oblong-elliptical, wall black to shiny black, single longitudinal opening slit with black, shelf-like zone along either side. *Ascomatal structure* typical of *Terriera* (see p. 37). *Paraphyses* 1·5 μm diam., increasing to 3–4 μm at irregularly swollen apex, branched once or twice near apex, forming dense epithecium 15–20 μm thick. *Asci* 80–95 × 7–8·5 μm, ± cylindrical, tapering slightly to rounded apex with undifferentiated wall, 8-spored, well-developed basal stalk at maturity with spores confined to upper 60–70 μm. *Ascospores* not seen released, 40–50 × 2–2·5 μm, 1 (–3)-septate, tapering slightly to ends.

Typification: Brazil: AMAZONAS: Plateau of Serra Araca, N side of North Mtn, cloud forest, 17–22 Feb. 1984, *G.J. Samuels* 335 (**NY**!, *holotypus* of *T. latiascus*; **PDD** 58076!, *isotypus*).

Hosts: ?*Euterpe* (*Arecaceae*); *Heliconia* (*Heliconiaceae*).

Distribution: Brazil, Trinidad.

Illustrations: Figs 77, 78.

Notes: This species is characterized by small ascomata, short, broad asci and broad ascospores. Asci are similar in size to those of *T. fuegiana*, but differences in host and distribution suggest that these are distinct species.

Fig. 78. *Terriera latiascus* (**PDD** 58076). Asci and apex of paraphyses (bar = 20 µm).

Although no substratum details are given on the GJS 335 specimen label, it appears to be part of a dried palm frond. The three collections are very similar in macroscopic appearance and in the shape and size of the hymenial elements.

Another collection in **NY** on leaves of *Heliconia* (**Colombia:** DPTO. CHOCÓ: Quibdó-Medellín Rd, on *Heliconia*, 10 Aug. 1976, *K.P. Dumont et al. s. nùm.*; **NY**) represents a different species with very narrow asci (90–100 × 4·5–5·5 µm) and additional collections are required to assess its taxonomic position.

Additional specimens examined: Trinidad: Sangre Grande, on ?*Euterpe*, 1912–1913, *R. Thaxter s. num.* (**FH**, as *Lophodermium* sp.); Cumuto, on *Heliconia*, 1912–1913, *R. Thaxter s. num.* (**FH**, as *Lophodermium* sp.).

Terriera longissima P.R. Johnst., **sp. nov.**

Etymology: Referring to the unusually long asci and ascospores.

A *T. cladophila* ascis 175–210 × 6–6·5 µm, ascosporis 120–130 µm longis differt.

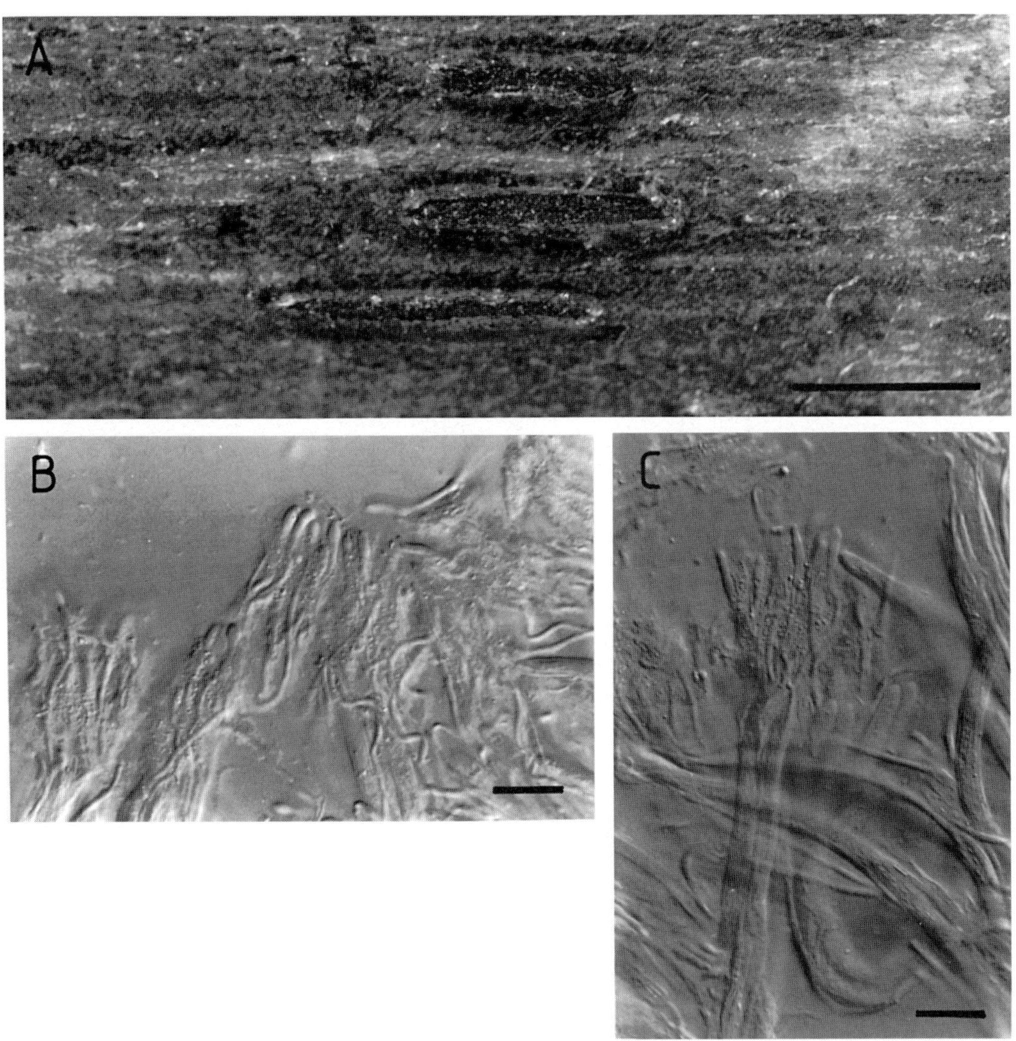

Fig. 79. *Terriera longissima* (**PDD** 58055). **A**, ascomata (bar = 1 mm); **B–C**, apex of asci and paraphyses, showing crystalline inclusions in epithecial gel (bar = 20 μm).

Infected areas on dead leaves, often slightly darker than surrounding host tissue, associated with numerous, somewhat diffuse zone lines, containing scattered ascomata, conidiomata not seen. *Ascomata* 1–2 × 0·3 mm, oblong to sublinear in outline, ends rounded, wall variable in colour, some quite pale, but most dark brown to black, single, longitudinal opening slit with black shelf-like zone along either side. *Ascomatal insertion* subepidermal. *Ascomatal structure* typical of *Terriera* (see p. 34). *Paraphyses* 1·5–2 μm diam., swelling to 4–6 μm diam. at ± clavate apex, branched once or twice near apex, embedded in clear gel with small crystalline inclusions, forming

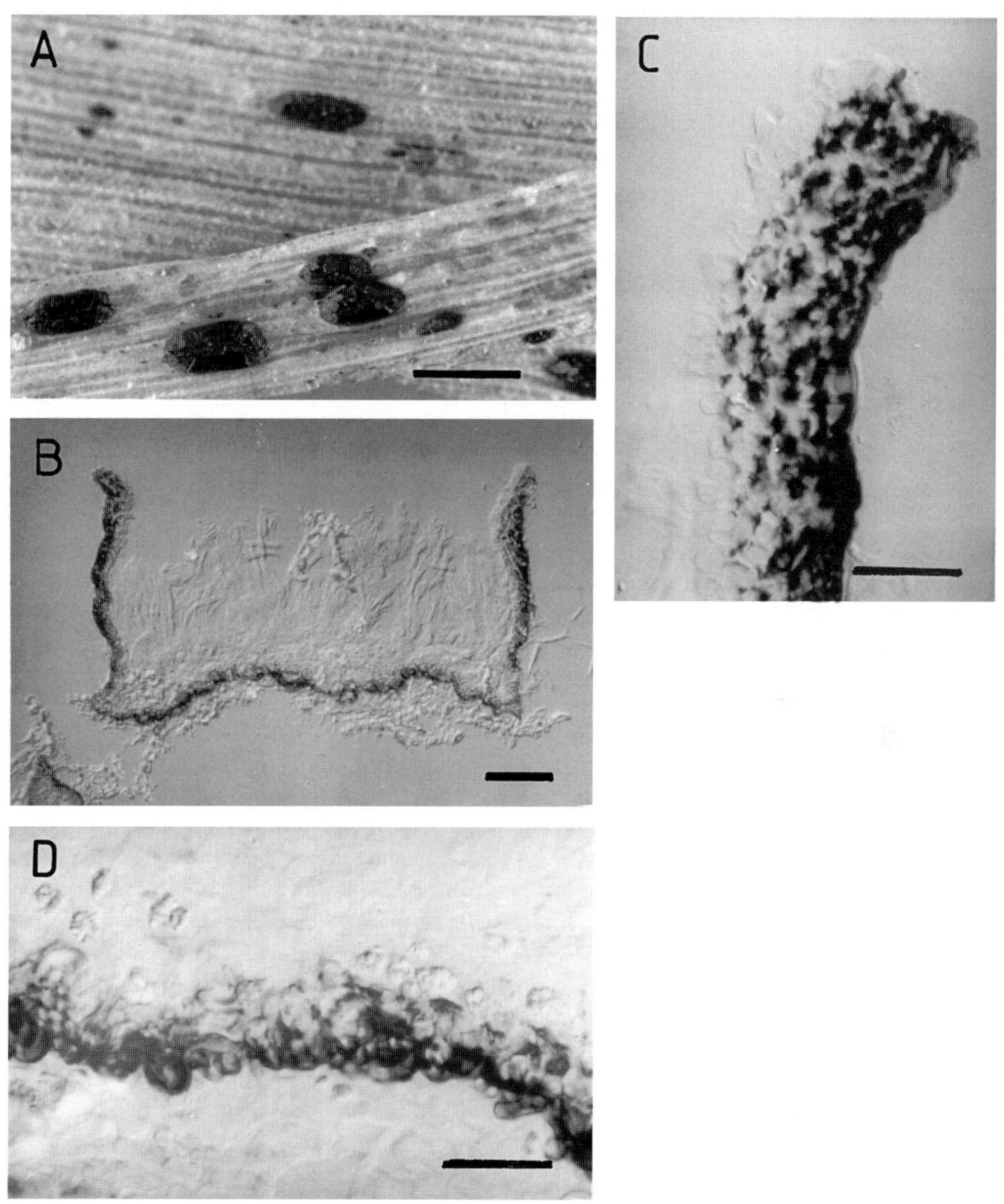

Fig. 80. *Lophodermium* cf. *luzulae* (**PDD** 60249). **A**, ascomata (bar = 1 mm); **B**, ascoma in vertical section (bar = 50 µm); **C**, upper wall of ascoma in vertical section; **D**, lower wall of ascoma in vertical section (bars = 20 µm).

dense epithecium 50 µm thick. *Asci* 175–210 × 6–6·5 µm, cylindrical, tapering slightly to rounded apex with undifferentiated wall, 8-spored, well-developed basal stalk with spores confined to upper 150–160 µm. *Ascospores* not seen released, approximately 120–130 µm long.

Typification: Guyana: POTARO-SIPARUNI REGION VII: Potaro Subregion: Mt Ayanganna, on ?*Bambusaceae* (as 'grass'), 11–12 Mar. 1987, *G.J. Samuels* 5113 (**NY**!, *holotypus*; **PDD**! 58055, **BPI**, **BRG**, *isotypi*).

Host: ?*Bambusaceae.*

Distribution: Guyana.

Illustration: Fig. 79.

Notes: *Terriera longissima* is distinguished by its very long asci and ascospores. Known from only two collections, the host appears to be the same species in both, although on the herbarium packets one is labelled as 'grass', the other as 'bamboo'.

The holotype collection also contains a second species of *Terriera* with shorter, broader ascomata, shorter and broader asci and possibly conidiomata, although it is in poor condition, with most ascomata over-mature.

Additional specimen examined: Guyana: Mt Wokomung, on bamboo, 11 Jul. 1989, *G.J. Samuels* 6573 (**NY**, as *Lophodermium* sp.).

Lophodermium *cf.* **luzulae** Hazsl., *Mat. természettund. Közl.* **21**: 186 (1886).

Infected areas on dead leaves, paler than surrounding host tissue, not associated with zone lines, containing groups of ascomata, conidiomata not seen. *Ascomata c.* 1 mm diam., broadly elliptical to ± round in outline, wall black, no paler zone along future line of opening, one or more irregular opening slits, with no lip cells, hymenium partly exposed in dry ascomata, orange-brown. *Ascomatal insertion* partly subcuticular, partly intra-epidermal, with variation both between and within ascomata. *Upper wall* 30–60 µm thick in vertical section, comprising two layers of approximately equal thickness, outer layer with several rows of angular to globose cells, 5–8 µm diam., with irregularly thickened walls, the cells adjacent to broken-down host epidermal cells or cuticle smaller and darker, inner layer comprising hyaline, thin-walled, cylindrical periphysoids embedded in gel. *Lower wall* 10–15 µm thick in vertical section, comprising several layers of dark-walled cells variable in shape and size, in squash mount darkened cells of lower wall 3–4 µm diam., narrow-cylindrical, irregularly bent or curved, arranged in two or three rows across base of ascoma. *Paraphyses* 1·5–2 µm diam., increasing to 4–6 µm diam. at clavate apex, forming regular palisade-like layer above asci. *Asci* 115–130 × 8·5–10 µm, clavate, tapering to small, truncate apex with undifferentiated wall, 8-spored, well-developed basal stalk with spores confined to upper 80–100 µm. *Ascospores* 50–70 × 1·5 (–2) µm, tapering slightly to base, 0–3-septate, ± straight when released, apical and basal gelatinous caps both small, globose.

Typification: 'Az itt jellemzett gomba nö a magas Tátrán Luzula maxima levelein, s elszakasztható a Pázsitok levelein termö töalaktól, tömlöi es spórái alapján mint *L. luzulae* nov. spec. Rajz 6. Tab. III' (HAZSLINSKY, 1886).

Hosts: *Calamagrostis* (*Poaceae*); *Carex* (*Cyperaceae*); *Luzula* (*Juncaceae*).

Distribution: Europe.

Illustrations: Figs 80, 82.

Notes: The type specimen of *L. luzulae* is not present in either of the two herbaria (**BP** and **RB**) listed in *Index Herbariorum* as containing Hazslinsky material. Many specimens from **BP**

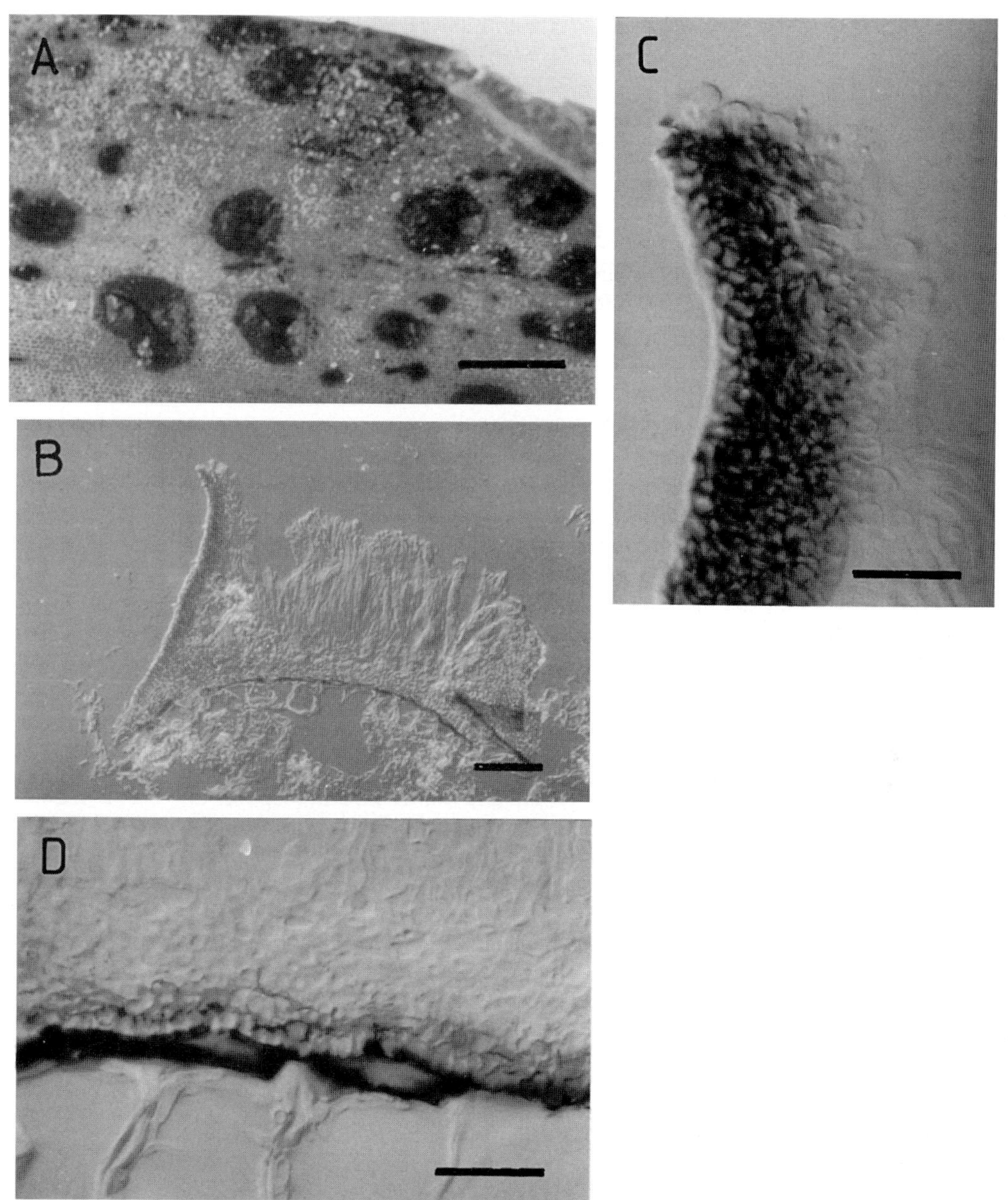

Fig. 81. *Lophodermium* cf. *luzulae* 'species 2' (*Müller & Cassagrande*, 22 Sep. 1965, **ZT**). **A**, ascomata (bar = 1 mm); **B**, ascoma in vertical section (bar = 50 µm); **C**, upper wall of ascoma in vertical section; **D**, lower wall of ascoma in vertical section (bars = 20 µm).

were destroyed during the Second World War (J. Gönczöl, pers. comm.). The description and illustrations provided by HAZSLINSKY (1886) are of little use in determining the distinguishing features of this species. Drawings labelled *L. luzulae* which match those published by HAZSLINSKY are present in **S**, but again there is no material of the fungus.

MÜLLER (1977) listed a number of collections from **ZT** as *L. luzulae*, and discussed the differences between *L. luzulae* and another species found on *Juncaceae*, *L. juncinum*. Collections deposited in **ZT** under both names were examined and, among those in which mature ascomata were found, four putative species have been recognized. All four are known from few collections, making the findings reported here tentative.

One of the species, found only on *Juncus*, is considered to represent *L. juncinum sensu* MÜLLER (1977) and is described fully under that name. It is characterized by the ascomata developing beneath the epidermis, but with the covering host tissue splitting, folding back and eroding as the ascomata mature, the dull grey-brown ascomata becoming more or less erumpent by the time the opening slit has developed.

The other three putative species have shiny black ascomata developing either between the cuticle and epidermis or within the epidermis, always remaining covered by some host tissue. All three have ascomata of similar size and shape, opening by a single slit or by several radiate slits, the opening not associated with any differentiated cells. When rehydrated the ascomata have the appearance of a species of *Coccomyces* such as *C. coronatus* or *C. tumidus*. These species can be distinguished by ascus and ascospore size and by paraphysis shape. One of them, with circinate paraphyses and a thick, pale brown upper ascomatal wall with a broad, loose periphysoid layer, is described fully under the name *L. tumidulum*.

The other two putative species have paraphyses swollen at the apex, forming a regular epithecium-like palisade above the developing asci, a thinner, darker upper wall and a compact periphysoid layer, usually embedded in gel. One of these is referred to *L.* cf. *luzulae* (see description above). The other is known from only two small collections (**Switzerland:** WALLIS: Aletschreservat, Gersternwald, on *Luzula silvatica*, 22 Sep. 1965, *E. Müller & F. Casagrande s. num.* (**ZT**); Aletschreservat, Längmoosweg, on *Juncus filiformes*, 8 Sep. 1970, *E. Müller s. num.* (**ZT**)). Although in ascomatal appearance and structure it is similar to *L.* cf. *luzulae*, it differs microscopically, with asci 120–150 × 10–12 µm and non-septate ascospores 30–50 × 2–2·5 µm; see **Figs 81**, **83**, as '*L.* cf. *luzulae* species 2'.

Coccomyces coronatus has the same kind of ascomatal structure as *L.* cf. *luzulae* (see **Fig. 84**) and the same distinctive orange-brown hymenium when dry. It differs in having wider asci (11–13 µm) and non-septate, wider ascospores (2–2·5 µm). See also Notes under *Rhytisma juncicola*, which is possibly conspecific with *C. coronatus*.

Several other species of *Rhytismataceae* found on *Juncaceae* and *Cyperaceae* have a similar macroscopic appearance to *L. tumidulum*, *L.* cf. *luzulae* and the other putative species discussed above. These include *Duplicaria antarctica*, *D. acuminata* Ellis & Everh. (see discussion under *D. antarctica*) and species of *Hypohelion* (JOHNSTON, 1990*c*). *Hypoderma alpinum* (SPOONER, 1981) and *Coccomyces insignis* P. Karst. (SHERWOOD, 1980; see discussion under *Lophium eriophori*), both described from *Carex*, are also similar in ascomatal appearance and structure. This similarity could be indicative either of common ancestry or of parallel adaptation to a similar habitat.

Additional specimens examined: ***Lophodermium* cf. *luzulae*: Switzerland:** GRAUBÜNDEN: Zuoz, rechte Talseite, on *Luzula sieberi*, 22 Jul. 1971, *E. Müller s. num.* (**ZT**); Unterengadin, Susch, on *Calamagrostis villosa*, 3 Aug. 1980, *L. Petrini s. num.* (**PDD** 60249, **ZT**); WALLIS: Aletschreservat, Moräneweg, on *Carex sempervirens*, 20 Sep. 1965, *E. Müller s. num.* (**K**).

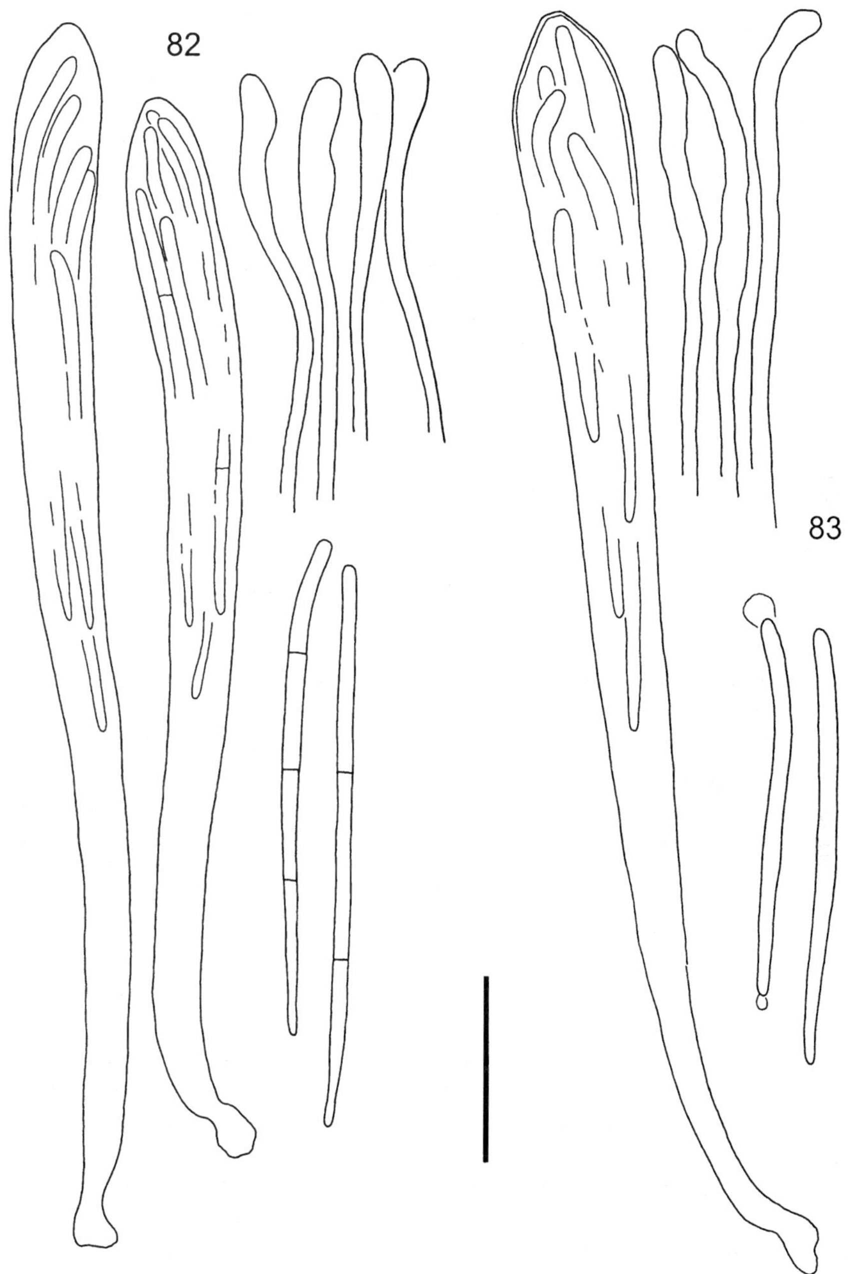

Figs 82–83. 82. *Lophodermium* cf. *luzulae* (*Müller*, 20 Sep. 1965, **K**). Asci, apex of paraphyses and ascospores. **83.** *Lophodermium* cf. *luzulae* 'species 2' (*Müller & Cassagrande*, 22 Sep. 1965, **ZT**). Asci, apex of paraphyses and ascospores (bar = 20 μm).

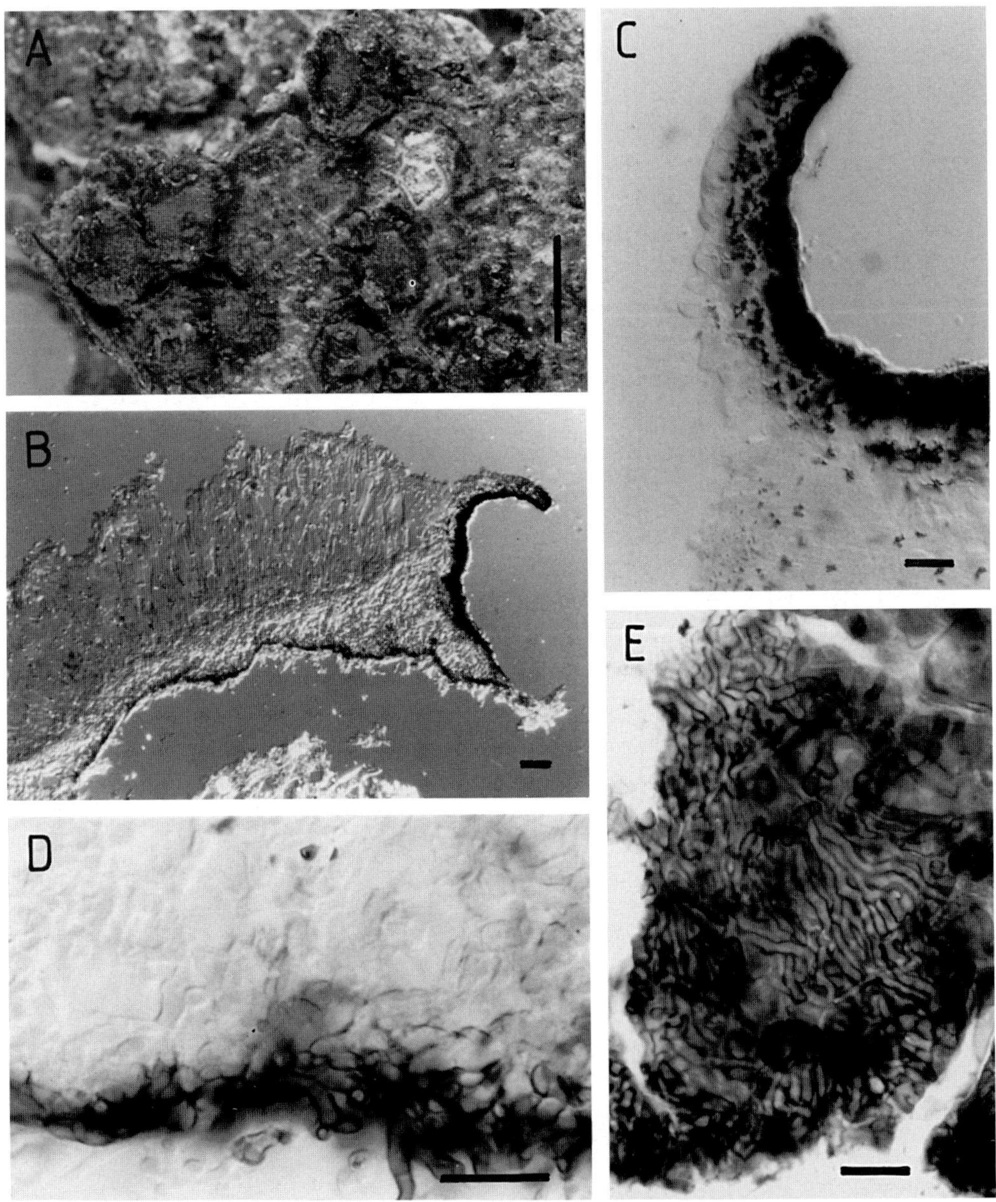

Fig. 84. *Coccomyces coronatus* (**PDD** 14713). **A**, ascomata (bar = 1 mm); **B**, one side of ascoma in vertical section (bar = 50 μm); **C**, upper wall of ascoma in vertical section; **D**, lower wall of ascoma in vertical section; **E**, lower wall of ascoma in squash mount (bars = 20 μm).

***Duplicaria acuminata*:** **USA:** WASHINGTON: Marysville, on *Juncus biformis*, Jul. 1928, *J.M. Grant s. num.* (**FH**); Skamania Co.: Mt St Helens, nr Timberline Camp, on *Juncus*, 22 Jul. 1951, *W.B. & V.G.C. Cooke s. num.* (**CUP** 52770); WYOMING: Albany Co.: Medicine Row Mtns, Lake Marie, on *Juncus drummondii*, 22 Jul. 1939, *W.G. Solheim & J.F. Brenckle s. num.* (Solheim & Cummins, *Mycoflora saximontanensis* no. 303; **CUP**).

***Coccomyces coronatus*:** **England:** YORKSHIRE: Hackfall, on *Quercus*, Nov. 1948, *S.J. Hughes & J. Webster s. num.* (**PDD** 14713). **USA:** NEW HAMPSHIRE: Carroll Co.: Redstone, on *Quercus*, 5 Sep. 1963, *H.E. & M.E. Bigelow s. num.* (**NY**).

Lophodermium minutum Hilitzer, *Věd. Spisy čsl. Akad. zeměd.* **3**: 152 (1929).

Typification: 'In vaginis Stipae pennatae (S. Joannis) a cl. Baudyš apud Doly prope Pragam lecta' (HILITZER, 1929).

Host: *Stipa pennata* (*Poaceae*).

Distribution: Europe.

Notes: HILITZER (1929) cited no individual specimen as the type. None of Hilitzer's material matching the location details given in the protologue could be found at **PRM** (Z. Pouzar, pers. comm.). Dr V. Antonín (pers. comm.) stated that he was unable to find the specimen 'not only in our herbarium **BRNM** but in other Baudyš's places of work in Brno', concluding that Hilitzer's material is probably lost.

Lophodermium miscanthi Tehon, *Illinois biol. Monogr.* **13**(4): 54 (1935), *nom. inval.*, *ICBN* Art. 36.1.

Infected areas on dead leaves, not associated with bleaching of the host tissue or zone lines, containing groups of ascomata, conidiomata not seen. *Ascomata* 0·7–2·5 × 0·3–0·4 mm, oblong-elliptical to sublinear in outline, ends ± rounded, unopened ascomata with wall pale grey and well-differentiated line along future line of opening, opened ascomata with wall dark grey to ± concolorous with surrounding host tissue and with narrow, black line around outer edge. *Ascomatal insertion* intrahypodermal, developing within hypodermal cells beneath host cuticle and thick-walled epidermal cells, some ascomata developing within the fibre bundles sometimes present beneath the epidermis, eventually splitting them. *Ascomatal structure* typical of *Terriera* (see p. 37), but in open ascomata characteristic dark periphysoid layer much reduced, with only scattered periphysoid-like cells evident along edge of opening slit, similar in appearance to very early stage of periphysoid development in most species of *Terriera* (see **Fig. 12**). *Paraphyses* 1·5 µm diam., often tangled and swollen to 2·5–4 µm diam. at apex, embedded in gel, extending 20–30 µm beyond asci as well-developed epithecium. *Asci* 130–175 × 7·5–8·5 (–9·0) µm, cylindrical, tapering slightly to ± broadly rounded apex with undifferentiated wall, 8-spored, basal stalk with spores confined to upper 100–140 µm. *Ascospores* not seen released, 2–2·5 µm wide, 100–120 µm long, gelatinous sheath not observed.

Typification: Philippines: LAGUNA: nr Los Baños, Mt Maquiling, on *Miscanthus japonicus*, Apr. 1914, *C.F. Baker* 155 (**DAR** 46902!; **NY**!, specimen on which the invalid *L. miscanthi* was based).

Host: *Miscanthus japonicus* (*Poaceae*).

Distribution: Philippines.

Illustrations: Figs 85, 86.

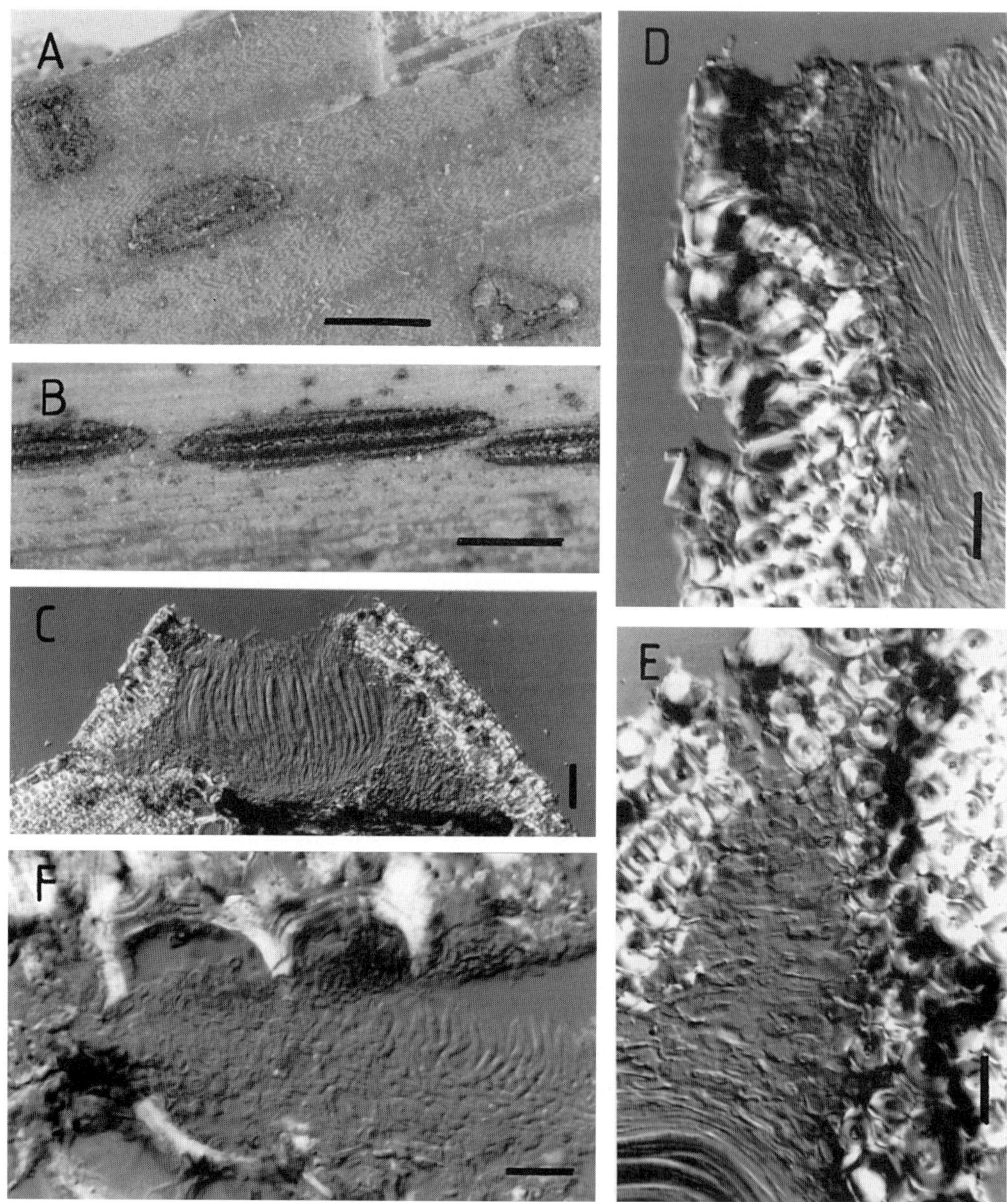

Fig. 85. *Lophodermium miscanthi* (**DAR** 46902). **A**, pale-walled ascomata (bar = 1 mm); **B**, dark-walled ascomata (bar = 1 mm); **C**, ascoma in vertical section (bar = 50 µm); **D**, detail of margin of ascomatal opening in vertical section (bar = 20 µm); **E**, detail of lower corner of ascoma in vertical section (bar = 20 µm); **F**, immature ascoma in vertical section, showing palisadic nature of primordium (bar = 50 µm).

Notes: *Lophodermium miscanthi* clearly belongs in *Terriera*, but the clypeus remains poorly developed, possibly because of the robust nature of the covering host tissue. The black, shelf-like projections which typically extend over the top of the hymenium in species of this genus are lacking; however, the macroscopic appearance of the unopened ascomata, the stroma comprising cylindrical cells arranged with a vertical orientation and the general aspect of the hymenial elements are all typical of *Terriera*. The macroscopic appearance of the ascomata in the *Baker* 155 specimen is variable in colour and outline: some are irregularly shaped and concolorous with the surrounding host tissue, others narrowly elliptical in outline with dark walls (see **Fig. 85**), but they do not differ microscopically. Ascomata in an earlier collection by Merrill are all dark-walled and narrowly elliptical.

Lophodermium miscanthi appears to match *Terriera arundinacea* in most respects (see Notes under the latter species). The differences in ascomatal structure may relate to differences in host leaf structure. Given the questions remaining concerning acceptable species limits for tropical species of *Terriera*, this epithet is neither validated here nor placed in synonymy at present.

The *Baker* 155 duplicate in **NY** is a mixed collection, with some leaves having another species of *Lophodermium* similar in ascomatal structure to *Lophodermium* Group E (see p. 43). The ascomata are paler and smaller (0·4–0·6 × 0·2 mm) than those of *L. miscanthi*, with paraphyses slightly swollen and unbranched, asci 80–90 × 8–9 μm, clavate with a broad-truncate apex and ascospores 45–50 × 2 μm, coiling on release. A new species is not proposed; additional, more recent collections are needed to describe this fungus properly.

Additional specimen examined: Philippines: LAGUNA PROV.: Luzon, Mt Maquiling, on *Miscanthus sinensis*, Mar. 1913, *E.D. Merrill* 8665 (**NY**, as *L. arundinaceum*).

Lophodermium montanum Ferraris, *Malpighia* **16**: 456 (1902).

Typification: 'Su foglie secche di *Tolfeldia calyculata*. Cormayeur. Settembre 1900' (FERRARIS, 1902).

Host: *Tofieldia calyculata* (*Liliaceae*).

Distribution: Italy.

Notes: No type material located.

Terriera nematoidea (P.R. Johnst.) P.R. Johnst., **comb. nov.**
Lophodermium nematoideum, P.R. Johnst., *N.Z. Jl Bot.* **29**: 401 (1991).

Infected areas on dead leaves, slightly paler than surrounding host tissue, associated with partial zone lines, containing ascomata and conidiomata. *Ascomata* 0·7–1·7 × 0·2–0·3 mm, elliptical to sublinear in outline, unopened ascomata wall black, with pale areas near both ends, and with faint pale zone along future line of opening, single longitudinal opening slit with black shelf-like zone along either side. *Ascomatal insertion* subepidermal, sometimes also beneath scattered host fibre bundles. *Ascomatal structure* typical of *Terriera*, but with darkened lower wall very well developed. *Paraphyses* 1–1·5 μm diam., undifferentiated at apex, embedded in gel, extending *c.* 20 μm beyond asci. *Asci* 70–80 × 5–6·5 μm, subclavate, tapering slightly to truncate apex with slightly thickened wall and small central pore, 8-spored, well-developed basal stalk with spores confined to upper 40–50 μm. *Ascospores* 30–35 × 1 μm, tapering slightly to ends, gently curved

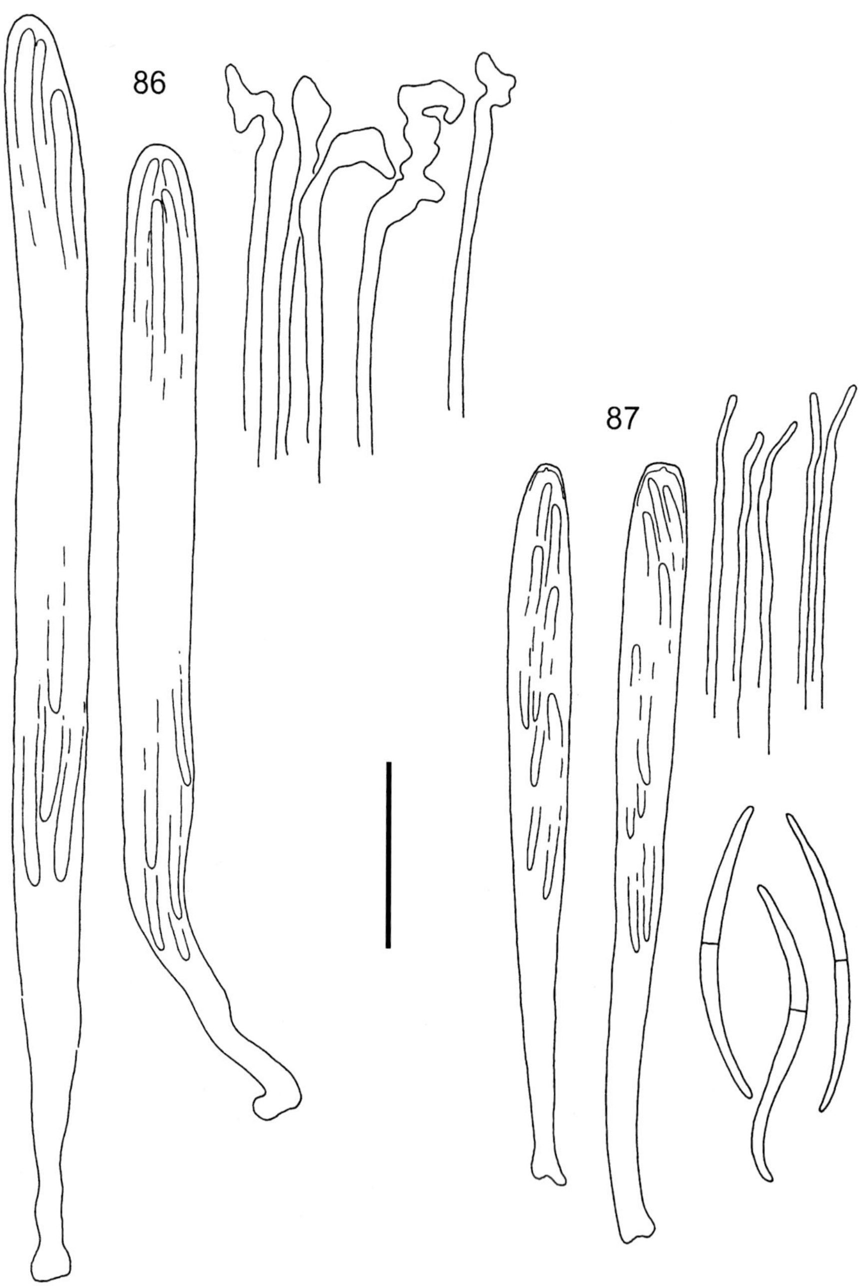

Figs 86–87. 86. *Lophodermium miscanthi* (*Baker* 155, **FH**). Asci and apex of paraphyses. **87.** *Terriera nematoidea* (**PDD** 54132). Asci, apex of paraphyses and ascospores (bar = 20 µm).

or sigmoid on release, 1-septate, gelatinous sheath lacking. *Conidiomata* 0·3 mm diam., round in outline, wall pale brown with darker line around outside edge, in vertical section upper and lower walls comprising one to three rows of brown to pale brown cells and lined with one or two rows of thin-walled, hyaline cells on which conidiogenous cells are held. *Conidiogenous cells* 8–12 × 1·5–2·5 µm, solitary, tapering to apex, proliferation sympodial, often with 2 developing conidia held at apex. *Conidia* 3–4 × 1 µm, oblong-elliptical with ends rounded, non-septate, hyaline.

Typification: New Zealand: NORTHLAND: Russell Forest, Punaruku Rd, Hori Wehi Wehi Track, on *Gahnia*, 11 Aug. 1988, *P.R. Johnston* R800 (**PDD** 54132!).

Host: *Gahnia* (*Cyperaceae*).

Distribution: New Zealand.

Illustrations: Fig. 87; also JOHNSTON (1991, figs 5, 9).

Notes: *Terriera nematoidea* is distinguished by its short, broad asci with a truncate apex and short ascospores. Unusually for a species of *Terriera*, the epithecium is poorly developed.

Lophodermium nitidum P.R. Johnst., **sp. nov.**

Etymology: Referring to the shiny black macroscopic appearance of the ascomata.

A *L. arundinaceo* ascomatibus atronitidis, margo eorum perdistincto, cellulis atrobrunneis parietis inferioris ascomatum angularibus vel cylindricalibus et in stratis pluribus dispositis differt.

Infected areas on dead leaves, not associated with bleaching of host tissue or zone lines, containing scattered ascomata, conidiomata not seen. *Ascomata* 0·5–0·7 mm long, broadly elliptical in outline, ends rounded, sometimes with small apiculus, wall shiny black, margin sharply defined, single, longitudinal opening slit lined with thick, pale lips, in some collections the lips becoming dark as ascomata mature. *Ascomatal insertion* subepidermal. *Covering layer* up to 50 µm thick in vertical section, comprising clypeus of dark-walled hyphae within epidermal cells, and upper wall of ascomatal stroma. *Upper wall* up to 30 µm thick in vertical section, comprising three or four rows of ± globose cells with walls irregularly encrusted with dark brown material, but with group of very dark cells adjacent to well-developed lips. *Lower wall* 15–20 µm thick in vertical section, comprising single row of pale-walled, angular cells below which are several rows of angular to globose cells, 4–6 µm diam., with thick, dark walls, in squash mount darkened cells of lower wall short-cylindrical to cylindrical, 4–6 µm diam., in two to four layers. *Paraphyses* 1·5–2 µm diam., circinate, often coiling at apex, sometimes also irregularly swollen. *Asci* 110–125 (–130) × (7·5–) 8·5–9·5 (–10) µm, subclavate, tapering gradually to small, rounded apex with undifferentiated wall, 8-spored, basal stalk developing at maturity with spores confined to upper 85–95 µm. *Ascospores* (50–) 55–65 (–75) × 1·5 (–2) µm, straight when released, tapering gradually to base, apical gelatinous cap globose, 3·5–5 µm diam., basal gelatinous cap cylindrical, 4–6 × 1·5–2 µm, gelatinous sheath *c.* 5 µm diam.

Typification: Switzerland: WALLIS: Lötschental, Ried, on *Calamagrostis villosa*, 26 Jul. 1979, *L. Petrini* 8 (**ZT**!, *holotypus*; **PDD**! 65719, *isotypus*)

Hosts: *Calamagrostis*, *Deschampsia*, *Festuca*, *Molinia*, *Nardus*, *Sesleria* (*Poaceae*); *Carex* (*Cyperaceae*).

Distribution: Europe, North America.

Illustrations: Figs 9C, D, 10B–D, 88, 90.

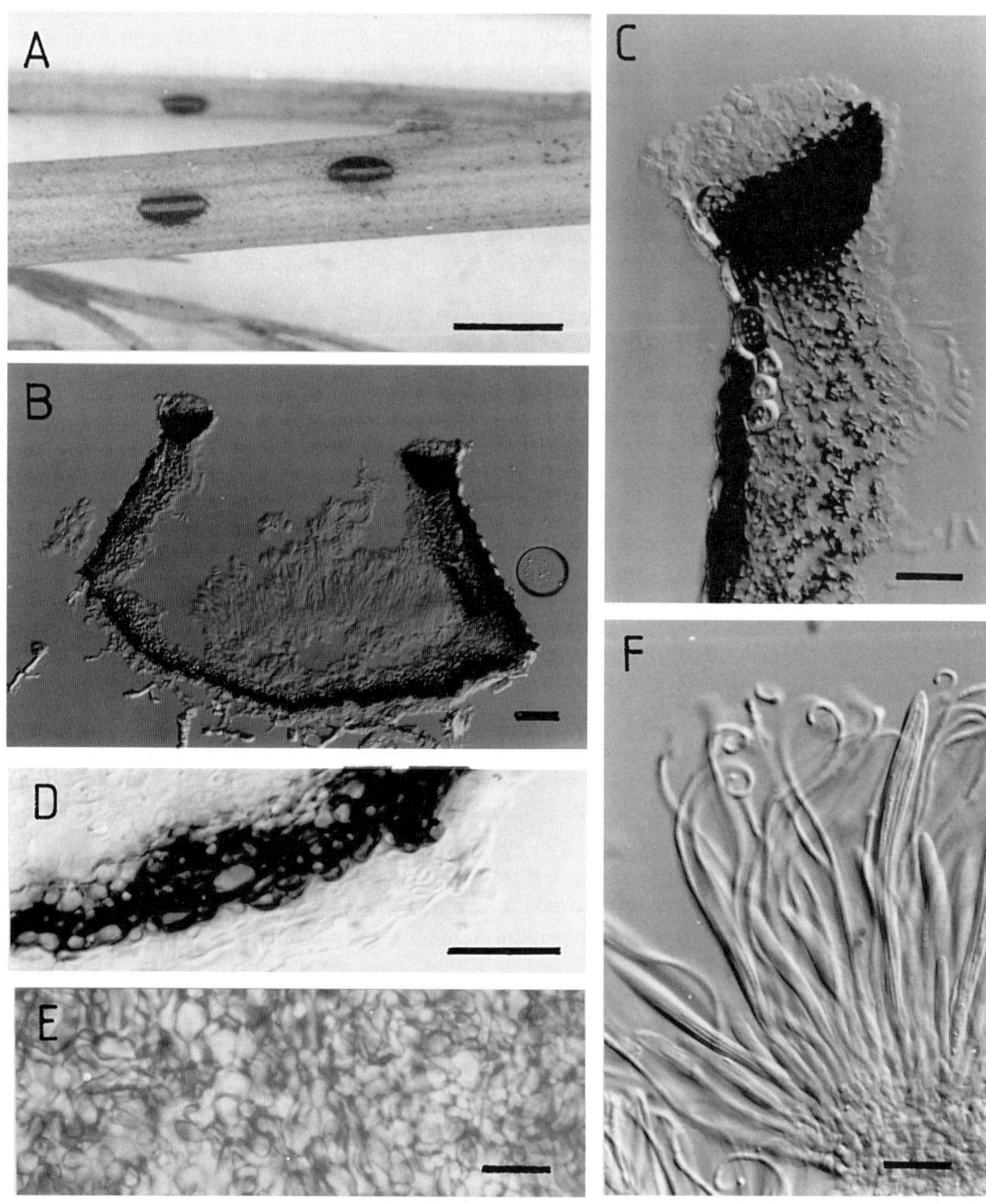

Fig. 88. *Lophodermium nitidum* (**PDD** 65286). **A**, ascomata (bar = 1 mm); **B**, ascoma in vertical section (bar = 50 μm); **C**, upper wall of ascoma in vertical section; **D**, lower wall of ascoma in vertical section; **E**, lower wall of ascoma in squash mount; **F**, asci and paraphyses (bars = 20 μm).

Notes: *Lophodermium nitidum* has an ascomatal structure typical of the *actinothyrium*-group of *Lophodermium* Group A (see p. 33). It can often be distinguished macroscopically from other species in this group by the shiny black ascomata with a sharply defined margin and thick lip cells.

This species is similar to *L. culmigenum* in ascus and ascospore dimensions, although both are slightly longer and wider in the latter. *Lophodermium culmigenum* is most readily distinguished by its generally larger ascomata, usually with poorly-defined margins, lower wall comprising narrow hyphae forming a *textura intricata* and paraphyses usually bent rather than coiling at the apex. There also appears to be some host specialization between the two species; see Notes under *L. culmigenum*.

Lophodermium actinothyrium is macroscopically similar to *L. nitidum* and is found on a similar range of hosts, but is distinguished by the lower wall of the ascomata comprising a single layer of angular/globose cells, by shorter and slightly narrower asci and shorter ascospores. Most collections of *L. actinothyrium* have ascomata with a small apiculus at the rounded ends, a feature generally lacking in *L. nitidum*.

Rehm, *Ascomyceten* no. 319b, discussed by TERRIER (1942) as *L.* '*non-alpinum*', appears typical of *L. nitidum*.

Several collections from North America which have been examined are similar to *L. nitidum*. The DiCosmo collection from Ontario matches the European collections closely, although there are several other groups of collections which appear to differ slightly in ascus and ascospore size and paraphysis shape (listed below as *L.* cf. *nitidum*, Groups 1–3). If the features used to distinguish the European species in *Lophodermium* Group A prove to be tenable, then these other groups may in future be considered distinct species. However, given the doubt at present surrounding appropriate species limits for extra-European collections of grass-inhabiting species of *Lophodermium* (see p. 36), no new species are proposed here.

Lophodermium cf. *nitidum* Group 1 has ascomata similar in appearance to *L. nitidum* and a similar lower wall structure. However, the asci (mostly 120–140 × 9·5–11 µm) and ascospores (65–80 × 2–2·5 µm) are larger.

Lophodermium cf. *nitidum* Group 2 is similar to *L. nitidum* macroscopically, in ascomatal structure and in paraphysis shape, but has asci (90–105 × 7·5–8·5 µm) and ascospores (50–65 × 1·5 µm) more similar in size to *L. actinothyrium*. The latter differs in the structure of the ascomal lower wall and the darkened cells being globose, forming a single layer across the base.

Lophodermium cf. *nitidum* Group 3 is a mixed collection with *L.* cf. *nitidum* Group 2. The lower wall structure is the same, but the margin of the ascomata is diffuse rather than sharp and the asci (90–105 × 9–10 µm, ± cylindrical, with rostrate apex), ascospores (65–75 × 2 µm) and paraphyses (undifferentiated to slightly swollen at the apex) all differ.

Additional specimens examined: Austria: SALZBURG: Pinzgau, National Park Hohe Tauern, between Finkau and Trisslalm, on *Calamagrostis villosa*, 23 Jul. 1992, *C. Scheuer s. num.* (**GZU**, as *L. apiculatum*); STEIERMARK: Wölzer Tauern, Planneralpe, on *Calamagrostis villosa*, 24 Jul. 1981, *D. Kores s. num.* (**GZU**, as *L. apiculatum*). **Canada:** ONTARIO: Bruce Peninsula, Hope Bay, on grass, 12 Aug. 1979, *F. DiCosmo s. num.* (**FH**, as *Lophodermium* sp.). **Czech Republic:** MORAVIA: Ùtěchov, Bílovice, on *Calamagrostis epigeios*, 24 Jun. 1973, *M. Svrček s. num.* (**PRM** 734438, as *L. apiculatum*); Ostrava, 'Halda' Lučina, on *Calamagrostis epigeios*, 24 Jul. 1970, *M. Svrček & J. Veselský s. num.* (**PRM** 714093, as *L. apiculatum*); Roždalovice, between Mlýnec and Křešice, on *Molinia coerulea*, 23 Jul. 1935, *A. Hilitzer s. num.* (**PRM** 820253, as *L. actinothyrium*). **France:** Chamonix, on grass leaves, 5 Jul. 1953, *L.E. Wehmeyer s. num.* (**DAOM** 123698). **Germany:** Schlesien, Schwarzbach, Isergebirge, on *Aira flexuosa*, 18 Jul. 1922, *H. Sydow s. num.* (Sydow, *Mycotheca germanica* no. 2147; **PDD** 55995, as *L. culmigenum*). **Latvia:** Ādaži, on *Molinia coerulea*, 8 Sep. 1929, *F. Smarods s. num.* (**K**, as *L. arundinaceum* var. *actinothyrium*). **Sweden:** UPPLAND: Dalby Parish: roadside *c.* 100 m S of Jerusalem, on *Deschampsia caespitosa*, 25 Jul. 1985, *K. & L. Holm* 3651 (**PDD** 59627, **ZT**). **Switzerland:** Gotthard Pass, on *Nardus stricta*, Nov. 1892, *H. Rehm s. num.* (Rehm, *Ascomyceten* no. 319b; **DAOM** 3852, as *L. arundinaceum* var. *alpinum*); GRAUBÜNDEN: Klosters, Schlappintobel, on *Calamagrostis villosa*, 6 Jul. 1980, *L. Petrini*

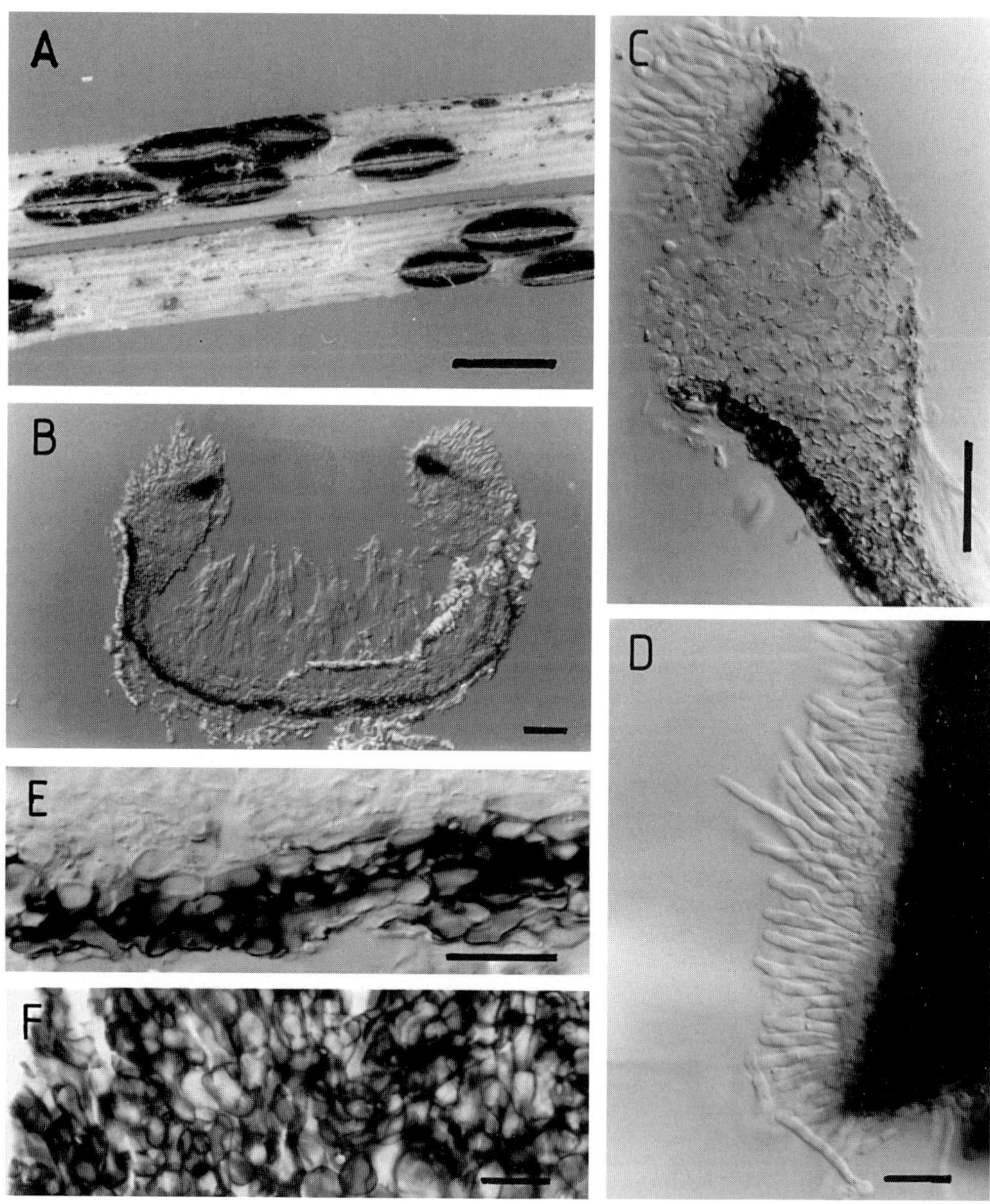

Fig. 89. *Lophodermium nonramosum* (**A**, *Scheuer* 812, **GZU**; **B–E**, **PDD** 64955; **F**, **PDD** 64956). **A**, ascomata (bar = 1 mm); **B**, ascoma in vertical section (bar = 50 μm); **C**, detail of upper wall of ascoma in vertical section; **D**, unbranched lip cells; **E**, detail of lower wall of ascoma in vertical section; **F**, lower wall in squash mount (bars = 20 μm).

50 (**PDD** 65717, **ZT**); Oberengadin, Zuoz, Crasta, on *Festuca*, 15 Jul. 1980, *L. Petrini* 110 (**PDD** 65718, **ZT**); Davos Dorf, on *Calamagrostis villosa*, 11 Jun. 1980, *L. Petrini* 40 & 42 (**PDD** 65286, **ZT**); Bergün, Val Tours, on *Calamagrostis villosa*, 29 Aug. 1980, *C. Scheuer s. num.* (**GZU**, as *L. apiculatum*); SCHAFFHAUSEN: Thayngen, Moos, on *Carex flacca*, 5 May 1980, *L. Petrini* 18 (**PDD** 65721, **ZT**); TICINO: San Gottardo, on grass, 21 Jul. 1980, *L. Petrini* 181 & 182 (**PDD** 65720, **ZT**); VALAIS: Zermatt, on grass, 7 Jul. 1953, *L.E. Wehmeyer s. num.* (**DAOM** 123669, as *L. gramineum*).

***Lophodermium* cf. *nitidum* Group 1: USA:** ALASKA: Glacier Bay National Monument, Bartlett Cove, road to lodge, on *Elymus arenarius* subsp. *mollis*, 20 Jun. 1980, *W.B. & V.G. Cooke* 58131 (**NY**, **DAOM** 179104, both as *L. arundinaceum*). **Canada:** BRITISH COLUMBIA: Vancouver I, Cordova Bay, on *Elymus mollis* subsp. *mollis*, 30 Jun. 1957, *W.G. Ziller s. num.* (**DAOM** 57491, as *L. arundinaceum*).

***Lophodermium* cf. *nitidum* Group 2: USA:** UTAH: Weber Co.: Wasatch Mtns, Snow Basin, on *Agropyron smithii*, 25 Jul. 1985, *C.T. Rogerson s. num.* (**NY**, as *L. arundinaceum, p.p.*).

***Lophodermium* cf. *nitidum* Group 3: USA:** UTAH: Weber Co.: Wasatch Mtns, Snow Basin, on *Agropyron smithii*, 25 Jul. 1985, *C.T. Rogerson s. num.* (**NY**, as *L. arundinaceum, p.p.*).

Lophodermium nivale (Maire) Arx & E. Müll., *Stud. Mycol.* 9: 133 (1975).

Hadotia nivalis Maire, *Bull. Séanc. Soc. Sci. Nancy* Sér. 3, **7**: 174 (1906).

Typification: 'Sur les chaumes desséchés d'*Alopecurus textilis* Boiss. dans la zone alpine du mont Argée (Erdjias-Dagh) au-dessus de Kaisserie, Cappadoce, 26/9, No. 1147' (MAIRE, 1906).

Host: *Alopecurus textilis* (*Poaceae*).

Distribution: North Africa.

Notes: The type has not been located. The exsiccata Maire, *Mycotheca boreali-africana* nos 1–400 was examined at **FH**, but this species is not included in the set.

ARX & MÜLLER (1975) stated that this species is 'close to *Lophodermium alpinum* Rehm and *L.* (non) *alpinum* Rehm *sensu* TERRIER (1942)'; it is the type species of the genus *Hadotia* Maire.

Lophodermium nonramosum P.R. Johnst., sp. nov.

Etymology: Referring to the unbranched lip cells, unusual in *Lophodermium s. str.*

A *L. arundinaceo* cellulis atrobrunneis parietis inferioris ascomatum late cylindricalibus vel globosis, cellulis labiorum non ramosis, ascis subclavatis vel subfusiformibus, 130–175 × 12·5–14 (–15) µm, ascosporis 80–105 × 2·5 µm differt.

Infected areas on dead leaves, not associated with bleaching of host tissue or zone lines, containing scattered ascomata, conidiomata not seen. *Ascomata* 0·5–1·5 × 0·3–0·5 mm, oblong-elliptical in outline, ends ± rounded but edges not sharply defined, wall pale to dark grey, with distinct pale zone along future line of opening, single longitudinal opening slit lined with broad layer of lip cells. *Ascomatal insertion* subepidermal. *Covering layer* up to 90–120 µm thick in vertical section, narrower towards base of wall, comprising clypeus of dark-walled hyphae within epidermal cells, and upper wall of ascomatal stroma. *Upper wall* up to 50–80 µm thick, pale, comprising mostly thin-walled, angular to globose cells, 8–10 µm diam., with walls slightly and irregularly encrusted, but with group of cells with thick and dark walls adjacent to well-developed lips, sometimes with scattered fibre cells embedded within wall, lip cells hyaline, thin-walled, cylindrical, rarely branched, in unopened ascomata upper wall lined with hyaline, short-cylindrical periphysoids, lost after ascomata open. *Lower wall* 15–30 µm thick in vertical section, comprising three or four rows of angular to globose cells, 5–8 µm diam., the lowermost one to three rows with slightly thickened, dark walls, in squash mount dark cells of lower wall globose to short-cylindrical, 8–11 µm diam., arranged in one to three rows across base of ascoma. *Paraphyses* 1·5–

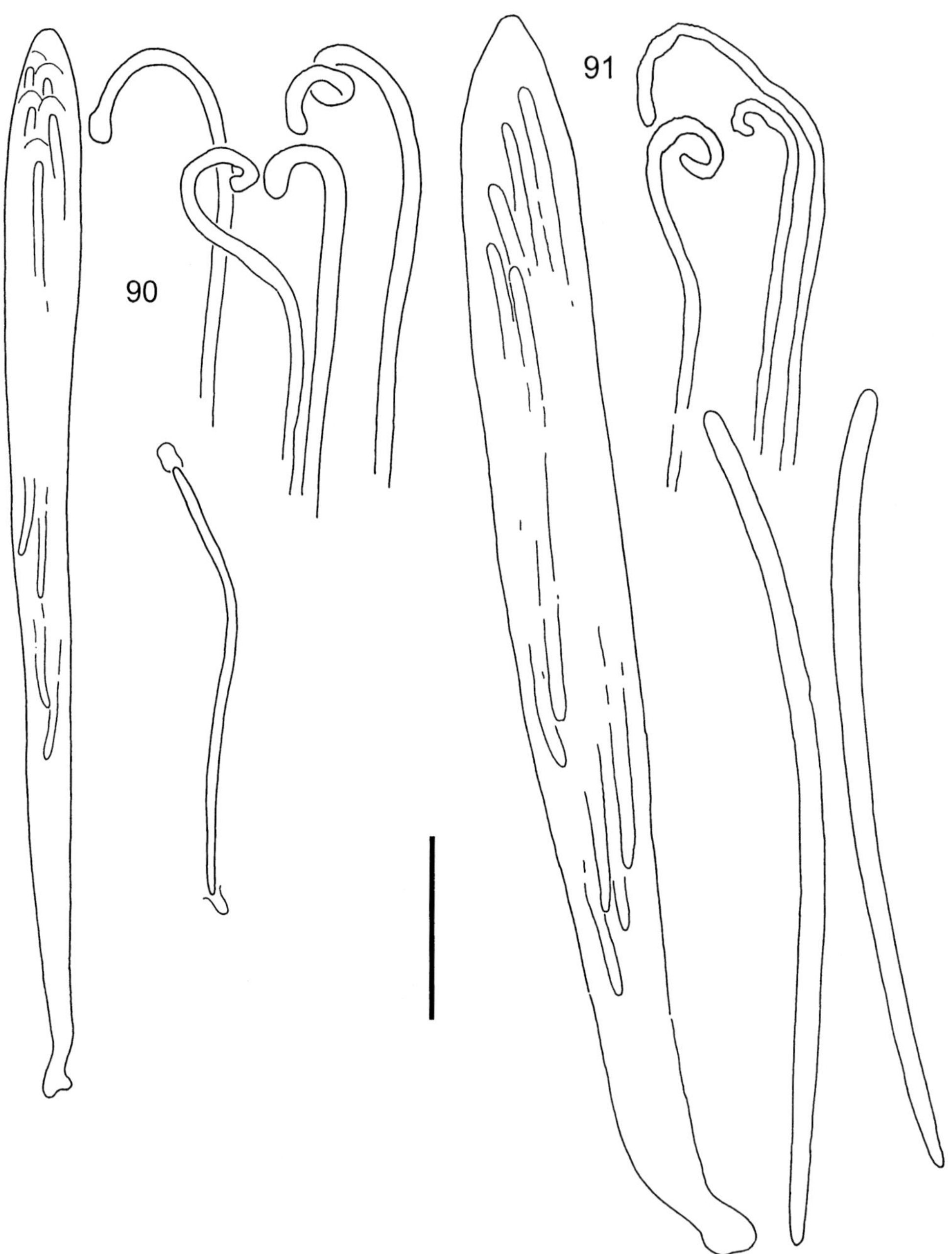

Figs 90–91. 90. *Lophodermium nitidum*. Ascus and apex of paraphyses (**PDD** 65286) and ascospore in water, loose, outer sheath lost (**PDD** 59627). **91.** *Lophodermium nonramosum* (**PDD** 64956). Ascus, apex of paraphyses and released ascospores, gelatinous caps and sheath lost (bar = 20 μm).

2 µm diam., circinate and coiling at apex. *Asci* 130–175 × 12·5–14 (–15) µm, subclavate to subfusoid, tapering gradually to truncate apex with undifferentiated wall, 8-spored, basal stalk at maturity with spores confined to upper 110–150 µm, spores widely spaced within ascus. *Ascospores* 80–105 × 2·5 µm, straight when released, tapering gradually to rounded base, non-septate, apical gelatinous cap globose, 5–6 µm diam., basal gelatinous cap broad-cylindrical, *c.* 4 µm diam., gelatinous sheath 7–8 µm thick.

Typification: Switzerland: TICINO: Airolo, Sassodella Boggia, on ?*Festuca*, 22 Jul. 1980, *L. Petrini* 170 (**ZT**!, *holotypus*; **PDD**! 64955, *isotypus*).

Hosts: *Festuca*, *Nardus* (*Poaceae*); *Carex* (*Cyperaceae*).

Distribution: Europe (Swiss Alps).

Illustrations: Figs 89, 91.

Notes: *Lophodermium nonramosum* is characteristic of the *actinothyrium*-group within *Lophodermium* Group A. It is distinguished in this group by its large asci and ascospores, very wide upper wall of the ascomata and unusual lip cells, which are only rarely branched; all other species in *Lophodermium* Group A have lip cells copiously branched.

Macroscopically, the collections on *Festuca* are reminiscent of *L. alpinum* and *L. gramineum*, their ascomata having grey to pale grey walls and a well-developed paler zone along the future line of opening, remaining evident along both sides of the opening slit in opened ascomata.

Additional specimens examined: Switzerland: GRAUBÜNDEN: Davos-Wolfgang, on *Festuca rubra*, 12 Jun. 1980, *L. Petrini* 39 (**PDD** 64954, **ZT**); Unterengaden, Prui-Laret, on *Nardus stricta*, 27 Aug. 1984, *O. Petrini* 135.2 (**PDD** 64956, **ZT**); Rhätische Alpen, Sertig-Tal, Davos, Wasserfall-Cheren, on *Carex sempervirens*, 24 Aug. 1980, *C. Scheuer* 812 (**GZU**).

Lophodermium occultum Petr., *Sydowia* **1**: 271 (1947).

Typification: Australia: NEW SOUTH WALES: Sydney area, Port Jackson, on *Spinifex hirsutus* (as *S. australis*), *s. dat.*, *Sieber s. num.*(Sieber, *Agrostotheca Novae Hollandiae* no. 62; **W**!, **MEL** 230308!).

Host: *Spinifex* (*Poaceae*).

Distribution: Australia.

Notes: No *Lophodermium* ascomata were seen on the type material examined.

The host cited on the label of the type specimen, and by PETRAK (1947), is *Spinifex australis* but, as noted by SAMUELS (1985), *S. australis* is not listed in *Index Kewensis*. There is a handwritten label with the isotype held at **MEL** annotated 'N 62 *Spinifex hirsutus*'; this is the only species of *Spinifex* known from Port Jackson (Tom May, **MEL**, pers. comm.).

Lophodermium oxyascum Speg., *Boln Acad. nac. Cienc. Córdoba* **11**: 251 (1888).
Hypoderma oxyascum (Speg.) Kuntze, *Revis. gen. pl.* **3**(3): 487 (1898).
Dermascia oxyasca (Speg.) Tehon, *Illinois biol. Monogr.* **13**(4): 70 (1935), *nom. inval.*, *ICBN* Art. 43.

Typification: Argentina: TIERRA DEL FUEGO: Staten I, on grass, 1882, *C. Spegazzini* (**LPS** 1004!).

Hosts: *Luzula* (*Juncaceae*); *Uncinia* (*Cyperaceae*); *Festuca*, *Poa* (*Poaceae*) (SPEGAZZINI, 1887).

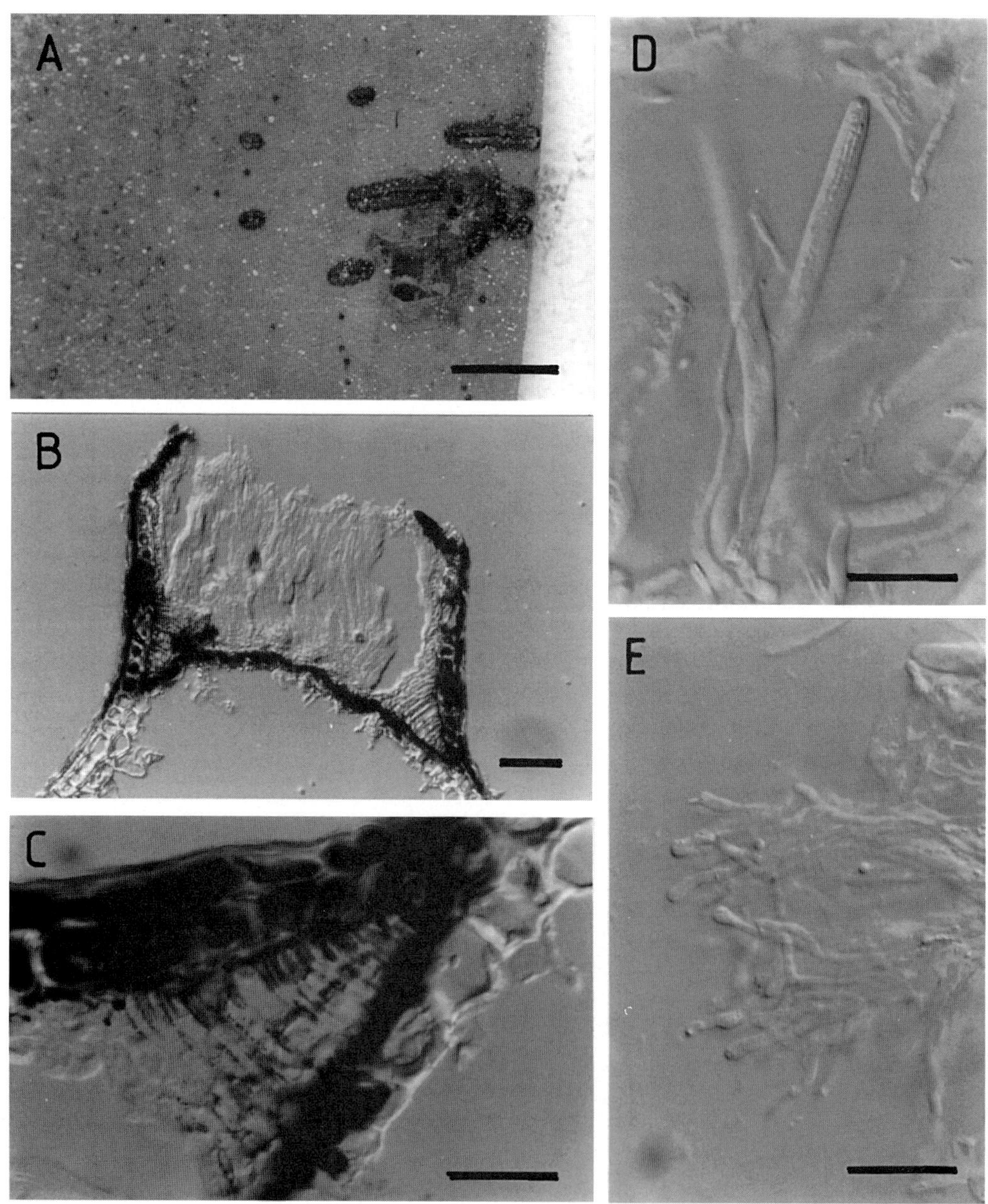

Fig. 92. *Terriera javanica* var. *pandani* (holotype, **PAD**). **A**, ascomata (bar = 1 mm); **B**, ascoma in vertical section (bar = 50 µm); **C**, detail of corner of ascoma in vertical section; **D**, apex of asci; **E**, apex of paraphyses (bars = 20 µm).

Distribution: Tierra del Fuego.

Notes: The type specimen comprises a tiny collection in poor condition from which neither macroscopic nor microscopic features could be determined. Although SPEGAZZINI (1887) stated that this species was found on several hosts on Staten I and Fuegia, only the type specimen was available at **LPS**.

Terriera pandani (Tehon) P.R. Johnst., **comb. nov.**
Clithris pandani Tehon, *Bot. Gaz.* **65**: 555 (1918).

Infected areas on dead leaves, not associated with bleaching of host tissue or zone lines, containing groups of ascomata, conidiomata not seen. *Ascomata* 0·4–1·2 × 0·2–0·3 mm, oblong to oblong-elliptical, ends rounded, wall black, well-developed paler zone along future line of opening, single, longitudinal opening slit with black shelf-like zone along either side. *Ascomatal insertion* intrahypodermal, beneath epidermis and uppermost layer of hypodermal cells. *Ascomatal structure* typical of *Terriera* (see p. 37), with covering layer comprising epidermal and hypodermal cells filled with dark brown, thick-walled hyphae of clypeus. *Paraphyses* swelling slightly to subclavate apex, branched two or three times near apex, embedded in gel, forming compact epithecium above asci. *Asci* 100–120 × 5–6 µm, cylindrical, tapering suddenly to small, rounded apex with undifferentiated or slightly thickened wall and broad central pore, 8-spored, short basal stalk with spores confined to upper 80–90 µm. *Ascospores* not seen released, 50–70 × 1–1·5 µm, no gelatinous sheath.

Typification: Puerto Rico: San Juan, on *Pandanus*, 16 Nov. 1913, *F.L. Stevens* 4090A (**ILL** 7426!).

Host: *Pandanus* (*Pandanaceae*).

Distribution: Puerto Rico.

Notes: *Terriera pandani* is similar in ascus size and paraphysis shape to *T. clithris*, which, however, has asci with a more or less truncate apex. It is clearly distinguished from *T. javanica* var. *pandani*, the latter having shorter asci and ascospores, and asci broadly rounded at the apex.

See also Notes under *T. javanica*.

Terriera javanica var. **pandani** (Penz. & Sacc.) P.R. Johnst., **comb. nov.**
Lophodermium javanicum var. *pandani* Penz. & Sacc., *Malpighia* **11**: 529 (1897).

Infected areas on dead leaves, paler than surrounding host tissue, not associated with zone lines, containing scattered ascomata, conidiomata not seen. *Ascomata* 0·7–1·3 × 0·3 mm, oblong to sublinear in outline, ends rounded, wall black, single longitudinal opening slit with black flattened zone along either side. *Ascomatal insertion* intrahypodermal, developing below epidermis and first layer of hypodermal cells. *Ascomatal structure* typical of *Terriera* (see p. 37), with covering layer comprising epidermal and hypodermal cells filled with dark brown, thick-walled hyphae of clypeus. *Paraphyses* 1·5 µm diam., irregularly swollen up to 4–5 µm at apex, often branching near apex, forming compact epithecium above asci. *Asci* 90–95 × 6–6·5 µm, cylindrical, broadly rounded apex with undifferentiated wall, 8-spored, short basal stalk at maturity with spores extending ± to base. *Ascospores* not seen released, 50–60 × 1·5 µm.

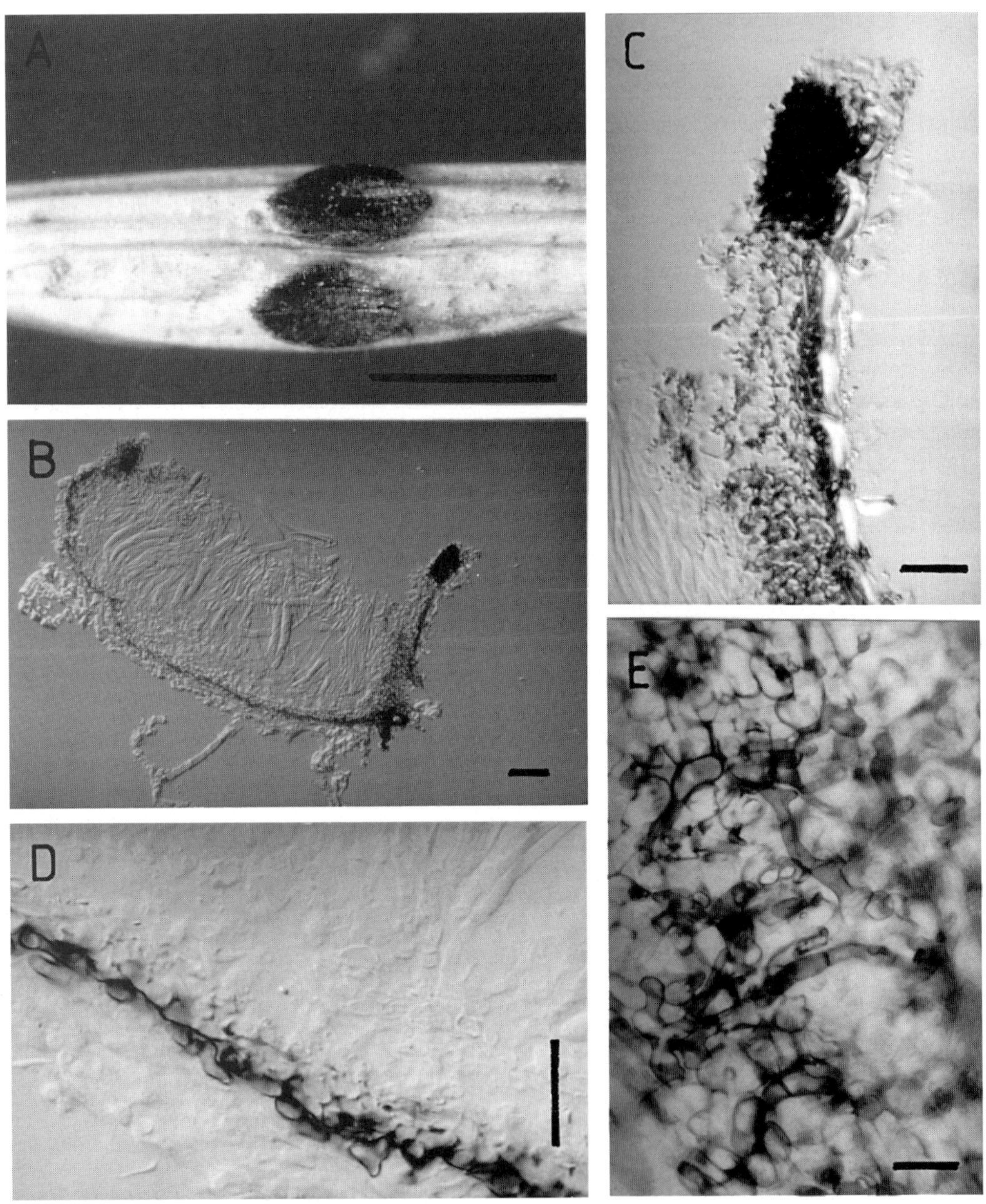

Fig. 93. *Lophodermium petriniae* (**A–D**, **PDD** 65383; **E**, **PDD** 65391). **A**, ascomata (bar = 1 mm); **B**, ascoma in vertical section (bar = 50 μm); **C**, upper wall of ascoma in vertical section; **D**, lower wall of ascoma in vertical section; **E**, lower wall of ascoma in squash mount (bars = 20 μm).

Typification: Indonesia: JAVA: Tjibodas, on *Pandanus*, 27 Feb. 1897, *O. Penzig* 185 (**PAD**!).

Host: *Pandanus* (*Pandanaceae*).

Distribution: Java.

Illustration: Fig. 92.

Notes: *Terriera javanica* var. *pandani* is characterized by its short ascospores and asci, the latter with a broadly rounded apex, and branched paraphyses. It is very similar to several other monocotyledon-inhabiting species of *Terriera*: see Notes under *L. andropogonis* and *T. sacchari* (Lyon) P.R. Johnst.

This species is also similar to *Lophodermium passiflorae* (Rehm) Tehon and *L. reyesianum* Rehm, both described from southern Asia on non-monocotyledonous hosts and typical of the genus *Terriera*. The relationships between many tropical species of *Terriera*, often known from type specimens alone, will remain unresolved until more material becomes available.

Lophodermium petriniae P.R. Johnst., **sp. nov.**

Etymology: In honour of Liliane Petrini, a collector and student of grass-inhabiting species of *Lophodermium*.

A *L. arundinaceo* cellulis labiorum debiliter evolutis, cellulis atrobrunneis parietis inferioris ascomatum late cylindricalibus et in strato unico, ascis plus minusve cylindricalibus, ascosporis (75–) 90–110 × (1·5–) 2 µm, conidiomatibus nigris differt.

Infected areas on dead leaves, slightly paler than surrounding host tissue, not associated with zone lines, containing groups of ascomata and conidiomata. *Ascomata* 0·6–1·2 × 0·3–0·5 mm, oblong-elliptical to elliptical in outline, ends ± acute, wall mostly dark grey, some with irregular pale grey patches, single, longitudinal opening slit, with indistinct lips visible as slit develops, but soon lost. *Ascomatal insertion* subepidermal. *Covering layer* up to 65 µm thick in vertical section, comprising clypeus of dark-walled hyphae within epidermal cells, and upper wall of ascomatal stroma. *Upper wall* in unopened ascomata comprising two layers, outer layer mostly of angular to globose cells, 5–8 µm diam., with slightly and irregularly thickened walls, but with groups of very dark-walled cells immediately to either side of future line of opening, inner layer up to 30 µm thick, comprising hyaline, thin-walled, cylindrical, dichotomously branching periphysoids, in opened ascomata comprising mostly angular to globose cells, 5–8 µm with walls slightly and irregularly encrusted with dark brown material and group of very dark cells adjacent to poorly-developed lips, host fibre cells sometimes embedded within wall. *Lower wall* 10–15 µm thick in vertical section, comprising two or three rows of ± globose cells, the lowermost row of cells with darkened wall (in addition there may be scattered, dark-walled, cylindrical cells across base of wall), in squash mount darkened cells of lower wall broad-cylindrical, mostly 6–8 µm diam., arranged in layer one or two cells wide across base of ascoma, with scattered, narrow, dark-walled hyphae across base of wall. *Paraphyses* 1·5–2 µm diam., circinate and coiling at apex. *Asci* 110–130 × 9–10 µm, ± cylindrical, tapering suddenly to small, subtruncate apex with undifferentiated wall, 8-spored, short basal stalk at maturity with spores confined to upper 90–110. *Ascospores* (75–) 90–110 × (1·5–) 2 µm, tapering slightly to base, apical gelatinous cap globose to broad-cylindrical, 3·5–5 µm diam., basal gelatinous cap narrow-cylindrical, 3–5 µm long, gelatinous sheath 5–6 µm diam. *Conidiomata* 0·2–0·3 mm diam., round in outline, wall dark grey to black, intra-epidermal. *Upper wall* up to 6 µm thick in vertical section, comprising one or two rows of

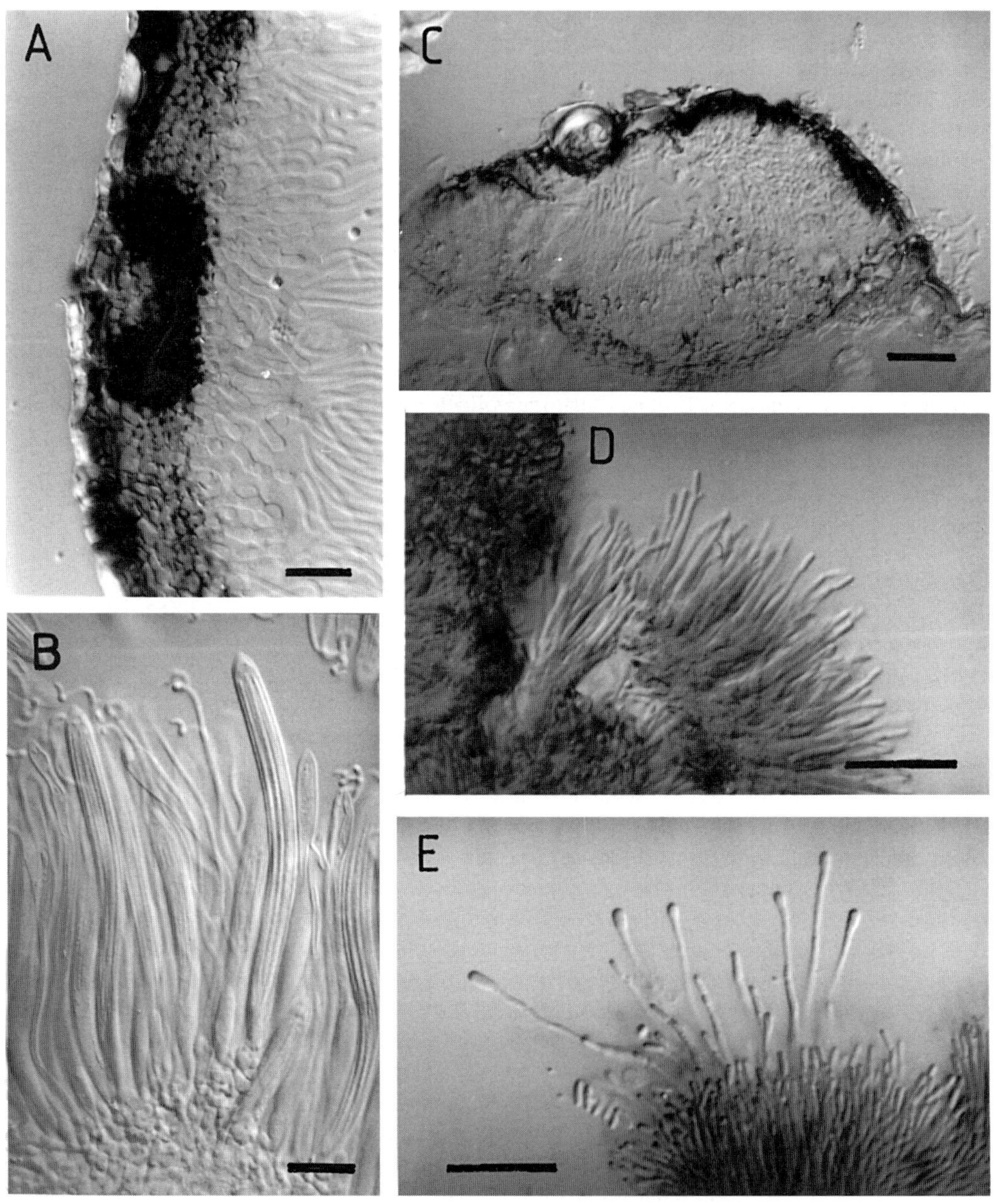

Fig. 94. *Lophodermium petriniae* (**A**, **B**, **PDD** 65388; **C–E**, **PDD** 65383). **A**, upper wall of unopened ascoma in vertical section; **B**, asci and paraphyses; **C**, conidioma in vertical section; **D**, conidiogenous cells and conidia; **E**, trichogynes (bars = 20 μm).

angular to globose cells with thick, dark walls. *Lower wall* with outer layer comprising pale brown, thin-walled hyphal cells, 4–6 μm wide, running ± parallel with surface of host leaf, inner layer comprising two or three rows of globose cells, 5–7 μm diam., with hyaline to pale brown walls, conidiogenous cells held on inner layer of cells. *Conidiogenous cells* 12–19 × 1·5–2·5 μm, cylindrical, tapering slightly to apex, proliferation sympodial, often with two developing conidia held at apex. *Conidia* 4–5 × 1–1·5 μm, cylindrical, hyaline, non-septate, filiform trichogynes present among conidiogenous cells.

Typification: Switzerland: GRAUBÜNDEN: Klosters, Schlappintobel, on *Agrostis gigantea*, 6 Jul. 1980, *L. Petrini* 48 (**ZT**!, *holotypus* of *L. petriniae*; **PDD** 65388!, *isotypus*).

Hosts: *Agrostis*, *Anthoxanthum*, *Calamagrostis* (*Poaceae*).

Distribution: Switzerland.

Illustrations: Figs 93, 94.

Notes: *Lophodermium petriniae* has an ascomatal structure typical of *Lophodermium* Group A, but cannot be placed unequivocally in one of the subgroups recognized within Group A (see p. 33). It has the well-developed periphysoid layer of the *alpinum*-group, but also has poorly-developed lip cells. It is characterized within Group A by its large, black conidiomata, asci close to being strictly cylindrical, long ascospores and well-developed periphysoids. It is known from several hosts, but only from the Swiss Alps. Unusually for *Rhytismataceae*, mature conidia and ascomata are found together; in most other species conidiomata are over-mature by the time ascomata are open.

Ascus shape in *L. petriniae* is more or less the same as in *L. sesleriae* and very similar to *L. gramineum*. These two last-named species both have an ascomatal structure typical of the *alpinum*-group. The former is distinguished by its lack of conidiomata and by its ascospores which taper to a more or less acute base, while the latter, also lacking conidiomata, has strictly cylindrical asci, and paraphyses undifferentiated or slightly swollen apically, never circinate.

Additional specimens examined: Switzerland: GLARUS: Braunwald, Gumen, on *Agrostis agrostiflora*, 26 Jul. 1986, *O. Petrini* 155 (**PDD** 65389, **ZT**); GRAUBÜNDEN: Bergün, on *Agrostis schleicheri*, 30 Aug. 1984, *O. Petrini* 124 (**PDD** 65383, **ZT**); Bergün, on *Anthoxanthum alpinum*, 30 Sep. 1984, *O. Petrini* 125 (**PDD** 65386, **ZT**); Bergün, Avers, below Cresta, on *Calamagrostis villosa*, 18 Aug. 1981, *B. Widler* 152 (**PDD** 65387, **ZT**); TICINO: Alpe di Piora, Mottone, on grass, 23 Jul. 1980, *L. Petrini* 192 (**PDD** 65390, **ZT**); TICINO: Alpe di Piora, Cadagno, on grass, 23 Jul. 1980, *L. Petrini* 207 (**PDD** 65391, **ZT**).

Lophodermium arundinaceum var. **phragmitis** Sacc. & Penz., in SACCARDO, *Michelia* **2**: 610 (1882).

Typification: 'in foliis vaginisque *Phragmitis communis*, Rhône, T.5728' (SACCARDO, 1882).

Host: *Phragmites communis* (*Poaceae*).

Distribution: France.

Notes: SACCARDO (1882) noted that 'T' in the collection number represented a collection made by J. Therry, Lyon. No material matching that cited by SACCARDO was located in **P**. From the host alone, it is likely that *L. arundinaceum* var. *phragmitis* represents *L. arundinaceum*.

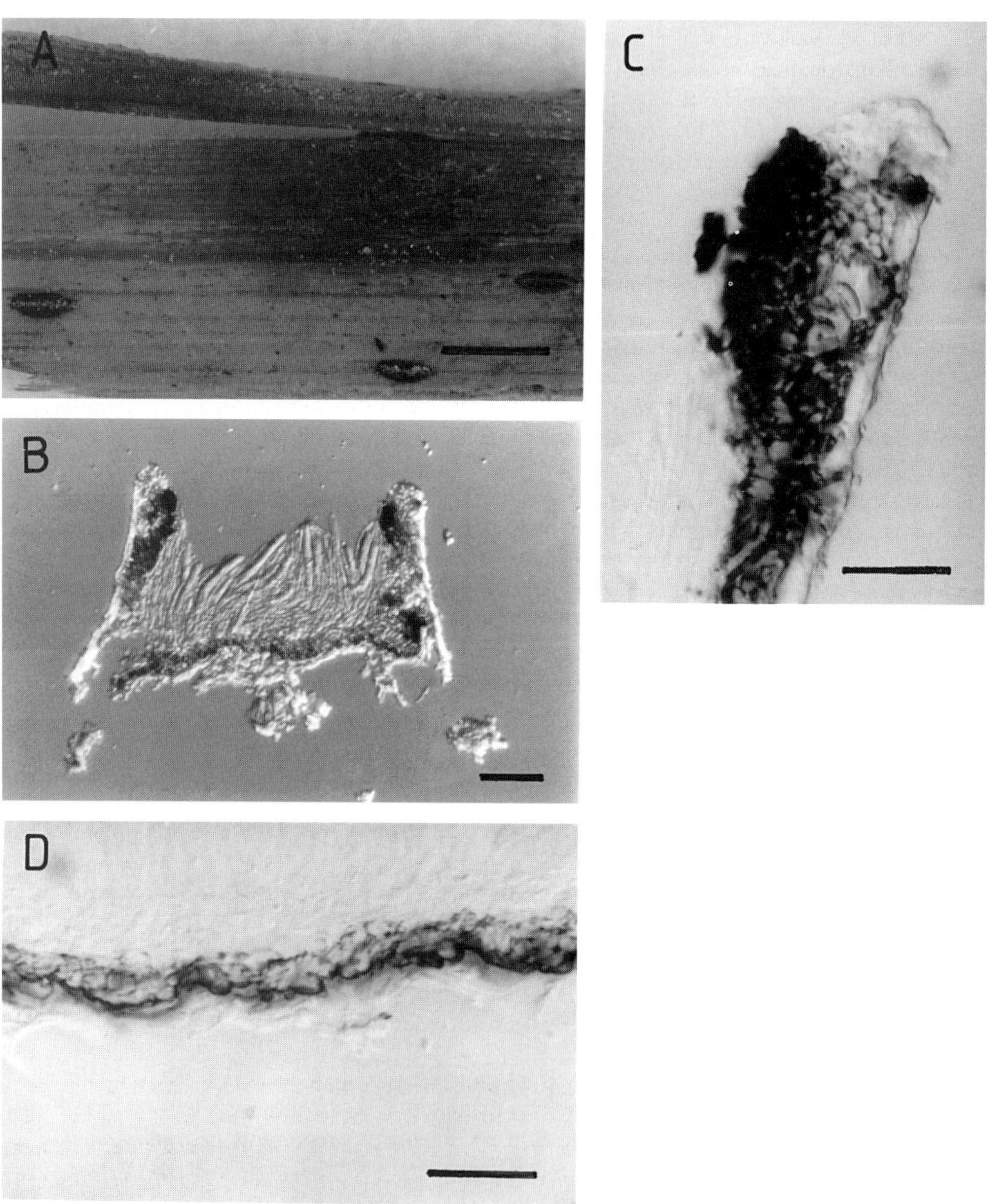

Fig. 95. *Lophodermium raapianum* (lectotype, **PAD**). **A**, ascomata (bar = 1 mm); **B**, ascoma in vertical section (bar = 50 µm); **C**, upper wall of ascoma in vertical section; **D**, lower wall of ascoma in vertical section (bars = 20 µm).

Lophodermium arundinaceum var. **piptatheri** Ranoj., *Annls mycol.* **8**: 354 (1910).

Typification: 'Auf trockenen Blättern von *Piptatherum paradoxum* P.B. Gradac bei Krepoljin im Juni 1908' (RANOJEVIČ, 1910).

Host: *Piptatherum paradoxum* (*Poaceae*).

Distribution: Europe.

Notes: No type material was located.

Lophodermium raapianum Penz. & Sacc., *Malpighia* **11**: 529 (1897).

Infected areas on dead leaves, slightly paler than surrounding host tissue, some groups of ascomata associated with broad, brown zone lines, but only where another species of *Rhytismataceae* (represented in this material by conidiomata alone) occurs on adjacent part of leaf, containing scattered ascomata. *Ascomata* 0·5–1 × 0·25–0·3 mm, oblong-elliptical in outline, unopened ascomata with wall pale to dark grey, irregular and mottled with distinct paler zone along future line of opening, opened ascomata with wall mottled, pale to dark grey and narrow black line marking outside edge, longitudinal opening slit lined with yellowish lips. *Ascomatal insertion* subepidermal. *Covering layer* in unopened ascomata (at stage when asci starting to elongate) up to 60 µm thick in vertical section, comprising clypeus of dark-walled hyphae within epidermal cells and upper wall of ascomatal stroma. *Upper wall* of unopened ascomata in vertical section up to 30 µm thick, comprising angular to globose cells with thick, darkened walls, especially in inner part, lined with poorly-developed layer of periphysoids, fibre bundles of host sometimes embedded, in vertical section opened ascomata with upper wall comprising mostly angular, very thick-walled cells, 4–6 µm diam., lip cells possibly unbranched. *Lower wall* in vertical section 15 µm thick, comprising three or four rows of angular to ± cylindrical cells, 5–8 µm diam., lowermost row with darkened, thickened walls. *Paraphyses* 1·5–2 µm diam., ± undifferentiated, sometimes slightly bent or swollen near apex. *Asci* 95–120 × 5·5–8 µm, cylindrical, tapering gradually to rounded apex with undifferentiated wall, 8-spored, spores extending ± to base. *Ascospores* 60–80 × 1·5–2 µm, tapering slightly to both ends, gelatinous caps and sheaths not observed.

Typification: Indonesia: SUMATRA: Batoe, Poeloe Tana Masa, on *Scirpus*, Sep. 1896, *H. Raap s. num.* (**PAD**!, *lectotypus*, selected here; **BO**!; **W**!, *isolectotypi*).

Host: *Scirpus* (*Cyperaceae*).

Distribution: Sumatra.

Illustrations: Figs 95, 97.

Notes: The relationship of *L. raapianum* to other monocotyledon-inhabiting species is uncertain. In ascomatal structure it appears most similar to *L. agathidis*: macroscopically distinct pale zone along future line of opening, subepidermal ascomata, upper wall with a broad layer of darkened cells, the unbranched lip cells tending to be lacking along the inner part of the opening slit, *cf.* **Figs 21, 95**, although the two species differ in ascus and ascospore size.

Lophodermium raapianum is also similar in some respects to *L. inclusum*, another *Cyperaceae*-inhabiting species. Both have darker cells in the upper wall of the ascoma than in most of the *Poaceae*- and *Cyperaceae*-inhabiting species, and host fibre bundles commonly embedded in the upper wall. This last feature, although to some extent related to host leaf structure, can also be fungus species-specific. For example, there are two species of *Lophodermium* on a single species of *Gahnia* in New Zealand, but only one has fibre bundles

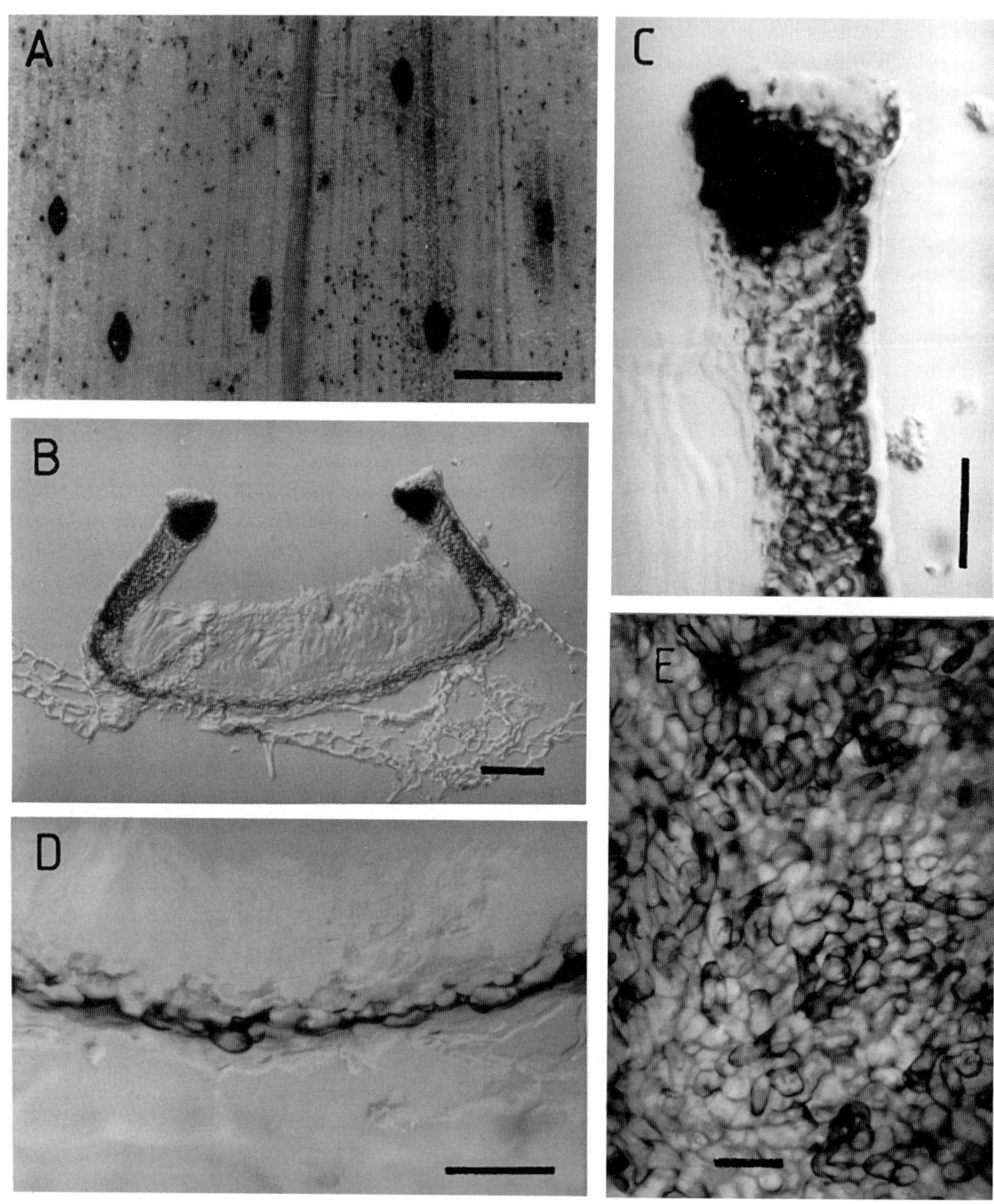

Fig. 96. *Lophodermium robergei* (Desmazières, *Pl. crypt. N. France* no. 169, **K**). **A**, ascomata (bar = 1 mm); **B**, ascoma in vertical section (bar = 50 μm); **C**, upper wall of ascoma in vertical section; **D**, lower wall of ascoma in vertical section; **E**, lower wall of ascoma in squash mount (bars = 20 μm).

embedded in its ascomatal walls. *Lophodermium inclusum* differs from *L. raapianum* in lacking lip cells.

Lophodermium robergei (Desm.) Sacc., *Syll. Fung.* **2**: 796 (1883).
Hysterium robergei Desm., *Mém. Soc. r. Sci. Agric. Arts Lille*, 1842: 140 (1843).
Hypoderma robergei (Desm.) Kuntze, *Revis. gen. pl.* **3**(3): 487 (1898).
Lophodermellina robergei (Desm.) Höhn., *Hedwigia* **62**: 79 (1920).
Lophodermium brachypodii Hilitzer, *Věd. Spisy čsl. Akad. zeměd.* **3**: 93 (1929).

Infected areas on dead leaves, not associated with bleaching of host tissue or zone lines, containing discrete patches of ascomata, conidiomata not seen. *Ascomata* 0·4–0·8 × 0·25–0·3 mm, elliptical in outline, ends rounded to ± acute, edges sharply defined, wall black, single longitudinal opening slit with lips variable in appearance, macroscopically distinct in some collections, dark and indistinct in others (see Notes below). *Ascomatal insertion* subepidermal. *Covering layer* up to 40–50 µm thick, comprising clypeus of dark brown, thick-walled hyphae, 2–3 µm diam., within epidermal cells, and upper wall of ascomatal stroma. *Upper wall* up to 25–40 µm thick in vertical section, comprising mostly globose cells, 4–6 µm diam., with walls irregularly encrusted with dark brown material, and with restricted group of very dark brown cells adjacent to poorly-developed lips. *Lower wall* 15 µm thick in vertical section, comprising two to four rows of ± globose to short-cylindrical cells, the lowermost row with darkened walls, in squash mount these short-cylindrical to irregular in shape, 5–8 µm diam., arranged in layer one or two cells wide across base of ascoma. *Paraphyses* 1–1·5 µm diam., circinate and coiling at apex. *Asci* 80–90 (–100) × (6·5–) 7–8 (–8·5) µm, subclavate to subfusoid, tapering gradually to small subtruncate apex with undifferentiated wall, 8-spored, well-defined basal stalk with spores confined to upper 50–60 µm. *Ascospores* (30–) 35–45 × 1·5 (–2) µm, tapering to base, apical gelatinous cap globose, 3–4 µm diam., basal gelatinous cap narrow-cylindrical, 3–4 × 1 µm, gelatinous sheath not seen.

Typification: France: NORMANDIE: Parc de Libisey, on *Bromus sylvaticus*, 31 Mar. 1842, *M. Roberge s. num.* (**S**!, *lectotypus* of *H. robergei*, selected here). **Czech Republic:** BOHEMIA: Dubá, on *Brachypodium pinnatum*, 6 Jul. 1927, *A. Hilitzer s. num.* (**PRM** 693137!, *lectotypus* of *L. brachypodii*, selected here).

Hosts: *Agropyron*, *Brachypodium*, *Bromus*, *Festuca*, *Oryzopsis* (*Poaceae*).

Distribution: Europe, Canada, Australia (see Notes below).

Illustrations: Figs 7B–E, 96, 98.

Notes: *Lophodermium robergei* has an ascomatal structure typical of the *actinothyrium*-group of *Lophodermium* Group A (see p. 33). It is distinguished from other species in the group by its short and narrow asci and short ascospores, the latter similar in size to those of *L. sieglingiae* and *L. caricinum*. The asci of *L. sieglingiae*, however, are wider and the ascospores have a characteristic very narrow, tail-like base. In *L. caricinum* the ascomata have a thicker upper wall lined with a characteristic broad, tangled layer of periphysoids and clavate asci with a long basal stalk.

No specimens were cited by DESMAZIÈRES (1843), the only information about the material examined being that the host was *Bromus sylvaticus* and that it was sent to him by Roberge in 1839 and 1842. The specimen here selected as lectotype fits this information perfectly. DUBY (1862) placed this species in synonymy with *L. arundinaceum* var. *gramineum*, and TEHON (1935) with *L. gramineum*. Although the type specimen of *L. robergei* is macroscopically similar to some

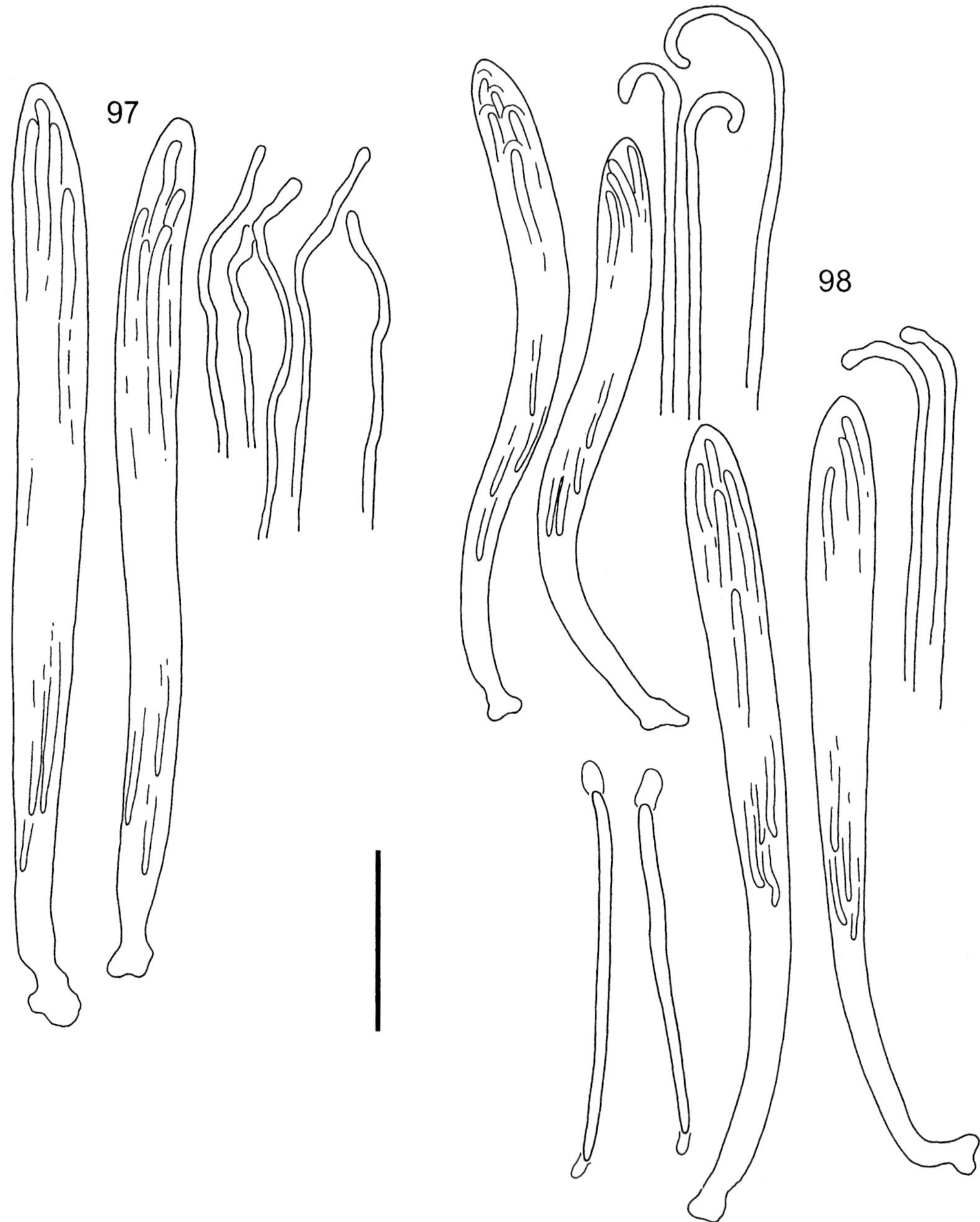

Figs 97–98. 97. *Lophodermium raapianum* (lectotype, **PAD**). Asci and apex of paraphyses. **98.** *Lophodermium robergei*. Asci and apex of paraphyses, above (*Roberge*, 31 Mar. 1842, **S**); asci, apex of paraphyses and ascospores, below (**PDD** 64470) (bar = 20 μm).

collections of *L. gramineum*, it is distinct both in ascomatal structure and microscopically.

Lophodermium brachypodii is microscopically indistinguishable from *L. robergei*, despite most collections examined from *Brachypodium* having lip cells well-developed compared to the poorly-differentiated ones seen in most collections from other hosts. Ascospore lengths of 70 µm cited by HILITZER (1929) and TEHON (1935) for *L. brachypodii* differ from those given above. However, all specimens from *Brachypodium* cited below, many of which were collected by Hilitzer, have the shorter ascospores described here.

There are two specimens of *Lophodermium* on *Brachypodium* in **PRM** collected by Hilitzer before 1929: **PRM** 693137 and 693153. Either or both could represent type material, and both specimens have been annotated by Z. Pouzar as 'certainly the type material'. However, **PRM** 693153 (**Czech Republic:** BOHEMIA: Hostín apud Srbsko, on *Brachypodium pinnatum*, 12 Jun. 1927, *A. Hilitzer s. num.*) matches *L. culmigenum* morphologically. **PRM** 693137 contains the same species as the other collections on *Brachypodium* deposited in **PRM** by Hilitzer, and is selected here as the lectotype.

The identity of the collections cited from Canada (**DAOM** 26555) and Australia (**PDD** 37490) remains equivocal. Ascospores of the Canadian collection are longer (45–55 × 1·5 µm) than typical, although other morphological features appear to fit well. The Australian collection has paraphyses bent rather than coiling at the apex and often slightly and irregularly swollen. A relationship between this collection and those from New Zealand which were discussed under *L. actinothyrium* should be considered.

Additional specimens examined: Australia: NEW SOUTH WALES: Tuggerah Lakes, on *Bromus uniloides*, 14 May 1969, *J. Walker s. num.* (**DAR** 17198, as *L. gramineum*). **Canada:** QUÉBEC: Wakefield, on *Oryzopsis asperifolia*, 15 Jun. 1951, *J.A. Parmelee & I.L. Conners s. num.* (**DAOM** 26555, as *L. arundinaceum*). **Czech Republic:** BOHEMIA: loco 'Skarman' ap. Domazlice, on *Brachypodium pinnatum*, 11 Sep. 1929, *A. Hilitzer s. num.* (**PRM** 693178); in colle 'Dubovy vrch' ap. Horsovsky Tyn, ap. locum 'Obora', on *Brachypodium pinnatum*, 31 Jul. 1931, *A. Hilitzer s. num.* (**PRM** 693179); Bohemia meridionalis, Sep. 1929, *A. Hilitzer* (**PRM** 693180); 'Sv. Markyta' ap. Jindrichuv Hradec, on *Brachypodium pinnatum*, *s. hosp.*, 12 Sep. 1929, *A. Hilitzer s. num.* (**PRM** 693181); Věznice, Nová Hut, Beroun, on *Brachypodium pinnatum*, 20 Jun. 1935, *A. Hilitzer s. num.* (**PRM** 820265); Žďár, Kaplice, Husí, on *Brachypodium pinnatum*, 2 Aug. 1970, *M. Svrček s. num.* (**PRM** 715267); Prudice, Tábor, on *Brachypodium silvatica*, 12 Aug. 1949, *M. Svrček s. num.* (**PRM** 756753); Dubichy, Velemiá, on *Brachypodium pinnatum*, 22 Sep. 1954, *M. Svrček s. num.* (**PRM** 771609); Srbsko, Karlštejn, Dřínová hora, on *Brachypodium pinnatum*, 7 Jun. 1972, *M. Svrček s. num.* (**PRM** 756752); Weisskirchen, Nordbahndamm, on *Brachypodium pinnatum*, 1 Mar. 1912, *F. Petrak s. num.* (Petrak, *Flora bohemiae et moraviae* no. 217; **K**; **FH**, as *L. arundinaceum* f. *apiculatum*). **England:** OXFORDSHIRE: *s. loc.*, on *Brachypodium*, 15 Sep. 1949, *Rev. Adams s. num.* (**IMI** 37302, as *L. gramineum*). **France:** *s. loc.*, on *Bromus sylvaticus*, *s. dat. nec coll.* (Desmazières, *Pl. crypt. N. France* Edn 2, Sér. 2, no. 169; **K**). **Italy:** CALABRIA: *s. loc.*, on *Agropyron ?repens*, 17 Jun. 1989, *R.W.G. Dennis s. num.* (**K**). **Germany:** ERFURT: Wittrodaer Forst, on *Brachypodium*, 20 Jul. 1907, *H. Diedicke s. num.* (Migula, *Kryptogamae Germaniae, Austriae et Helvetiae exsiccatae* no. 240; **K**; **S**; **NY**, as *L. arundinaceum*); SAXONY: *s. loc.*, on *Brachypodium sylvaticum*, 24 Jun. 1900, *W. Krieger s. num.* (Krieger, *Fungi saxonici* no. 1577; K, as *L. arundinaceum*). **Switzerland:** AARGAU: Geissberg, Besserstein, *s. hosp.*, 15 May 1980, *L. Petrini s. num.*(**PDD** 64471, **ZT**); Geissberg, on *Brachypodium sylvatica*, 15 May 1980, *L. Petrini* 20 (**PDD** 65285, **ZT**); SCHAFFHAUSEN: Bargen, Mülihalde, on *Bromus erectus*, 25 May 1980, *L. Petrini s. num.* (**PDD** 64472, **ZT**). **Wales:** Llanberis, Vivian Quarry Walk, on *?Festuca*, 28 Aug. 1986, *P.R. Johnston & E.M. Gibellini s. num.* (**PDD** 64470).

Lophodermium rottboelliae Sawada, *Rep. Govt. Res. Inst. Formosa* **87**: 3 (1944).

Typification: Taiwan: *s. loc.*, on *Rottboellia* [as '*Rottoboellia*'] *compressa*, *s. dat. nec coll.* (**TNS** F218983!, holotype).

Host: *Rottboellia* (*Poaceae*).

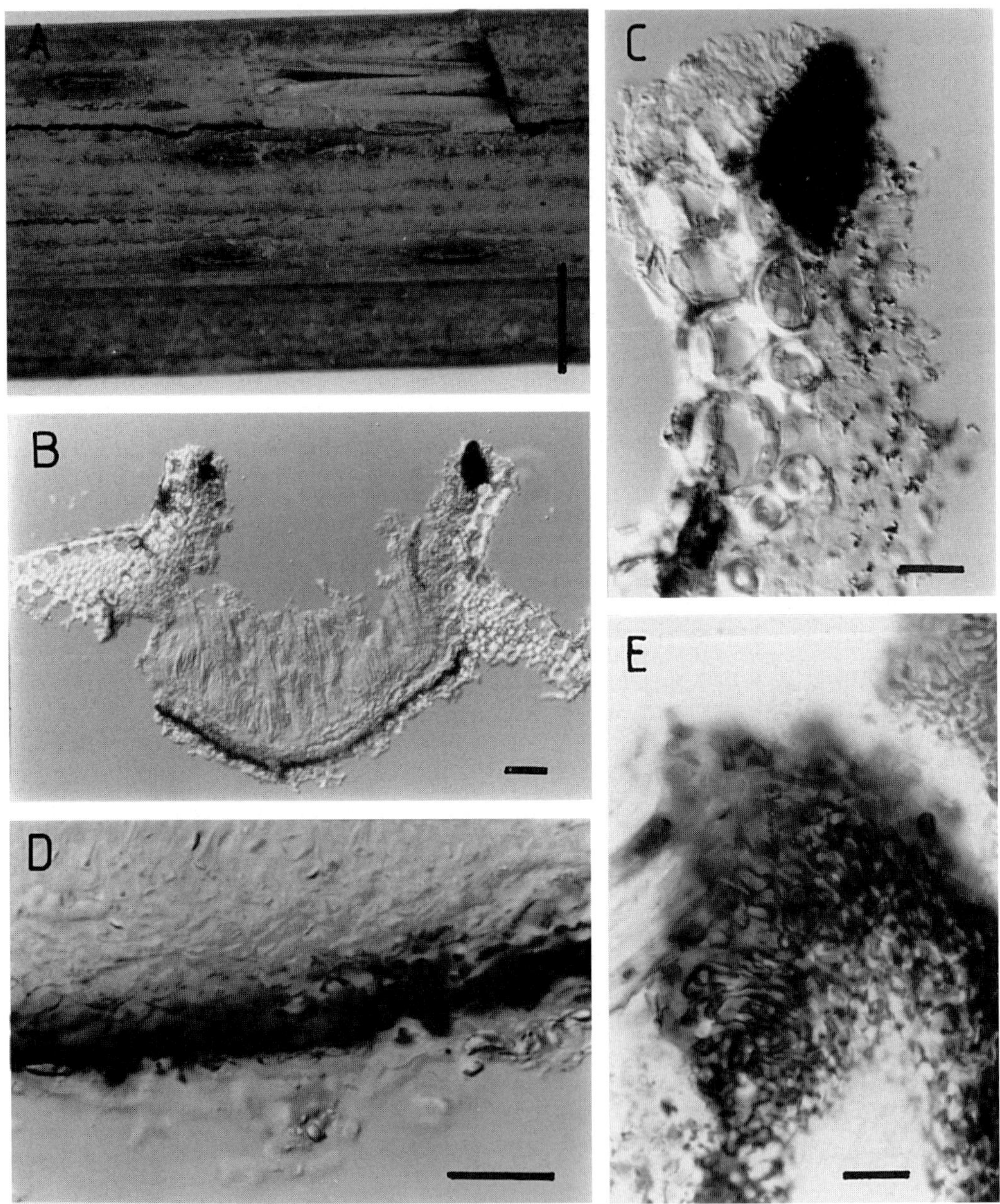

Fig. 99. *Lophodermium rubrum* (**PDD** 46154). **A**, ascomata (bar = 1 mm); **B**, ascoma in vertical section (bar = 50 μm); **C**, upper wall of ascoma in vertical section; **D**, lower wall of ascoma in vertical section; **E**, lower wall of ascoma in horizontal section (bars = 20 μm).

Distribution: Taiwan.

Notes: The original publication has not been seen. A note in the *Index of Fungi* **2**: 387 states that the diagnosis was in Japanese, which suggests that the name is invalid (*ICBN* Art. 36.1).

Macroscopically, the ascomata were typical in appearance of the *Poaceae*-inhabiting species of *Lophodermium s. str.*, although all ascomata examined were either over-mature or had lost all hymenial elements during storage.

Lophodermium rubrum P.R. Johnst., *N.Z. Jl Bot.* **27**: 267 (1989).

Infected areas on dead leaves, host tissue sometimes with pinkish tinge, not associated with zone lines, containing ascomata, conidiomata not seen. *Ascomata* 0·5–0·8 × 0·3 mm, ± elliptical in outline, ends rounded, ascomata deeply immersed with margin not sharply defined, wall pale grey to ± concolorous with surrounding host tissue toward edge of ascoma, dark grey around single longitudinal opening slit, lined with pale lip cells. *Ascomatal insertion* subepidermal and beneath bundles of host fibre cells below epidermis. *Covering layer* 80 µm thick in vertical section, comprising mostly ± intact epidermal and fibre cells of host and upper wall of ascomatal stroma, clypeus very poorly developed or lacking. *Upper wall* comprising mostly angular to globose cells, 8–12 µm diam., with walls slightly and irregularly encrusted with dark brown material and with group of very dark, thick-walled cells adjacent to well-developed lips. *Lower wall* 15–25 µm thick in vertical section, comprising five or six rows of cells, those in outer rows irregular in shape with dark, thick walls while those in inner rows angular with thin, pale brown walls, in squash mount darkened cells of lower wall long-cylindrical to hyphal, 3–4 µm diam., tangled, forming *textura intricata*. *Paraphyses* 1·5 µm diam., undifferentiated to slightly swollen at apex. *Asci* 120–165 × 9·5–10·5 µm, initially subfusoid, becoming subclavate with maturity, tapering gradually to small, subtruncate apex with undifferentiated wall, 8-spored, basal stalk at maturity with spores confined to upper 80–95 µm. *Ascospores* (60–) 70–85 × 1·5–2 µm, tapering gradually to base, apical gelatinous cap broad-cylindrical, rounded at apex, 2·5 × 4–4·5 µm, basal gelatinous cap tiny, 1–1·5 µm diam., gelatinous sheath sometimes present, narrow, 2·5–3 µm diam.

Typification: New Zealand: NORTHLAND: Poor Knights Is: Aorangi, Fraser Landing, on *Cortaderia*, 30 Aug. 1984, *R.E. Beever s. num.* (**PDD** 46154!, holotype).

Host: *Cortaderia* (*Poaceae*).

Distribution: New Zealand.

Illustrations: Figs 99, 100.

Notes: *Lophodermium rubrum* has ascomata typical in structure of the *arundinaceum*-group of *Lophodermium* Group A. Among these species it is characterized by the bright orange-red mycelium developing within the host tissue around the ascomata, and long asci. See also Notes under *L. culmigenum*.

Terriera sacchari (Lyon) P.R. Johnst., **comb. nov.**
Lophodermium sacchari Lyon, *Hawaii. Plrs' Rec.* **9**: 601 (1913).

Infected areas on dead leaves and leaf bases, not associated with bleaching of host tissue or zone lines, containing ascomata, conidiomata not seen. *Ascomata* 0·5–2 × 0·2–0·3 mm, narrow-oblong to sublinear in outline, ends rounded, wall black, unopened ascomata with well-developed pale

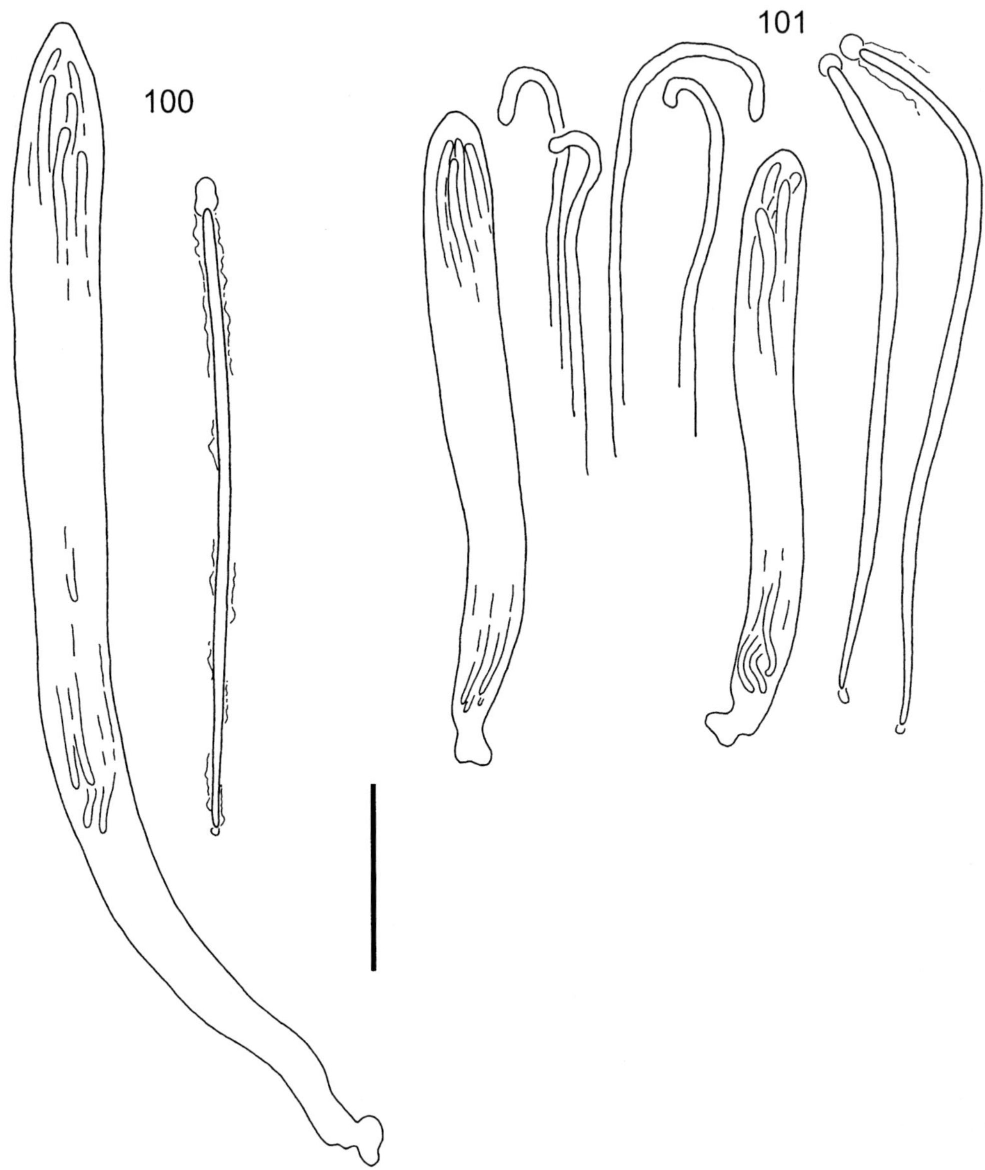

Figs 100–101. 100. *Lophodermium rubrum* (**PDD** 46154). Ascus, and ascospore in water. **101.** *Lophodermium seriatum* (*Eliasson*, 6 Sep. 1892, **K**). Asci, apex of paraphyses and ascospores (bar = 20 µm).

zone along future line of opening, single, longitudinal opening slit with narrow, black, flattened, shelf-like margin on either side. *Ascomatal insertion* subepidermal and often below host fibre cells. *Ascomatal structure* typical of *Terriera* (see p. 37). *Paraphyses* 1·5–2 µm diam., swollen to 2·5–4 µm at ± clavate apex, rarely branched, extending 20–30 µm beyond asci as well-developed epithecium. *Asci* 90–100 × 5–6 (–7) µm, cylindrical, ± broadly rounded at apex with undifferentiated wall, 8-spored, narrower basal stalk with spores confined to upper 50–70 µm. *Ascospores* not seen released, 1·5 µm wide, 50–60 µm long, gelatinous sheath not observed.

Typification: USA: HAWAII: Hawaii I, Ola'a, on *Saccharum officinarum*, 5 Mar. 1913, *M. Larsen s. num.* (**BISH** 487733!, holotype; **BISH** 487837!, isotype).

Host: *Saccharum officinarum* (*Poaceae*).

Distribution: USA (Hawaii).

Illustration: Fig. 102.

Notes: *Terriera sacchari* is distinguished within the genus by its unbranched paraphyses, narrow asci and more or less broadly rounded ascus apex. It is microscopically very similar to *T. javanica* var. *pandani* which, however, has paraphyses branched typically near the apex. *Terriera stevensii* P.R. Johnst., also described from a monocotyledon from Hawaii, differs in having longer asci and ascospores, and paraphyses embedded in thick gel.

Additional specimen examined: USA: HAWAII: Molokai, Wailau Valley, on *Saccharum officinarum*, 27 Dec. 1948, *Lohman s. num.* (**BISH** 603311, *p.p.*).

Terriera samuelsii P.R. Johnst., **sp. nov.**

Etymology: In honour of the collector of the type specimen, Gary Samuels.

A *T. arundinacea* pariete ascomatis atro, lineis atris stromaticis maculas pallidas cingentibus, conidiomatibus dilute brunneis deinde denigricantibus differt.

Infected areas on dead leaves, paler than surrounding host tissue, associated with incomplete, narrow, dark zone lines, containing ascomata and conidiomata. *Ascomata* 0·7–3 × 0·3 mm, oblong to sublinear in outline, ends rounded, wall black, with indistinct paler zone along future line of opening, single longitudinal slit with black, shelf-like area lining either side. *Ascomatal structure* typical of *Terriera* (see p. 37). *Paraphyses* 1·5–2 µm diam., undifferentiated to slightly swollen at apex, unbranched, embedded in hyaline to yellowish gel with numerous crystalline inclusions, extending 10–20 µm beyond asci as epithecium. *Asci* 125–140 × 7–8 µm, cylindrical, broadly rounded at apex with undifferentiated wall, 8-spored, well-developed basal stalk at maturity with spores confined to upper 80–100 µm. *Ascospores* (65–) 75–90 × 2 µm, ± straight when released, tapering slightly to ends, 1-septate, with no gelatinous sheath. *Conidiomata* 0·2–0·3 mm diam., round in outline, wall initially pale brown, becoming black. *Conidiogenous cells* 7·5–10 × 3–5 µm, solitary, ± cylindrical, proliferation sympodial, often with two conidia held at apex. *Conidia* 4·5–6 × 1·5 µm, short-cylindrical, straight, ends rounded, non-septate, hyaline.

Host: Unidentified monocotyledon.

Distribution: Brazil, Guyana.

Illustration: Fig. 103.

Notes: *Terriera samuelsii* is similar to the Asian *T. arundinacea* in ascus and ascospore size, and shape of the ascus apex and paraphyses. The latter species differs in having paler ascomata, lacking conidiomata and not being associated with zone lines.

Although the host remains unidentified, it appears to be the same in all collections examined.

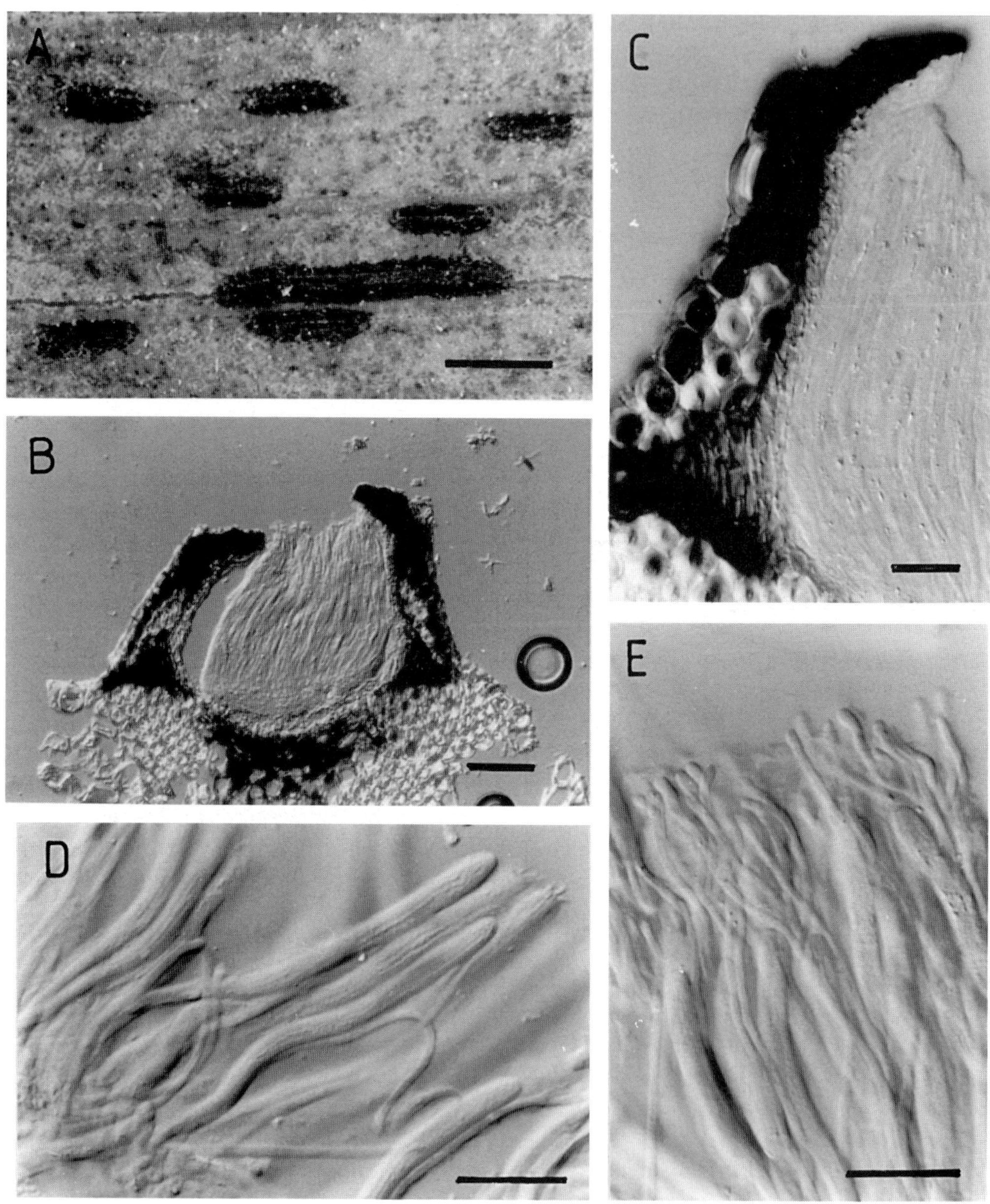

Fig. 102. *Terriera sacchari* (**BISH** 487733). **A**, ascomata (bar = 1 mm); **B**, ascoma in vertical section (bar = 50 μm); **C**, detail of upper wall of ascoma in vertical section; **D**, asci and paraphyses; **E**, apex of paraphyses (bars = 20 μm).

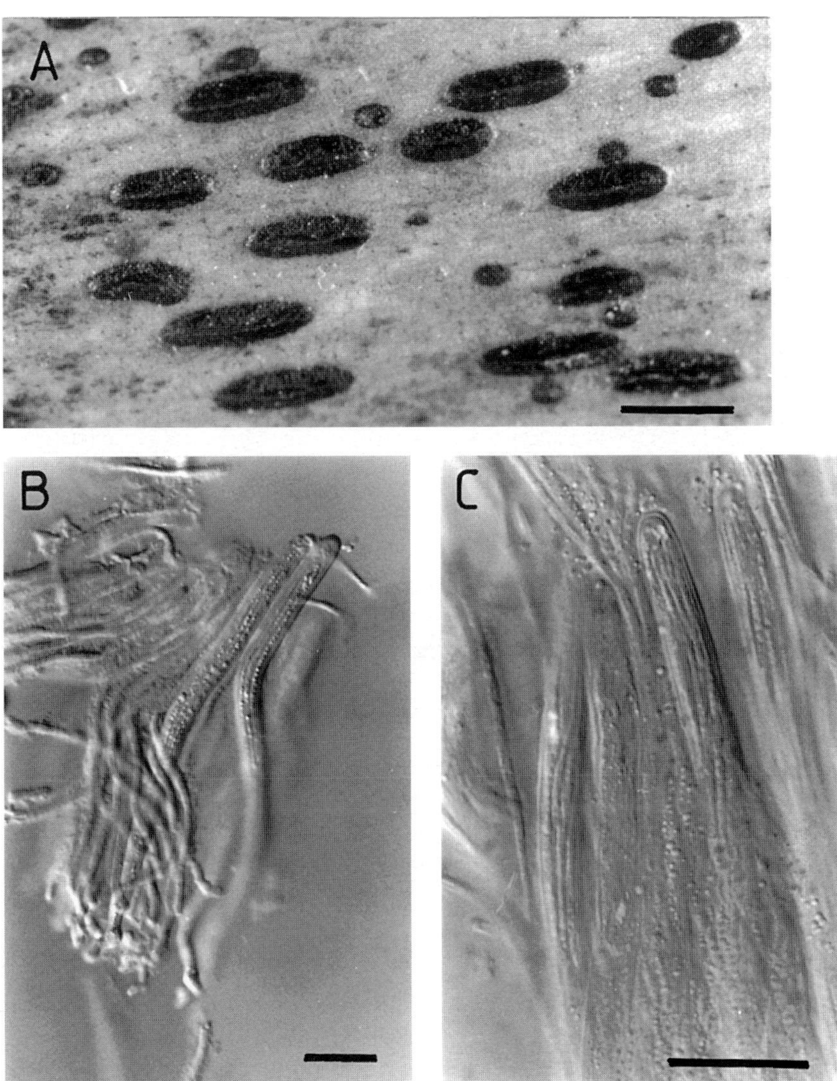

Fig. 103. *Terriera samuelsii* (**PDD** 58074). **A**, ascomata (bar = 1 mm); **B**, asci; **C**, apex of asci and paraphyses (bars 20 µm).

Additional specimens examined: Brazil: AMAZONAS: Serra Araca Plateau, 14–16 Feb. 1984, *G.J. Samuels* 278 (**NY**; **PDD** 58073); Serra Araca summit, Gallery Forest, 12 Feb. 1984, *G.J. Samuels* 224 (**NY**; **PDD** 58072). **Guyana:** POTARO-SIPARUNI: Potaro: Mt Ayanganna, eastern side, on grass, 11–12 Mar. 1987, *G.J. Samuels* 5113 (**NY**).

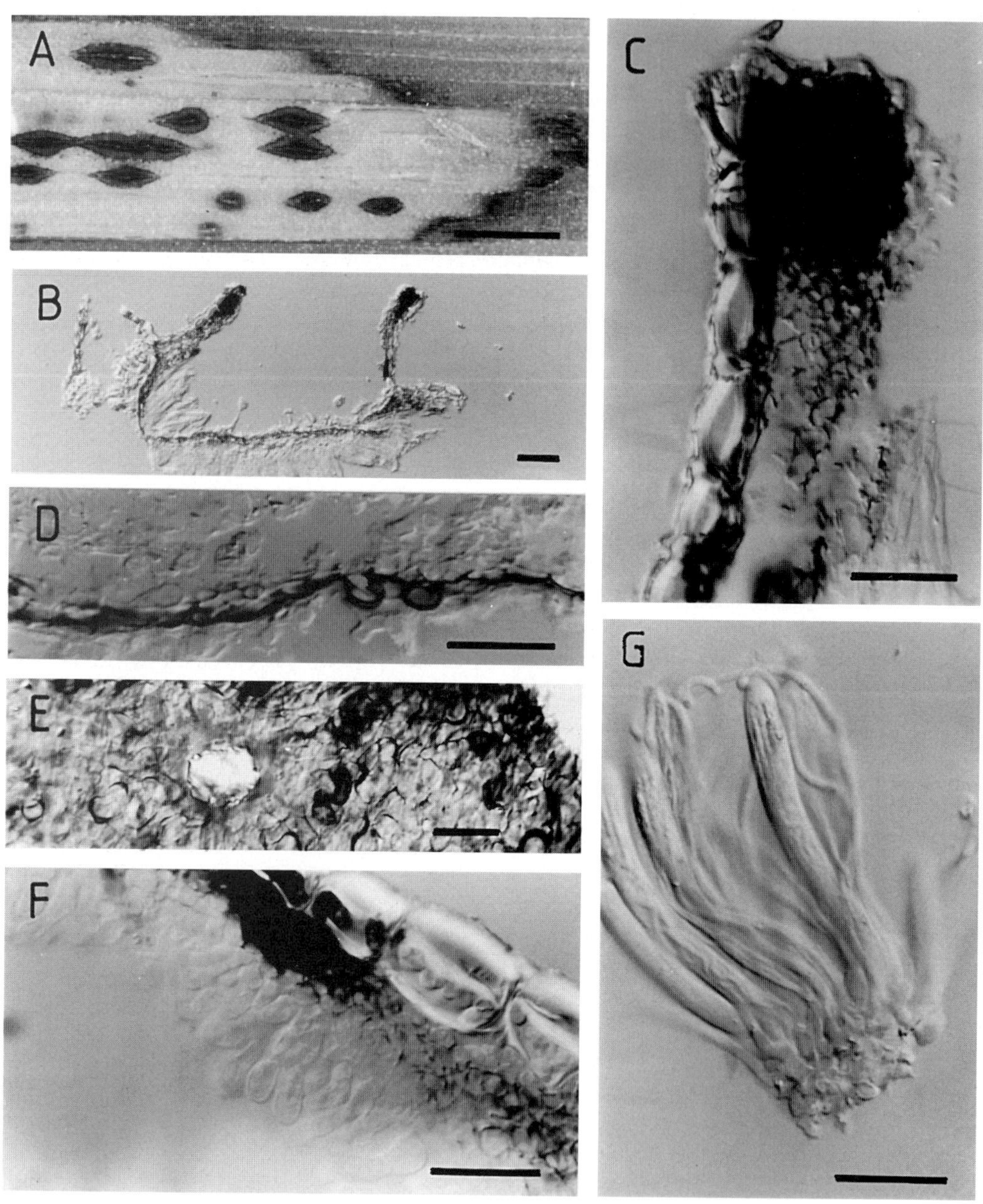

Fig. 104. *Lophodermium seriatum* (**A**, Fuckel, *Fungi rhenani* no. 741, **FH**; **B–E**, Libert, *Pl. crypt. Arduenna* no. 371, **K**). **A**, ascomata (bar = 1 mm); **B**, opened ascoma in vertical section (bar = 50 μm); **C**, upper wall of ascoma in vertical section; **D**, lower wall of ascoma in vertical section; **E**, lower wall of ascoma in squash mount; **F**, upper wall of unopened ascoma in vertical section; **G**, asci and paraphyses (bars = 20 μm).

Hypohelion scirpinum (DC.) P.R. Johnst., *Mycotaxon* **39**: 221 (1990).
Hypoderma scirpinum DC., in LAMARCK & CANDOLLE, *Fl. franç.* **6**: 166 (1815).
Hysterium scirpinum (DC.) Fr., *Bih. K. svenska VetenskAkad. Hand.* **40**: 95 (1819).
Lophodermium scirpinum (DC.) Chevall., *Fl. gén. env. Paris*: 436 (1826).
Hypodermopsis scirpinum (DC.) Kuntze, *Revis. gen. pl.* **3**(3): 487 (1898).

Notes: See JOHNSTON (1990*c*) for description, illustration, specimens examined and discussion.

Another species described from *Cyperaceae*, *Hypoderma alpinum* Spooner, also probably belongs in *Hypohelion*. Type material of *H. alpinum* has not been examined, but a collection from **GZU** (**Austria:** STEIERMARK: Wölzer Tauern, on *Carex paupercula*, 6 Sep. 1980, *C. Scheuer s. num.*) closely matches the description of *H. alpinum* provided by SPOONER (1981), except that the tiny ascomata often appear slightly irregular in outline. The type specimen of *Hypohelion parvum*, described by JOHNSTON (1990*c*), also closely matches the Austrian collection. If examination of the type specimen of *H. alpinum* confirms its identity with the Austrian collection, then *H. parvum* should be placed in synonymy with *H. alpinum* in the genus *Hypohelion*.

Lophodermium seriatum (Lib.) De Not., *G. Bot. ital.* **2**(2): 47 (1847).
Hysterium seriatum Lib., *Pl. crypt. Arduenna* no. 371 (1837).
Lophodermium arundinaceum var. *seriatum* (Lib.) Fuckel, *Symb. mycol.*: 257 (1870).
Hypoderma seriatum (Lib.) Kuntze, *Revis. gen. pl.* **3**(3): 487 (1898).

Infected areas initially forming necrotic patches, 1–4 × 0·2 mm, with darkened margin on otherwise green leaves which, after leaf has died, appear slightly paler than surrounding host tissue, not associated with zone lines, scattered ascomata visible, initially, within lesions on green leaves, continue development after leaf has died, conidiomata not seen. *Ascomata* 0·5–1 × 0·2–0·3 mm, elliptical in outline, ends acute, unopened ascomata with wall grey to black, if grey then outside edge and line to either side of centre of ascomata blackened, opened ascomata with wall dark brown to black, darker around outside edge and adjacent to opening, single, longitudinal slit with no lips. *Ascomatal insertion* subepidermal. *Covering layer c.* 40 µm thick in vertical section, comprising clypeus of dark brown, thick-walled hyphae, 3 µm diam., within epidermal cells, and poorly-developed upper wall of ascomatal stroma. *Upper wall* in unopened ascomata comprising two layers, outer layer mostly of angular to globose, pale-walled cells, 4–6 µm diam., with two groups of dark-walled cells immediately to either side of centre of ascoma, inner layer not extending across centre of ascoma, but otherwise 15–40 µm thick, comprising downward-projecting, hyaline, thin-walled periphysoids, in opened ascomata upper wall mostly of ± globose cells, 4–8 µm diam., with walls irregularly encrusted with dark material, but also patch of cells with thickly encrusted walls adjacent to ascomatal opening, periphysoids ± lost, lips lacking. *Lower wall* 10–15 µm thick in vertical section, comprising two or three rows of cells, with outer wall of lowermost row darkened, in squash mount darkened cells of lower wall globose to irregularly shaped, 8–15 µm diam., arranged in one or two rows across base of ascoma. *Paraphyses* 1·5–2 µm diam., circinate and coiling. *Asci* 75–100 × 6–7 µm, ± cylindrical, tapering suddenly to small subtruncate apex with undifferentiated wall, 8-spored, short basal stalk developing as matures with spores extending ± to base of ascus. *Ascospores* 70–80 × 1·5 µm, tapering gradually to narrow base, apical gelatinous cap globose, 3 µm diam., basal gelatinous cap globose, *c.* 1·5 µm diam., gelatinous sheath not observed.

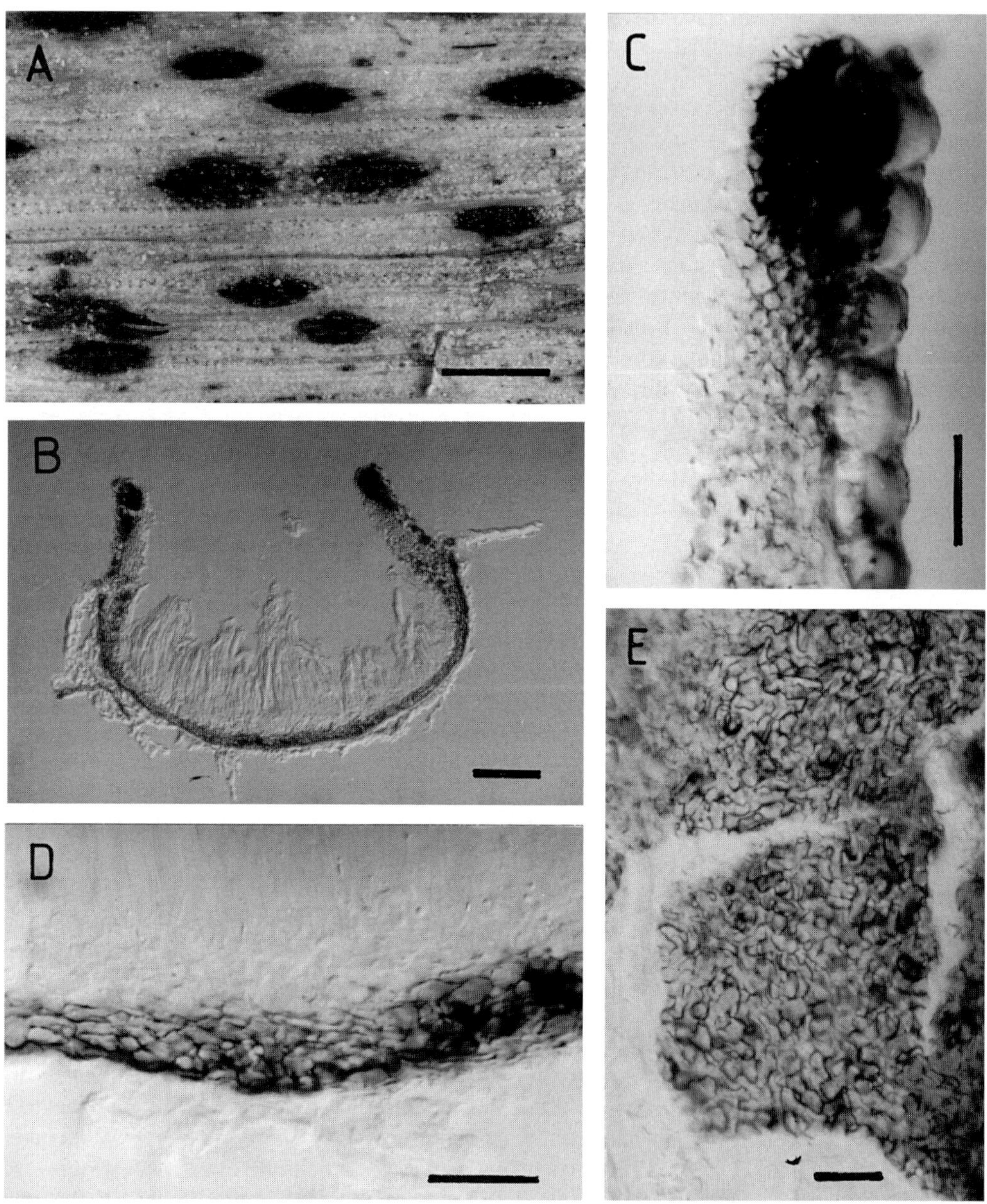

Fig. 105. *Lophodermium sesleriae* (**A**, **PRM** 693195; **B–D**, **PRM** 756746; **E**, Libert, *Pl. crypt. Arduenna* no. 73, **K**). **A**, ascomata (bar = 1 mm); **B**, open ascoma in vertical section (bar = 50 µm); **C**, upper wall of ascoma in vertical section; **D**, lower wall of ascoma in vertical section; **E**, lower wall of ascoma, squash mount (bars = 20 µm).

Typification: France: *s. loc.*, on *Festuca sylvatica* (Libert, *Pl. crypt. Arduenna* no. 371; **K**!, *lectotypus*, selected here; **FH**!, *isolectotypus*, selected here).

Host: *Festuca* (*Poaceae*).

Distribution: Europe.

Illustrations: Figs 8B–E, 101, 104.

Notes: *Lophodermium seriatum* has an ascomatal structure typical of the *alpinum*-group of *Lophodermium* Group A (see p. 33). Although known from few collections, it is distinctive biologically as the only grass-inhabiting species in which the ascomata begin development within necrotic lesions on otherwise green leaves. It is distinguished from other species in this group by its biology, narrow asci and circinate paraphyses.

FUCKEL (1873: 50) mentioned a *Leptostroma* state on *Calamagrostis* which he considered to be the anamorph of *L. arundinaceum* var. *seriatum* (citing Fuckel, *Fungi rhenani* Edn I, no. 2558). The duplicate of this exsiccatum in **K** does not match the type of *L. seriatum*, although it has what appear to be fruiting bodies, of about the correct size and shape for this species, developing within lesions on living leaves. However, these 'fruiting bodies' are stromatic structures comprising a solid tissue, *c.* 100 µm thick, of brown to pale brown, slightly thick-walled, angular cells of very uniform appearance, with the outermost layer of cells on all sides slightly thicker-walled and darker.

Additional specimens examined: Germany: RHEINLAND-PFALZ: Nassau, on *Festuca sylvatica, Fuckel s. num.* (Fuckel, *Fungi rhenani* no. 741; **FH**; **K**; **DAOM**, as *L. arundinaceum* e. *seriatum*). **Luxembourg:** between Neuport and Daverdisse, Merwart, on *Festuca sylvatica, Crepin s. num.* (Westendorp, *Herb. crypt.* no. 1046; **FH**; **K**). **Sweden:** VÄSTERGÖTLAND: Västra Tunhem Parish: Mt Halleberg, on *Festuca sylvatica*, 6 Sep. 1892, *A.G. Eliasson s. num.* (*Fungi suecici praesertim upsaliensis* no. 3490; **K**).

Lophodermium sesleriae Hilitzer, *Věd. Spisy čsl. Akad. zeměd.* **3**: 91 (1929).
Lophodermellina sesleriae (Hilitzer) Tehon, *Illinois biol. Monogr.* **13**(4): 81 (1935).

Infected areas on dead leaves, not associated with pale areas of host tissue or zone lines, containing scattered ascomata, conidiomata not seen. *Ascomata* 0·8–1·5 × 0·3–0·4 mm, elliptical in outline, ends acute, unopened ascomata with broad, quite well-defined paler zone along future line of opening, open ascomata with wall pale to dark grey, if pale then black line marking edge of ascoma, single longitudinal opening slit with no lips. *Ascomatal insertion* subepidermal. *Covering layer* up to 45 µm thick in vertical section, comprising clypeus of dark brown hyphae, 2–3 µm diam., within epidermal cells, and upper wall of ascomatal stroma. *Upper wall* in unopened ascomata with two layers of ± equal thickness, outer layer mostly of angular to ± globose cells, 4–6 µm diam., with irregularly thickened walls, but with patch of dark, very thick-walled cells immediately to either side of centre of ascoma, inner layer of ± monilioid periphysoids, not extending across centre line of ascoma, in opened ascomata upper wall with well-developed patch of very dark and very thick-walled cells adjacent to opening slit, periphysoids lost, lip cells lacking. *Lower wall* 20–25 µm thick in vertical section, comprising several rows of angular cells, 4–6 µm diam., lowermost two or three rows with darkened and slightly thickened walls, in squash mount darkened cells of lower wall 5–8 µm diam., irregularly cylindrical to angular-globose, in several rows across base of ascoma. *Paraphyses* 1·5–2 µm diam., circinate, tightly coiling at apex. *Asci* 100–125 × 7·5–9·5 µm, ± cylindrical, slightly narrower near base, tapering suddenly to small, subtruncate apex with undifferentiated wall, 8-spored, short basal stalk with spores confined to upper 90–100 µm. *Ascospores* 80–105 × 2–

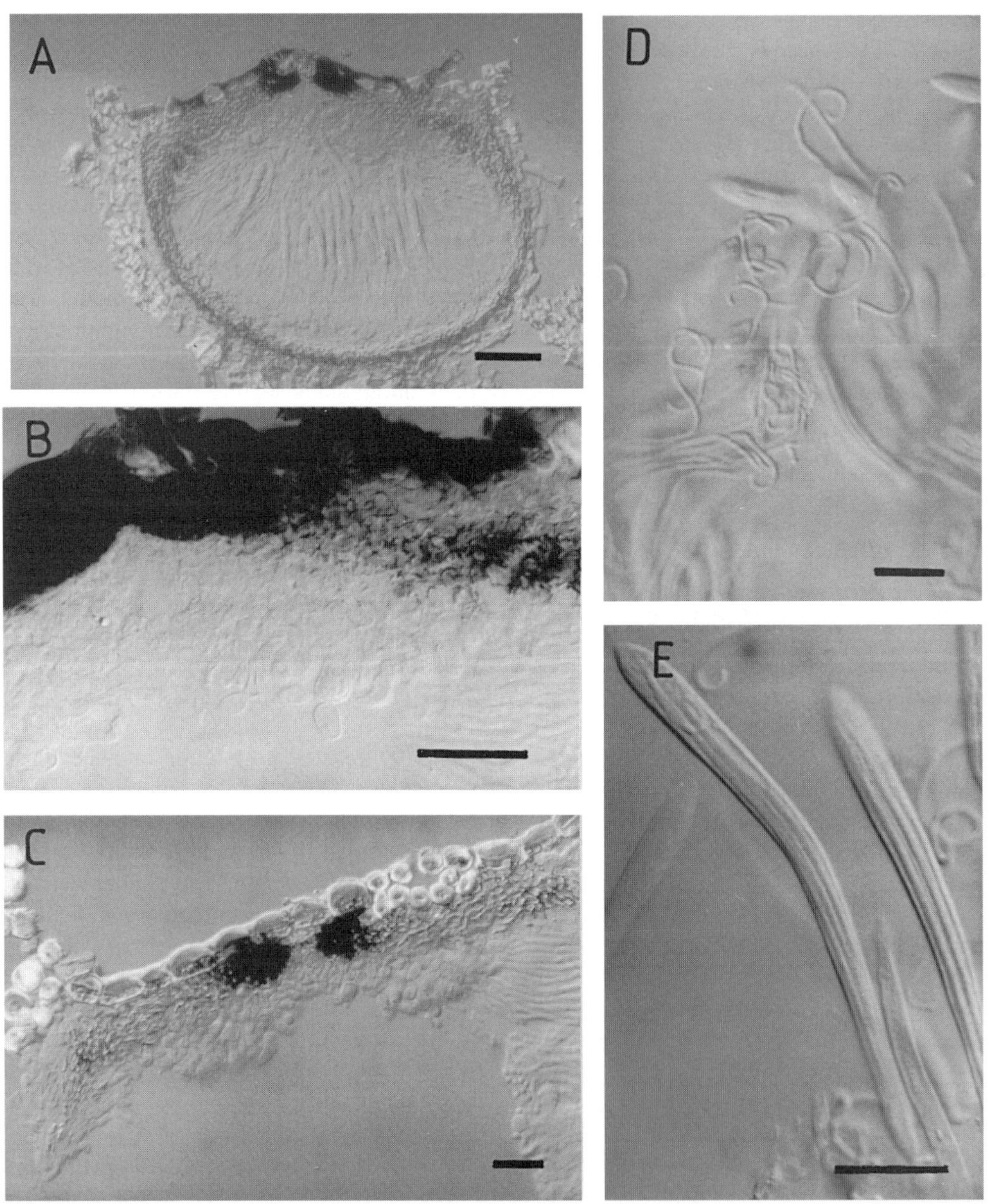

Fig. 106. *Lophodermium sesleriae* (**A**, **B**, **PRM** 756746; **C**, **D**, Libert, *Pl. crypt. Arduenna* no. 73, **K**; **E**, **PRM** 819810). **A**, unopened ascoma in vertical section (bar = 50 μm); **B**, upper wall of unopened ascoma in vertical section; **C**, upper wall of unopened ascoma in vertical section; **D**, paraphyses; **E**, ascus (bars = 20 μm).

2·5 μm, tapering to very fine, ± acute base, spores straight when released, non-septate, apical gelatinous cap globose, 2·5–3 μm diam., basal gelatinous cap tiny.

Typification: Czech Republic: BOHEMIA: Závist, on *Sesleria coerulea*, 13 Oct. 1927, *A. Hilitzer s. num.* (**PRM** 705175!, *lectotypus*, selected here).

Hosts: *Calamagrostis*, *Deschampsia*, *Sesleria* (*Poaceae*).

Distribution: Europe.

Illustrations: Figs 8A, 105, 106.

Notes: *Lophodermium sesleriae* is typical in ascomatal structure of the *alpinum*-group of *Lophodermium* Group A (see p. 33). It is distinguished from other species in this group by its circinate paraphyses, ascus and ascospore size and by having ascospores tapering to a more or less acute base.

Lophodermium sesleriae is macroscopically similar to two other species: *L. grandialpinum* also has the characteristic very long ascospores which taper to an acute base, but differs in having much wider, clavate asci, while *L. petriniae* differs in being associated with large, dark conidiomata and slightly narrower ascospores which do not taper to an acute base. In addition, in some ascomata at least, *L. petriniae* has poorly-developed lip cells as well as the broad layer of periphysoids typical of the *alpinum*-group.

HILITZER (1929) did not choose a type specimen, citing 'In foliis siccis Sesleriae coeruleae in Bohemia centrali dispersa: S. Prokop, Radotin, Zavist, Hostin' in the protologue. Two specimens in **PRM** collected by Hilitzer from *Sesleria* in 1927 have been examined, one from Hostín, the other from Závist. Each collection contains a distinct species of *Lophodermium*: one has small ascomata with macroscopically distinct lip cells and matches *L. sieglingiae* microscopically; the second has larger, elliptical ascomata with acute ends and no obvious lips, matching the description of this species given by HILITZER (1929) and is here regarded as the fungus referred to as *L. sesleriae* by him. The collection from Závist (**PRM** 705175) is in good condition and is here selected as the lectotype.

GRANITI (1952) provided a description for *L. sesleriae* based on Italian material from *Festuca*. He appears to have described a species with well-developed lip cells and ascospores shorter than *L. sesleriae* in the sense accepted here.

Additional specimens examined: Czech Republic: BOHEMIA: Hostín ap. Srbsko, on *Sesleria coerulea*, 22 Jun. 1927, *A. Hilitzer s. num.* (**PRM** 693195); Písek distr.: prope Cerhonice, Obora, on *Deschampsia caespitosa*, 19 Jul. 1973, *M. Svrček s. num.* (**PRM** 734660 as *Lophodermium* sp.); Tábor, Padolí, on *Calamagrostis epigeios*, 19 Aug. 1949, *M. Svrček s. num.* (**PRM** 756746, as *L. apiculatum*); Čimelice, in valle rivi Skalice, on *Calamagrostis epigeios*, 18 Aug. 1977, *M. Svrček s. num.* (**PRM** 819810, as *L. apiculatum*); *s. loc.*, on *Calamagrostis epigeios* (as *Arundis epigei*), *s. dat. nec coll.* (Libert, *Pl. crypt. Arduenna* no. 73; **HBG**; **K**). **Switzerland:** GRAUBÜNDEN: *s. loc.*, on *Sesleria coerulea*, 15 Sep. 1902, *A. Volkor s. num.* (**HBG**, as *L. arundinaceum*); *s. loc.*, on *Sesleria coerulea*, 16 Sep. 1902, *A. Volkor s. num.* (**HBG**, as *L. arundinaceum* var. *culmigenum*).

Lophodermium sieglingiae Hilitzer, *Věd. Spisy čsl. Akad. zeměd.* **3**: 92 (1929).
Lophodermium koeleriae Hilitzer, *Věd. Spisy čsl. Akad. zeměd.* **3**: 93 (1929).

Infected areas on dead leaves, slightly paler than surrounding host tissue, not associated with zone lines, containing ascomata, conidiomata not seen. *Ascomata* small, 0·3–0·5 × 0·2–0·3 mm, elliptical, ends rounded, wall uniformly black, single longitudinal opening slit with pale, yellowish lips. *Ascomatal insertion* subepidermal. *Covering layer* up to 50 μm thick in vertical section, comprising clypeus of dark brown, thick-walled hyphae, 2–3 μm diam., within epidermal cells, and upper wall of ascomatal stroma. *Upper wall* 30–35 μm thick, comprising mostly pale brown,

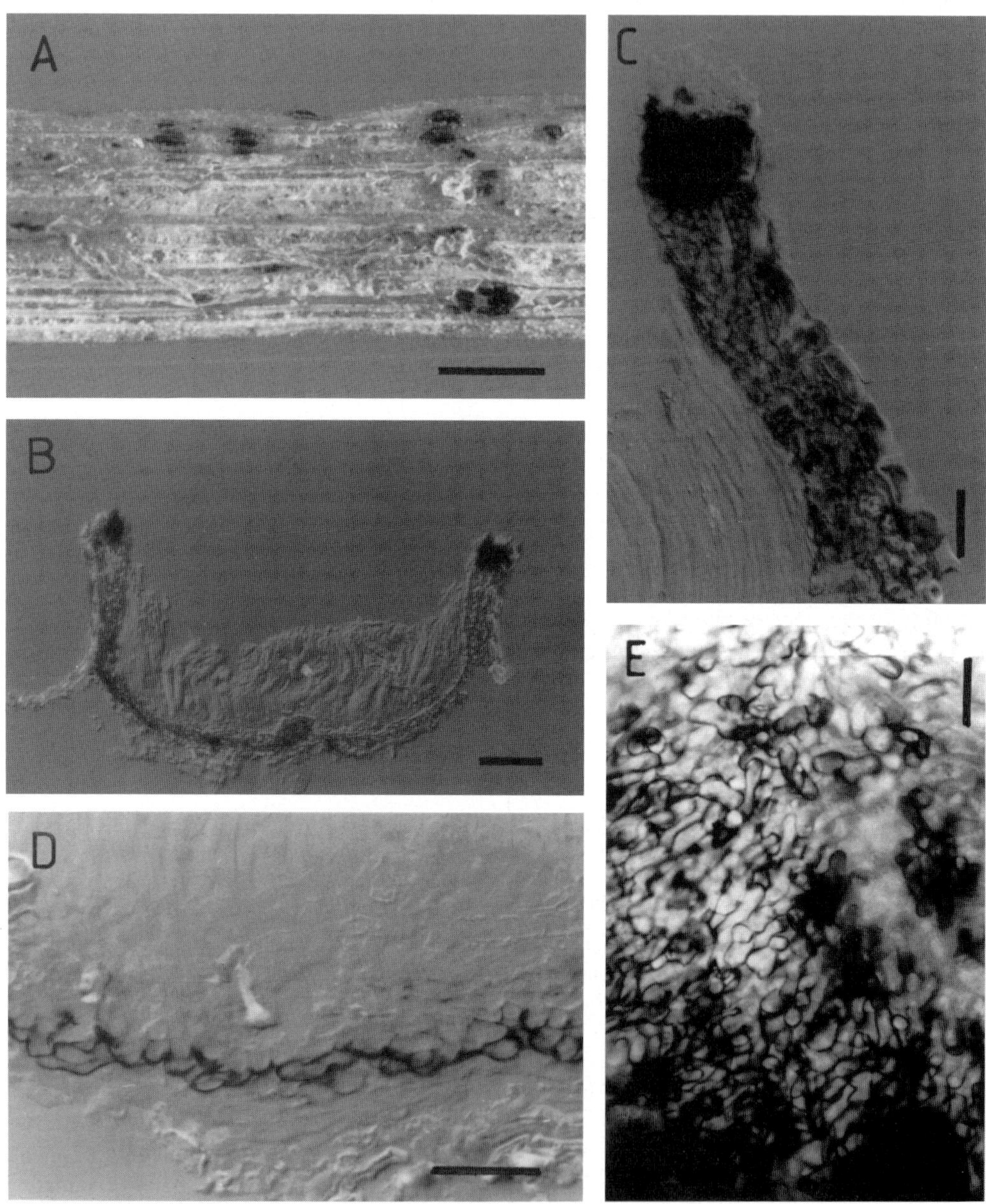

Fig. 107. *Lophodermium sieglingiae* (**A**, **C**, **PRM** 714793; **B**, **PRM** 693134; **D**, **E**, **PRM** 756741). **A**, ascomata (bar = 1 mm); **B**, ascoma in vertical section (bar = 50 µm); **C**, upper wall of ascoma in vertical section; **D**, lower wall of ascoma in vertical section; **E**, lower wall of ascoma in squash mount (bars = 20 µm).

angular to globose cells, 6–10 µm diam., with irregularly encrusted walls and patch of very dark-walled cells adjacent to well-developed lips. *Lower wall* up to 15 µm thick in vertical section, comprising two or three rows of globose cells, 5–8 µm diam., lowermost row with darkened walls or only its outermost wall darkened, inner rows hyaline to very pale brown, in squash mount darkened cells of lower wall irregularly cylindrical, 5–7 µm diam., sometimes with narrower, cylindrical cells scattered across base of stroma. *Paraphyses* (1·5–) 2 µm diam., circinate, sometimes also irregularly swollen to 3–4 µm diam. at apex. *Asci* 70–95 (–110) × 7·5–9 (–9·5) µm, cylindrical to subfusoid, tapering gradually to small rounded apex with undifferentiated wall, 8-spored, short basal stalk developing as matures with spores confined to upper 60–70 µm. *Ascospores* 35–55 × 2 µm, tapering suddenly to very narrow base, apical gelatinous cap 2–3 µm diam. within ascus, basal gelatinous cap 1–1·5 µm diam., gelatinous sheath not seen.

Typification: Czech Republic: BOHEMIA: Habr apud Mnichovice, on *Danthonia* (as *Sieglingia*) *decumbens*, 22 Oct. 1927, *A. Hilitzer s. num.* (**PRM** 693134!, *lectotypus* of *L. sieglingiae*, selected here); *idem loc.*, on *Koeleria cristata*, 22 Oct. 1927, *A. Hilitzer s. num.* (**PRM** 705158!, *lectotypus* of *L. koeleriae*, selected here).

Hosts: *Danthonia, Koeleria, Rytidosperma* (*Poaceae*).

Distribution: Europe, New Zealand.

Illustrations: Figs 10A, 107, 108.

Notes: *Lophodermium sieglingiae* has an ascomatal structure typical of the *actinothyrium*-group of *Lophodermium* Group A (see p. 33). It is characterized by small ascomata, short asci, and ascospores which suddenly narrow to an almost tail-like base. Macroscopically similar species of *Lophodermium* occurring on the same host are distinguished by ascospore shape.

New Zealand collections from *Rytidosperma* differ slightly from the European material in having two or three rows of darkened cells across the base of the ascomatal lower wall. Although EDGAR & CONNOR (1983) and CONNOR & EDGAR (1987) accepted *Rytidosperma* as a distinct genus, they noted JACOBS's (1982) suggestion that it may be congeneric with *Danthonia*, a common host for this fungus in Europe.

HILITZER (1929) did not designate a type specimen for *L. sieglingiae*, citing only locality details: 'In foliis et vaginis Koeleriae cristatae apud Habr prope Mnichovice.' The specimen selected above as lectotype matches these details and was annotated on the herbarium packet as 'holotype'. *Lophodermium koeleriae* is indistinguishable, both macroscopically and microscopically, from *L. sieglingiae*. HILITZER, and TEHON (1935), separated the two species primarily on the basis of ascomatal size, said to be larger for *L. sieglingiae*. The collections examined here, however, showed no differences in ascomatal size.

The existence of *L. sieglingiae* on closely related grasses in Europe and New Zealand suggests that this species probably occurs elsewhere. *Lophodermium danthoniae*, described on *Danthonia* from North America, may be conspecific; however, it is known only from the type collection, and as spores were not seen released from the asci, the characteristic ascospore shape could not be confirmed. Additional specimens on *Danthonia* from North America need to be examined.

Additional specimens examined: Czech Republic: BOHEMIA: Sv. Vít apud Trebon, on *Danthonia* (as *Sieglingia*) *decumbens*, 2 Oct. 1929, *A. Hilitzer s. num.* (**PRM** 705176); in sylvis boreo a vicum Stara Boleslav, on *Danthonia* (as *Sieglingia*) *decumbens*, 22 Feb. 1932, *A. Hilitzer s. num.* (**PRM** 705178); Těptín, Jílové, on *Danthonia decumbens*, 30 Sep. 1954, *M. Svrček s. num.* (**PRM** 756741); Čimelice, in silva Chlum, on *Danthonia* (as *Sieglingia*) *decumbens*, 15 Jul. 1970, *M. Svrček s. num.* (**PRM** 714973). **Finland:** Regio aboënsis: Västanfjärd, Söderrundvik, Grundsund, Norrtorp, on *Danthonia* (as *Sieglingia*) *decumbens*, 1 Sep. 1965, *H. Roivainen s. num.* (**DAOM** 119438, as *L. arundinaceum*). **Germany:** BERLIN: Krug, on *Danthonia* (as *Sieglingia*) *decumbens*, 6 Sep. 1875, *s. coll.* (**HBG**, as '*L. arundinaceum* var.').

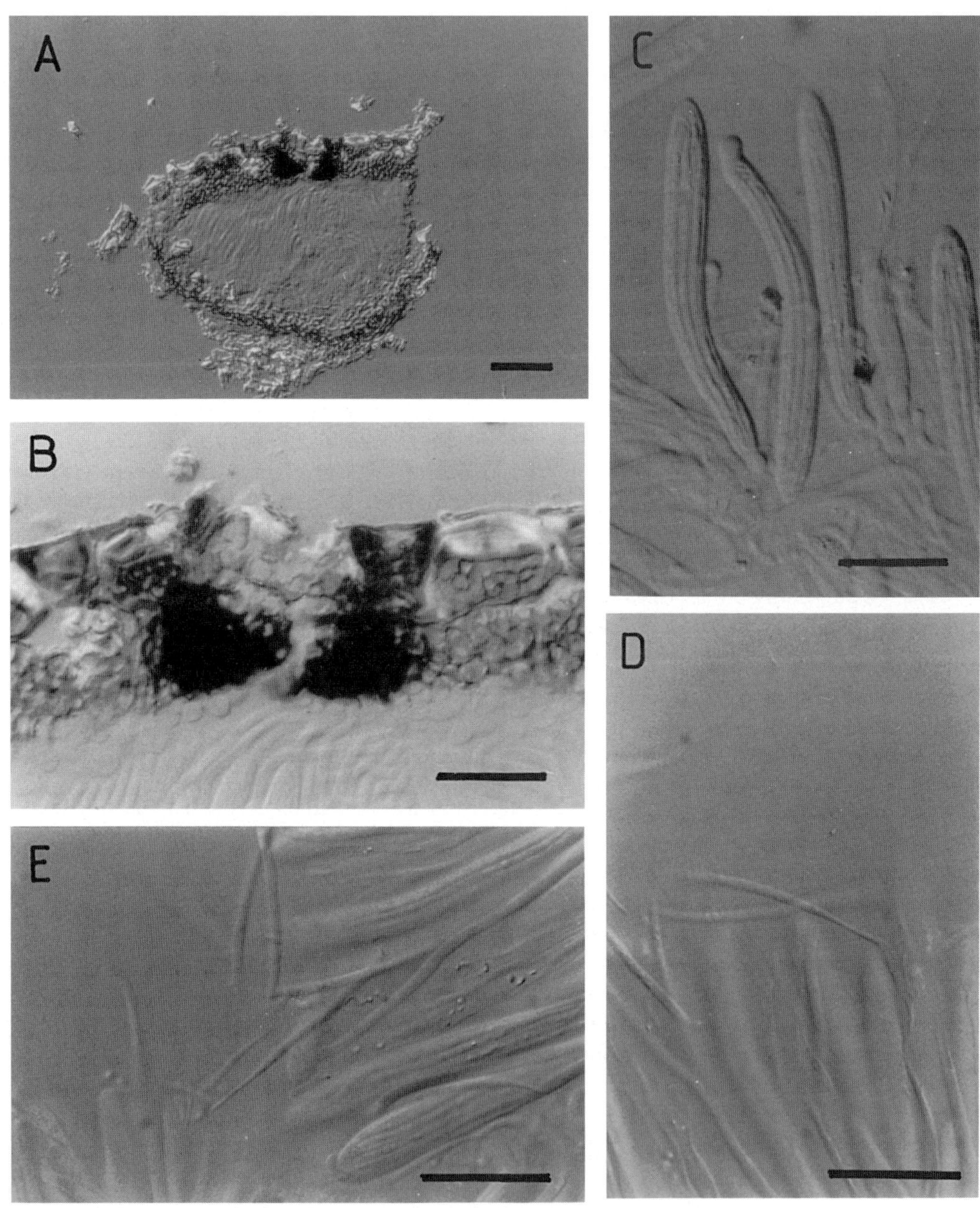

Fig. 108. *Lophodermium sieglingiae* (**A**, **B**, **PRM** 705176; **C**, **PRM** 693134; **D**, **E**, **PRM** 756741). **A**, unopened ascoma in vertical section (bar = 50 μm); **B**, upper wall of unopened ascoma in vertical section; **C**, asci (bar = 20 μm); **D–E**, released ascospores, showing sudden taper to the tail-like base (bars = 20 μm).

Scotland: *s. loc.*, on *Danthonia* (as *Sieglingia*), 12 Aug. 1990, *P.F. Cannon s. num.* (**IMI** 346118). **New Zealand:** TAUPO: Tongariro National Park, Turoa-Mangaturuturu Hut Track, on *Rytidosperma*, 23 Mar. 1984, *P.R. Johnston* R453 (**PDD** 49364); *idem loc.*, on *Rytidosperma*, 23 Mar. 1984, *P.R. Johnston* R498 (**PDD** 49363).

Terriera stevensii P.R. Johnst., **sp. nov.**

Lophodermellina stevensii Tehon, *Illinois biol. Monogr.* **13**(4): 81 (1935), *nom. inval.*, *ICBN* Art. 36.1.

Lophodermium stevensii Terrier, *Beitr. Krypt.-Fl. Schweiz* **9**(2): 32 (1942), *nom. inval.*, *ICBN* Art. 36.1.

A *T. cladophila* ascis 100–125 × 5–6 µm, ascosporis 60–80 × 1·5–2 µm, conidiomatibus differt.

Infected areas on dead leaves, slightly paler than surrounding host tissue, associated with zone lines only when other *Rhytismataceae* present on same leaf, containing scattered ascomata and, in some collections, conidiomata. *Ascomata* 0·5–1·2 × 0·2–0·3 mm, oblong in outline, ends rounded, wall black, with broad, well-developed pale zone along future line of opening, single longitudinal opening slit with black, flat, shelf-like zone along either side. *Ascomatal insertion* subepidermal to intrahypodermal. *Ascomatal structure* typical of *Terriera* (see p. 37). *Paraphyses* 1 µm diam., swelling to 2–4 µm diam. at ± clavate apex, embedded in gel, extending as well-developed epithecium 15–20 µm beyond asci. *Asci* 100–125 × 5–6 µm, cylindrical, broadly rounded at apex with undifferentiated wall, 8-spored, short basal stalk at maturity with spores extending ± to base. *Ascospores* not seen released, 60–80 × 1·5 (–2) µm, gelatinous caps or sheaths not observed. *Conidiomata* 0·2 mm diam., round in outline, lenticular, pale brown to concolorous with surrounding host tissue, with small, black, round, central ostiole. *Conidiogenous cells* cylindrical, 12–15 × 1·5–2 µm, proliferation appears to be both sympodial and percurrent within single fruit-body. *Conidia* 3–4 × 1 µm, oblong-elliptical with rounded ends, non-septate, hyaline.

Typification: USA: HAWAII: Hawaii I, Tantalus, on *Vincentia angustifolia*, 22 Jun. 1921, *F.L. Stevens* 622 (**BISH** 145937, *holotypus*; **BISH** 602324!; **ILL** 7572!, *isotypi*).

Host: *Vincentia* (*Cyperaceae*).

Distribution: Hawaii.

Illustrations: Figs 109, 115.

Notes: *Terriera stevensii* is characterized by ascus size, swollen unbranched paraphyses and the presence of conidiomata. STEVENS (1925) identified the collections as *L. arundinaceum*.

Additional specimens examined: USA: HAWAII: Hawaii I, Wahiawa, on *Vincentia angustifolia*, 3 Jun. 1921, *F.L. Stevens* 246 (**BISH** 145935 & 602323, **ILL** 7573); Hawaii I, Tantalus, on *Vincentia angustifolia*, 22 Jun. 1921, *F.L. Stevens* 652 (**BISH** 145936 & 602325, **ILL** 7569); Oahu, Mt Olympus, on *Vincentia angustifolia*, *F.L. Stevens* 727, 24 Jun. 1921 (**BISH** 145934 & 602326, **ILL**).

Lophodermium tumidulum Sacc., E. Bommer & M. Rousseau, in BOMMER & ROUSSEAU, *Bull. Soc. r. Bot. Belg.* **29**: 240 (1890).

Hypoderma tumidulum (Sacc., E. Bommer & M. Rousseau) Kuntze, *Revis. gen. pl.* **3**(3): 487 (1898).

Lophodermium inflatum Nograsek & Matzer, *Nova Hedwigia* **58**: 23 (1994).

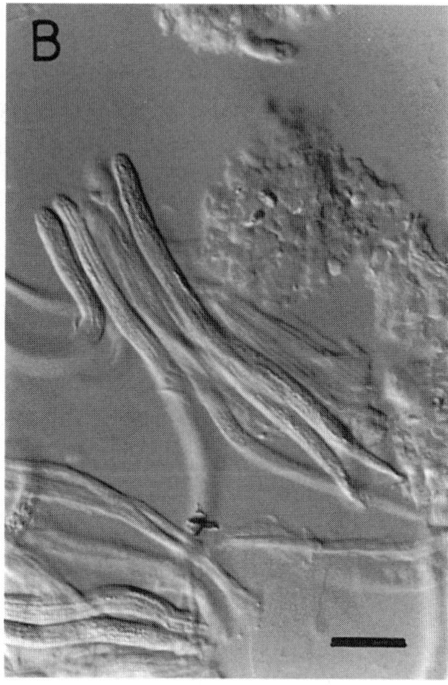

Fig. 109. *Terriera stevensii* (**BISH** 602324). **A**, ascomata (bar = 1 mm); **B**, asci (bar = 20 µm).

Infected areas on dead, partially decomposed leaves, not associated with bleaching of host tissue or zone lines, containing scattered ascomata, conidiomata not seen. *Ascomata* 0·5–0·8 × 0·3–0·4 mm, broadly elliptical to ovate in outline, prominent above leaf surface, appearing ± superficial, wall black, with no paler zone along future line of opening, opening by single slit or several radiate slits, lip cells lacking. *Ascomatal insertion* subcuticular. *Covering layer* in unopened ascomata *c.* 50 µm thick in vertical section, comprising mostly upper wall of ascomatal stroma. *Upper wall* mostly of globose cells, 5–8 µm diam., with walls slightly and irregularly encrusted with brown material, outermost row of cells darker and more thickly encrusted, inner part of wall comprising broad-cylindrical, hyaline, thin-walled, tangled periphysoids, in opened ascomata upper wall up to 100 µm thick in vertical section, similar in structure to that described for unopened ascomata, but cells with encrusted walls forming wider layer. *Lower wall* 10–15 µm thick in vertical section, comprising three or four rows of angular to globose cells, in squash mount darkened cells of lower wall short-cylindrical to irregularly shaped, 3–4 µm diam., arranged in three or four rows across base of ascoma. *Paraphyses* 2·5 µm diam., circinate and coiling at apex, some with short branches. *Asci* 140–180 × 11–14 µm, clavate, tapering to subtruncate apex with undifferentiated wall, 8-spored, basal stalk with spores confined to upper 100–120 µm. *Ascospores* 60–90 × 2–3 µm, straight when released, tapering slightly to base, non-septate, with small, globose, apical gelatinous cap.

Typification: Belgium: Westmalle, on *Scirpus caespitosus*, Sep. 1886, *M. Rousseau* K7215 (**BR** 9215-00!, holotype of *L. tumidulum*). **Austria:** STEIERMARK: Koralpe, Grillitschhütte, on *Sesleria varia*, 1 Jul. 1986, *M. Matzer* 1304 (**GZU**!, holotype of *L. inflatum*). **Sweden:** Torne Lappmark, Jukkasjärvi, on *Sesleria varia*, 14 Jul. 1986, *A. Nograsek s. num.* (**GZU**!, paratype of

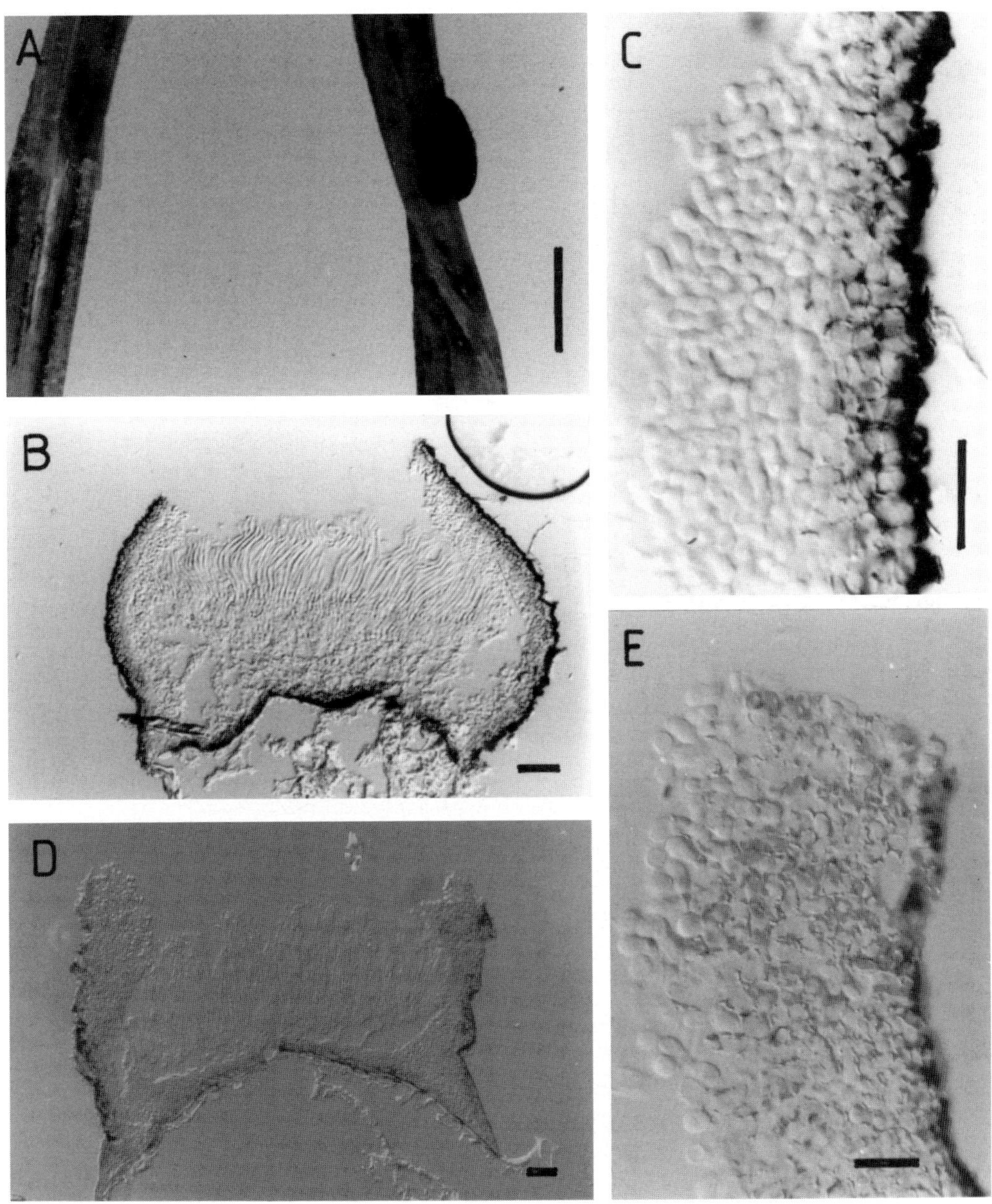

Fig. 110. *Lophodermium tumidulum* (**A–C**, **BR** 9215-00; **D**, **E**, *Müller*, 7 Aug. 1972, **ZT**). **A**, ascomata (bar = 1 mm); **B**, unopened ascoma in vertical section, wall of ascoma broken during sectioning (bar = 50 μm); **C**, upper wall of unopened ascoma (bar = 20 μm); **D**, opened ascoma in vertical section (bar = 50 μm); **E**, upper wall of opened (bar = 20 μm).

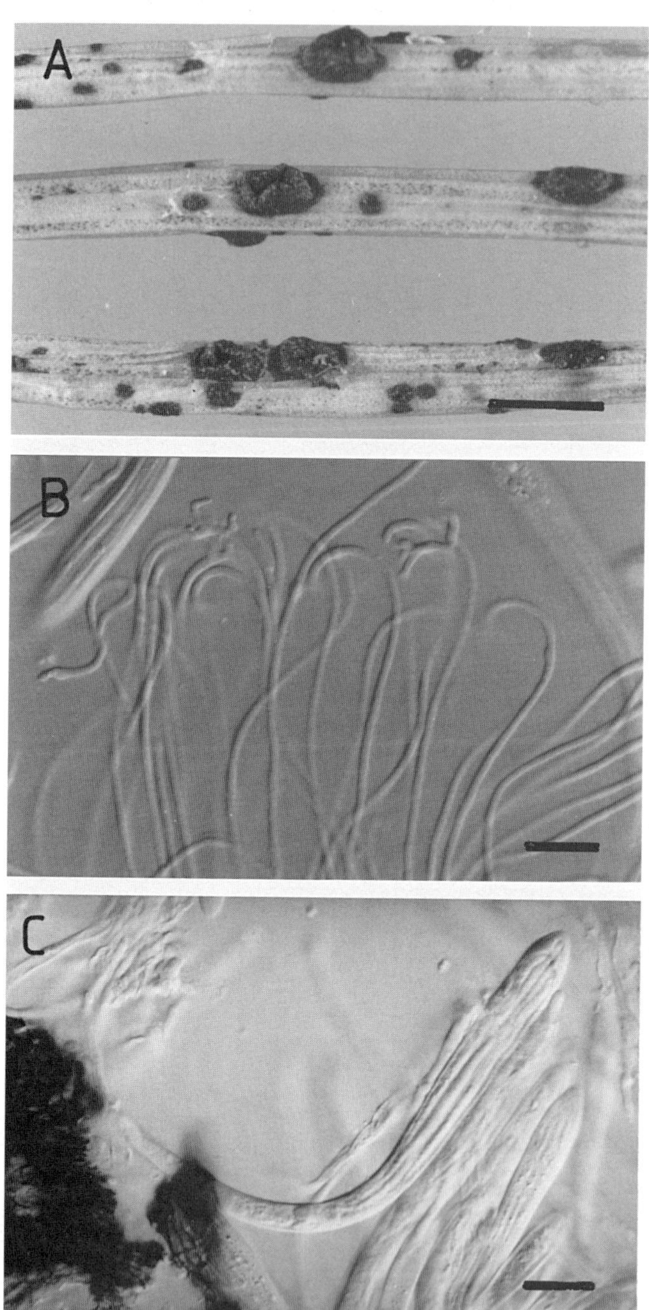

Fig. 111. *Lophodermium tumidulum* (**A**, **B**, *Dennis*, 9 Sep. 1970, **K**; **C**, *Müller*, 7 Aug. 1972, **ZT**). **A**, ascomata (bar = 1 mm); **B**, paraphyses; **C**, ascus (bars = 20 μm).

L. inflatum).

Hosts: *Scirpus* (*Cyperaceae*); *Juncus*, *Luzula* (*Juncaceae*); *Sesleria* (*Poaceae*).**Distribution:** Europe.

Illustrations: Figs 110, 111.

Notes: *Lophodermium tumidulum* is characteristic of *Lophodermium* Group C and is distinguished from the other species in this group, *L.* cf. *juncinum*, primarily by the depth of insertion of the ascomata. The type material of *L. tumidulum* is immature, the description of the hymenial elements being based on the other collections cited. These collections have an ascomatal structure and macroscopic appearance matching the type and, in size of asci and ascospores and shape of the paraphyses, they match the original description.

Lophodermium inflatum has been placed in synonymy with *L. tumidulum* despite the hymenium of *L. inflatum* not having been examined. Both the holotype and paratype collections are small and most ascomata are unopened. The description provided by NOGRASEK & MATZER (1994) gave ascus and ascospore dimensions more or less matching *L. tumidulum*, although the characteristic circinate and branched paraphyses were neither described nor illustrated there. Ascomatal structure of the two species is very similar (**Fig. 112**), as are the dimensions of the hymenial elements. Macroscopically, the ascomata of the type of *L. inflatum* differ slightly from the other specimens in being more highly arched above the leaf surface. However, this difference is not considered significant at the species level.

Lophodermium tumidulum is very similar to *Coccomyces tumidus* both macroscopically and in vertical section (see **Fig. 113**). The two species are almost indistinguishable in vertical section, although the lower wall of the ascomata of *C. tumidus* comprises narrow-cylindrical cells forming *textura intricata*. Microscopically, they both have circinate paraphyses and large clavate asci, but the ascospores of *C. tumidus* are shorter, 35–45 × 3·5–4 μm.

Additional specimens examined: ***Lophodermium tumidulum*:** **Austria:** Vorarlberg, Rätikon, Schruns, on *Carex*, 27 Jul. 1986, *M. & H. Mayrhofer* 7058 (**GZU**). **Switzerland:** GRAUBÜNDEN: Klosters ob Kuenisboden, on *Luzula sieberi*, 7 Aug. 1972, *E. Müller s. num.* (**ZT**); WALLIS: Aletschreservat bei Brig, on *Juncus trifidus*, 12 Sep. 1968, *J. Harr s. num.* (**ZT**); Aletschwald, Brig, on *Juncus trifidus*, 11 Sep. 1968, *J. Poelt* 13478 (**GZU**, as *L. juncinum*); Aletschreservat, Gersternwald, on *Luzula silvatica*, 22 Sep. 1965, *E. Müller & F. Casagrande s. num.* (**ZT**, *p.p.*, that represented by the unopened ascomata); Reidsalp, Bel Alp path, on *Scirpus caespitosus*, 9 Sep. 1970, *R.W.G. Dennis s. num.* (**K**).

***Coccomyces tumidus*:** **Germany:** BAVARIA: *s. loc.*, on *Acer pseudoplatanus*, 1 Sep. 1990, *D.W. Minter & P.R. Johnston s. num.* (**IMI** 341776, as *C. coronatus*); POTSDAM: Brandenburg, Schlachtensee, on *Quercus pedunculata*, 4 Oct. 1916, *H. Sydow s. num.* (Sydow, *Mycotheca germanica* no. 1611; **PDD** 42352, as *C. coronatus*). **USA:** MASSACHUSETTS: Concord, Estabrook Woods, on *Quercus rubra*, 27 Aug. 1978, *D.H. Pfister s. num.* (**FH**); Concord, Estabrook Woods, on *Quercus rubra*, 17 Sep. 1978, *D.H. Pfister s. num.* (**FH**).

Lophodermium typhinum (Fr.: Fr.) Lambotte, *Fl. mycol. belge* **2**: 452 (1880).
Hysterium typhinum Fr., *Syst. mycol.* **2**: 590 (1823).
Hypoderma typhinum (Fr.: Fr.) Kuntze, *Revis. gen. pl.* **3**(3): 487 (1898).

Infected areas on dead leaves, slightly paler than surrounding host tissue, not associated with zone lines, containing scattered ascomata, conidiomata not seen. *Ascomata* 0·4–0·7 × 0·3–0·4 mm, broadly elliptical in outline, ends rounded, wall black, single longitudinal opening slit, with well-developed whitish to yellowish lip cells. *Ascomatal insertion* subepidermal. *Covering layer* 50–65 μm thick in vertical section, comprising clypeus of dark brown, thick-walled hyphae, 3–4 μm diam., within ± intact epidermal cells and upper wall of ascomatal stroma. *Upper wall* comprising mostly angular cells, 7–10 μm diam., with walls irregularly encrusted with dark brown material

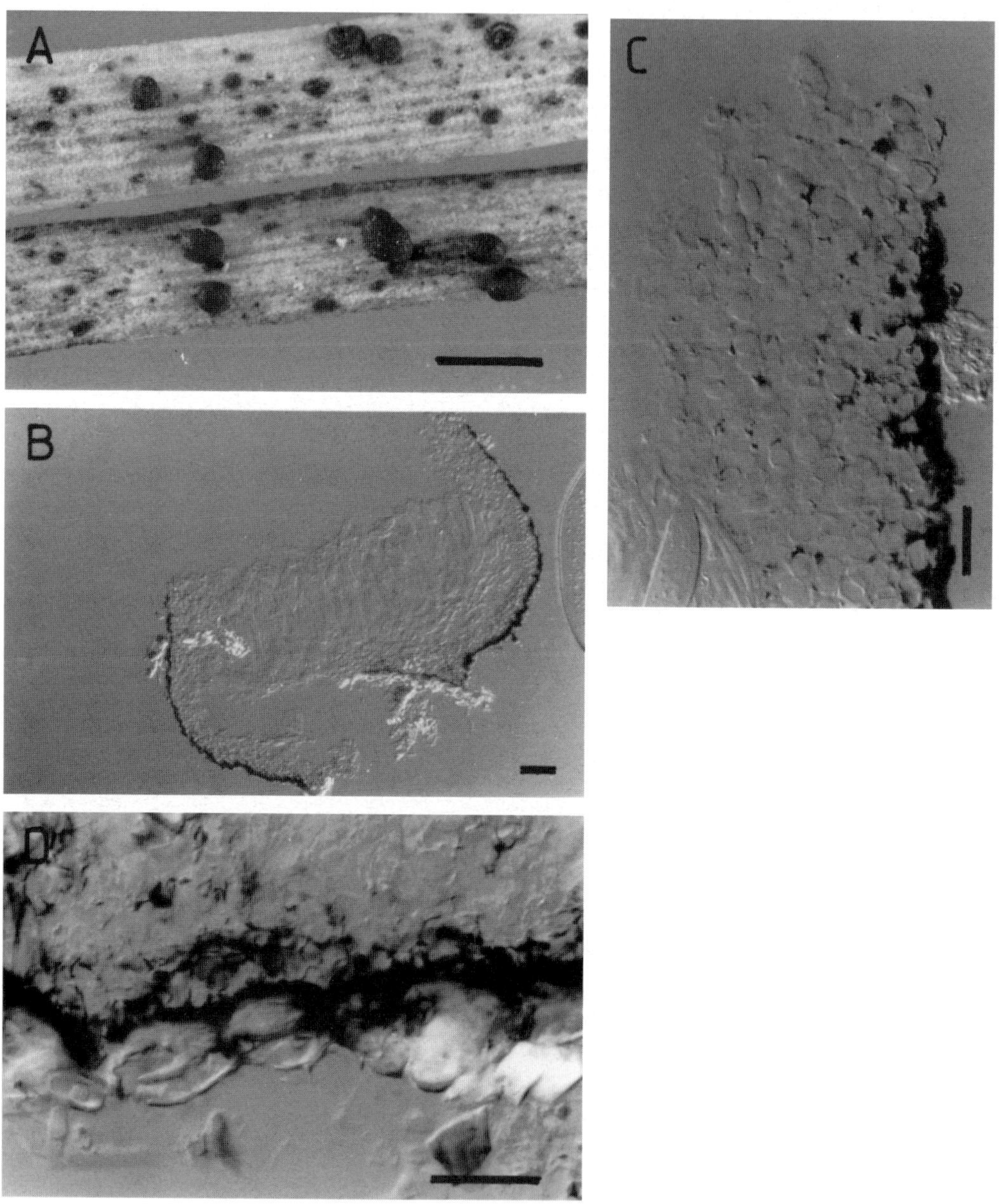

Fig. 112. *Lophodermium inflatum* (*Nograsek*, 14 Jul. 1986, **GZU**). **A**, ascomata (bar = 1 mm); **B**, ascoma in vertical section (bar = 50 μm); **C**, upper wall of ascoma in vertical section; **D**, lower wall of ascoma in vertical section (bars = 20 μm).

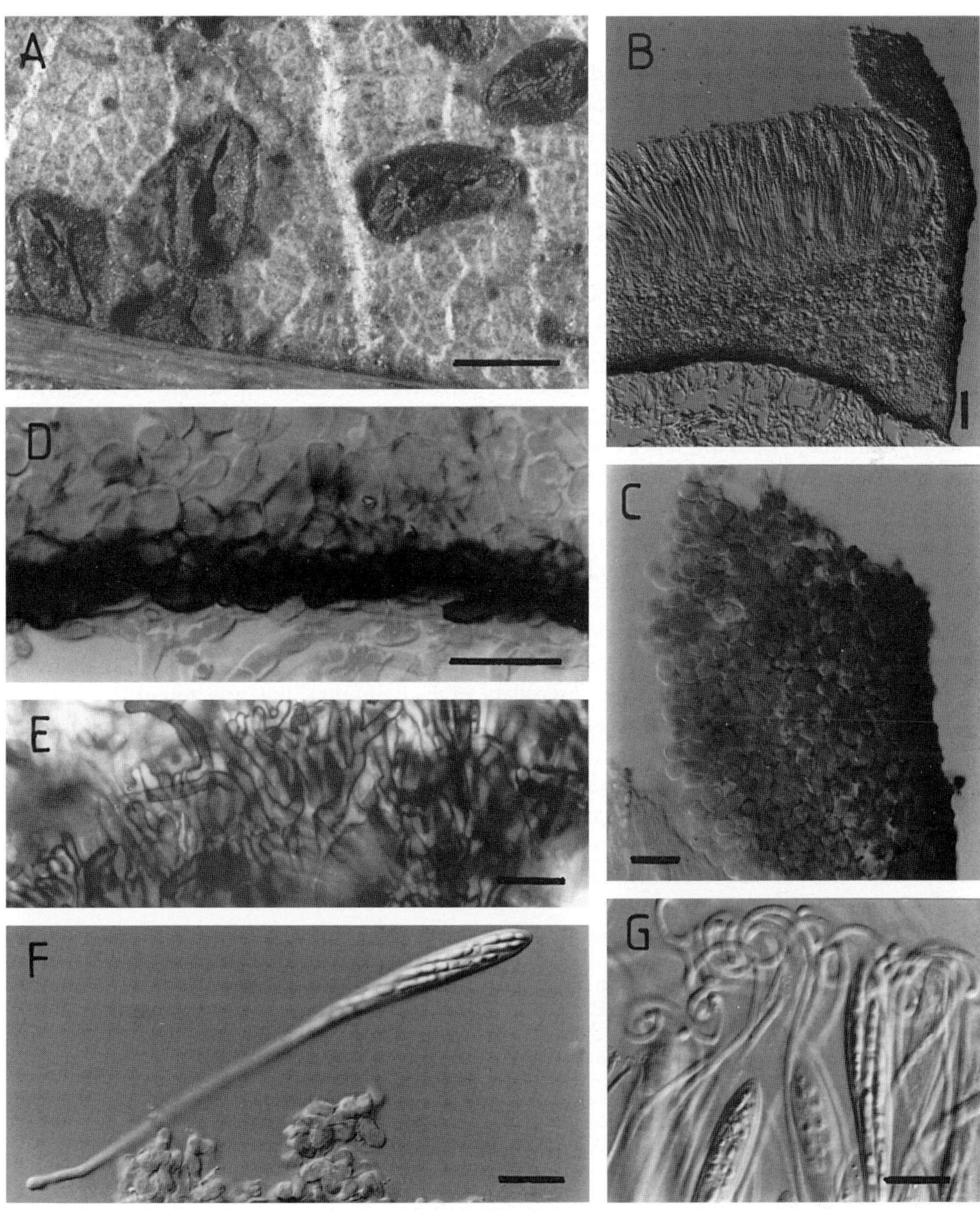

Fig. 113. *Coccomyces tumidus* (**A–D**, **IMI** 341776; **E–G**, *Pfister*, 27 Aug. 1978, **FH**). **A**, ascomata, hysterothecial form (bar = 1 mm); **B**, ascoma in vertical section, one side only (bar = 50 µm); **C**, upper wall of ascoma in vertical section; **D**, lower wall of ascoma in vertical section; **E**, lower wall of ascoma in squash mount; **F**, ascus; **G**, paraphyses (bars = 20 µm).

and with group of dark, thickly-encrusted cells adjacent to well-developed lips, groups of host fibre cells sometimes embedded within upper wall. *Lower wall* 10–20 µm thick in vertical section, comprising two to four rows of angular to globose cells, 4–6 µm diam., outer one or two rows with dark brown walls, inner rows hyaline and thin-walled, in squash mount darkened cells of lower wall angular to globose, 6–8 µm diam., arranged in one or two layers across base of ascoma. *Paraphyses* 2 µm diam., circinate and coiling at apex. *Asci* 85–110 × 8–9·5 µm, subclavate to subfusoid, tapering to small, subtruncate apex with undifferentiated wall, 8-spored, short basal stalk at maturity with spores confined to upper 65–75 µm. *Ascospores* 45–60 (–65) × 1·5–2 µm, tapering slightly to base, apical gelatinous cap globose, 2·5–3 µm diam., basal cap cylindrical, 2·5 × 1·5 µm, gelatinous sheath 4–5 µm diam.

Typification: No specimen cited by FRIES.

Host: *Typha* (*Typhaceae*).

Distribution: Europe, North America.

Illustrations: Figs 4B, 9A, B, 114, 116.

Notes: No type specimen was cited by FRIES and there is no material of *Lophodermium* in **UPS** collected from *Typha* by him. A neotype is, therefore, required, ideally selected from material collected in the vicinity of Lund, none of which was available for study.

Lophodermium typhinum has an ascomatal structure typical of the *actinothyrium*-group of *Lophodermium* Group A and is very similar morphologically to *L. actinothyrium*. The two differ macroscopically, *L. actinothyrium* typically having ascomata rounded at the ends with a small apiculus. These species may be conspecific, but are here retained as distinct for historical and practical reasons: combining them would require adoption of the name *L. typhinum* for one of the most common European *Poaceae*-inhabiting species of *Lophodermium*.

One collection examined from *Typha*, labelled as *L. typhinum* (**Sweden:** UPPLAND: Värmdö, Gustafsberg, on *Typha*, Aug. 1890, *H. Kugelberg s. num.*; **S**!), comprised dark fruiting bodies of a similar shape and size to unopened ascomata of *L. typhinum*, but was found to contain a *Leptostroma*-like anamorph. Conidiogenous cells (with sympodial proliferation) and conidia were very similar to those found in rhytismataceous anamorphs but, in comparison with these anamorphs, the conidiomata are unusually large and a layer of conidiogenous cells line the upper wall as well as the lower. No ascomata of *Lophodermium* were seen in this collection and no conidiomata were noted in any collection of *L. typhinum* examined.

Lophodermium culmigenum also occurs on *Typha*.

Specimens examined: Canada: ONTARIO: York Co.: Nashville, on *Typha latifolia*, 21 Jun. 1953, *R.F. Cain s. num.* (**TRTC** 24514, **P**); York Co.: Marsh River, on *Typha*, 10 Jun. 1936, *H.S. Jackson s. num.* (**DAOM** 86505, **TRTC** 12805). **England:** NORFOLK: Kings Lynn, on *Typha*, Feb. 1870, *Plowright s. num.* (**S**). SUFFOLK: *s. loc.*, on *Typha*, 17 Apr. 1980, *M.B. & J.P. Ellis s. num.* (**IMI** 252976). KENT: *s. loc.*, on *Typha*, 16 May 1987, *D.W. Minter s. num.* (**IMI** 316218). **France:** *s. loc.*, on *Typha latifolia*, Jul. 1899, *F. Fautrey* 2941 (*Herbier Cryptogamie de la Côte d'Or*, **P**). **USA:** NEW YORK: Orleans Co.: Lyndonville, on *Typha latifolia*, 13 Jun. 1890, *C.E. Fairman* 862 (**CUP** 15-131).

Lophodermium unciniae P.R. Johnst., *N.Z. Jl Bot.* **27**: 273 (1989).

Infected areas on dead leaves, paler than surrounding host tissue, often associated with broad, brown zone lines, containing ascomata and conidiomata. *Ascomata* 0·7–1·2 × 0·2–0·3 mm, oblong in outline, ends broadly rounded, wall black, unopened ascomata with distinct paler zone along the future line of opening, opening by longitudinal slit, often with short side-slits near ends of ascomata, hymenium covered when ascomata dry, upper wall appearing to have collapsed inwards.

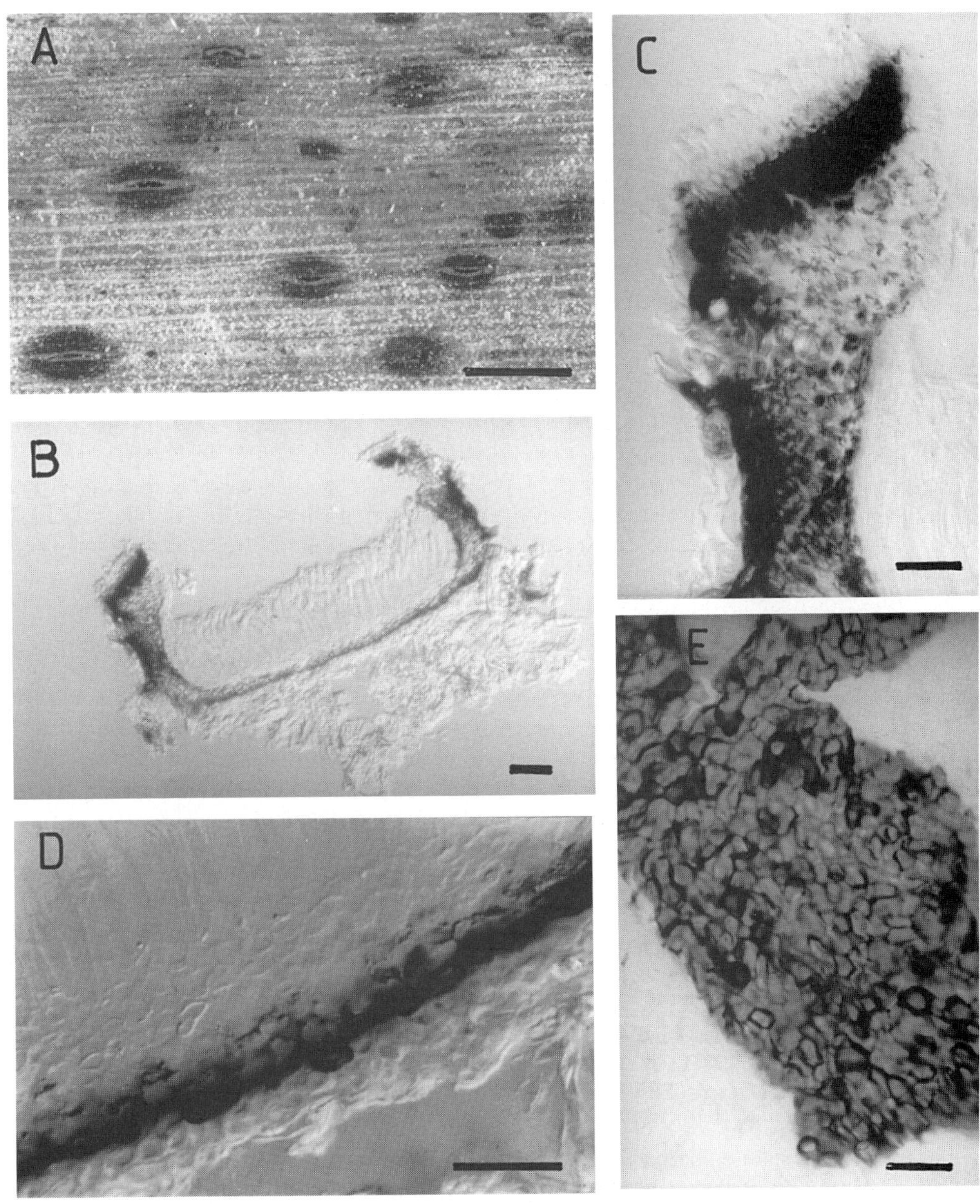

Fig. 114. *Lophodermium typhinum* (**A**, **E**, **TRTC** 24514; **B–D**, **DAOM** 86505). **A**, ascomata (bar = 1 mm); **B**, ascoma in vertical section (bar = 50 μm); **C**, upper wall of ascoma in vertical section; **D**, lower wall of ascoma in vertical section; **E**, lower wall of ascoma, squash mount (bars = 20 μm).

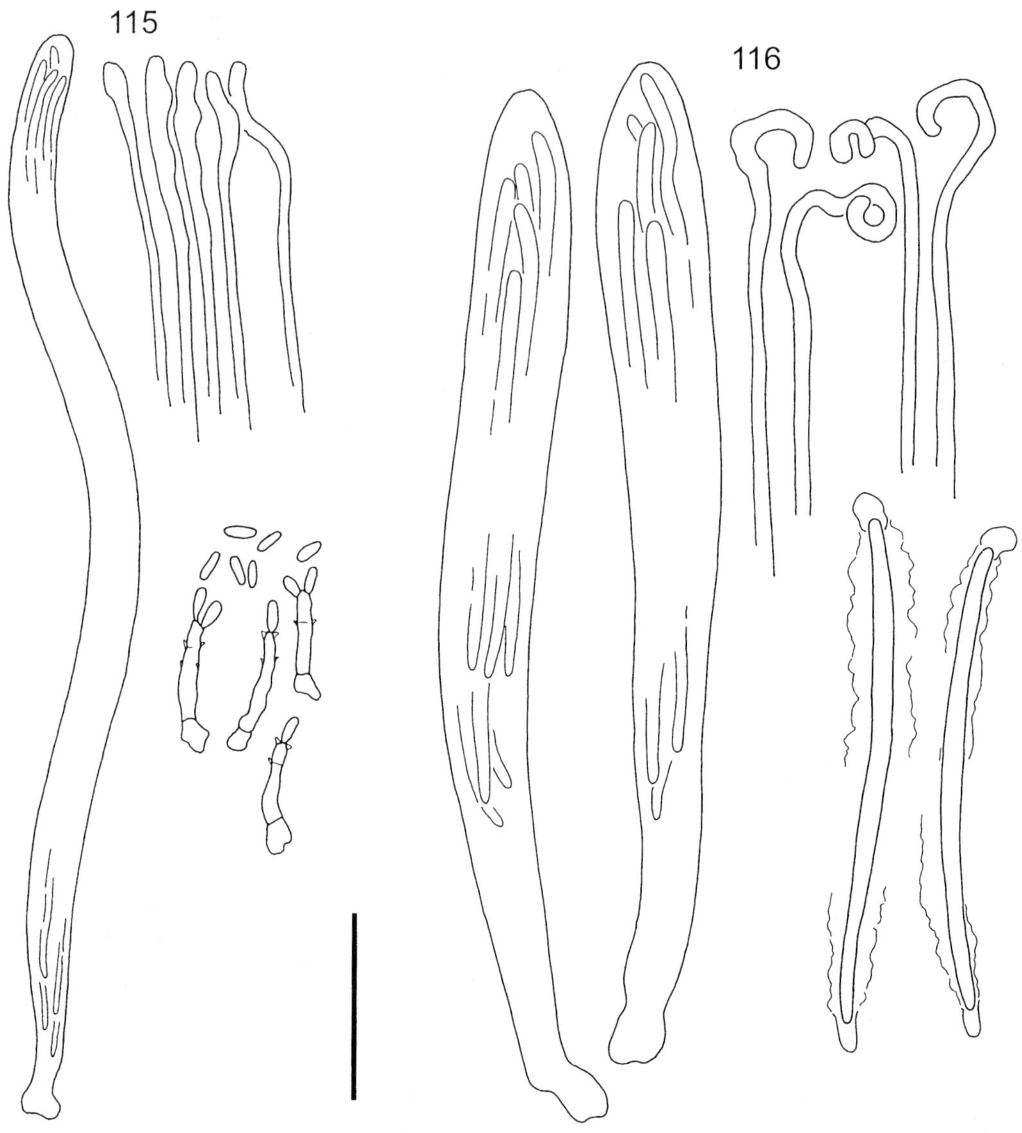

Figs 115–116. 115. *Terriera stevensii.* Ascus and apex of paraphyses (**BISH** 602324) and conidiogenous cells and conidia (*Stevens* 727, **ILL**). **116.** *Lophodermium typhinum.* Asci and apex of paraphyses (**CUP** 15-131) and ascospores in water (**TRTC** 24514) (bar = 20 µm).

Ascomatal insertion intra-epidermal. *Covering layer c.* 20 µm thick in vertical section, comprising host cuticle, broken epidermal cells, clypeus of narrow layer of dark-walled hyphae and upper wall of ascomatal stroma. *Upper wall* comprising several layers of angular, dark-walled cells, 4–6 µm diam., small group of lip cells along edge of opening slit, lips encrusted with dark brown material. *Lower wall* 20–25 µm thick in vertical section, comprising several rows of angular to globose cells with dark walls, lowermost two or three rows with very thick walls, in squash mount darkened cells of lower wall long cylindrical, arranged in several rows across base of ascoma. *Paraphyses* 1·5 µm diam., swelling to 2·5–4·5 µm diam. at apex where usually branched once or twice, embedded in gel, forming regular palisade 20–30 µm above asci. *Asci* 125–160 × 4·5–6·5 µm, cylindrical, tapering suddenly to rounded to subtruncate apex with thickened wall possessing broad central pore, 8-spored, empty basal stalk at maturity with spores confined to upper 120–130 µm. *Ascospores* 85–115 × 1·5 µm, filiform, tapering slightly to ends, bending to 120° *c.* one-third of distance from base, with small gelatinous, barb-like appendage at outer angle. *Conidiomata* 0·2 mm diam., round in outline, pale brown or concolorous with surrounding host tissue, with darker line around outside edge, intra-epidermal, upper wall lacking, lower wall comprising two or three rows of angular cells with pale brown, thin walls. *Conidiogenous cells* forming palisade-like layer across lower wall, cylindrical, 6·5–9 × 2·5–3·5 µm, proliferation sympodial. *Conidia* 3·5–4 × 1 µm, cylindrical, hyaline, non-septate.

Typification: New Zealand: COROMANDEL: Moehau, Te Hope Stream Track, 200 m, on *Uncinia*, 28 Aug. 1984, *P.R. Johnston* R538 (**PDD** 46156!, holotype).

Hosts: *Uncinia*, *Carex* (*Cyperaceae*); *Juncus* (*Juncaceae*).

Distribution: New Zealand.

Illustrations: Figs 13, 14A, B, 117.

Notes: In ascomatal structure *L. unciniae* is typical of Group E (see p. 43). It is very similar to the other *Cyperaceae*-inhabiting species in this group from New Zealand, *L. hauturuanum*. The two species are distinguished by host, macroscopic appearance of the ascomata, small differences in paraphysis shape and in the arrangement of the ascospores following release (JOHNSTON, 1994). See also Notes under *L. hauturuanum*.

Additional specimens examined: New Zealand: AUCKLAND: Waitakere Ranges, Scenic Drive, on *Uncinia*, 8 Oct. 1984, *P.R. Johnston* R584 (**PDD** 46155); Waitakere Ranges, Destruction Gully Track, on *Uncinia uncinata*, 14 Nov. 1983, *P.R. Johnston* 378A (**PDD** 49359); BULLER: Paparoa Range, Croesus Track, between road and Garden Gully Hut, on *Uncinia uncinata*, 4 May 1985, *P.R. Johnston* R684 (**PDD** 49024); Buller Gorge, Scenic Loop Rd, nr Murchison, on *Uncinia*, 17 Apr. 1983, *P.R. Johnston* R180 (**PDD** 49354); Lewis Pass, nr Springs Junction, on *Uncinia uncinata*, 8 Feb. 1989, *P.R. Johnston s. num.* (**PDD** 57138); COROMANDEL: Mt Kaitarakihi Track, on *Uncinia*, 28 Apr. 1988, *P.R. Johnston s. num.* (**PDD** 56852); Little Barrier I, Shag Track, on *Uncinia*, 15 Jun. 1984, *P.R. Johnston* LB9 (**PDD** 49357); NORTH CANTERBURY: Mt Thomas Forest, Richardson Track, on *Uncinia*, 21 Mar. 1991, *P.R. Johnston s. num.* (**PDD** 58593); SOUTHLAND: Bluff Hill, nr Glory Track, on *Uncinia*, 3 May 1984, *P.R. Johnston* R523 (**PDD** 49355); TAUPO: Kaimanawa Forest Park, on *Uncinia*, 6 May 1987, *P.R. Johnston s. num.* (**PDD** 54534); WELLINGTON: Tararua Ranges, nr Dundas Hut, Hut Ridge Track, on *Uncinia*, 7 Feb. 1985, *P.R. Johnston* R696 (**PDD** 49356); WESTLAND: Fox Glacier, Lake Gault Track, on *Uncinia*, 10 Apr. 1983, *P.R. Johnston* R173 (**PDD** 49358).

Lophodermium vrieseae Rehm, *Hedwigia* **39**: 212 (1900).

Infected areas on dead leaves, paler than surrounding host tissue, associated with incomplete, narrow zone lines, containing ascomata and structures with appearance of conidiomata of *Rhytismataceae*. *Ascomata* 0·3–0·5 × 0·2 mm, oblong-elliptical in outline, wall dark grey to black, with no paler zone along future line of opening, single longitudinal opening slit, lacking obvious lips macroscopically, few ascomata triangular in outline, with tri-radiate opening slit. *Ascomatal insertion* subepidermal. *Covering layer* 25–30 µm thick in vertical section, comprising clypeus within epidermal cells and upper wall of ascomatal stroma. *Upper wall* in unopened ascomata comprising mostly very dark brown, very thick-walled, angular cells, but with group of slightly paler, thinner-walled cells extending through wall near centre of ascoma, along future line of opening, lined with poorly-developed layer of periphysoids, in opened ascomata upper wall

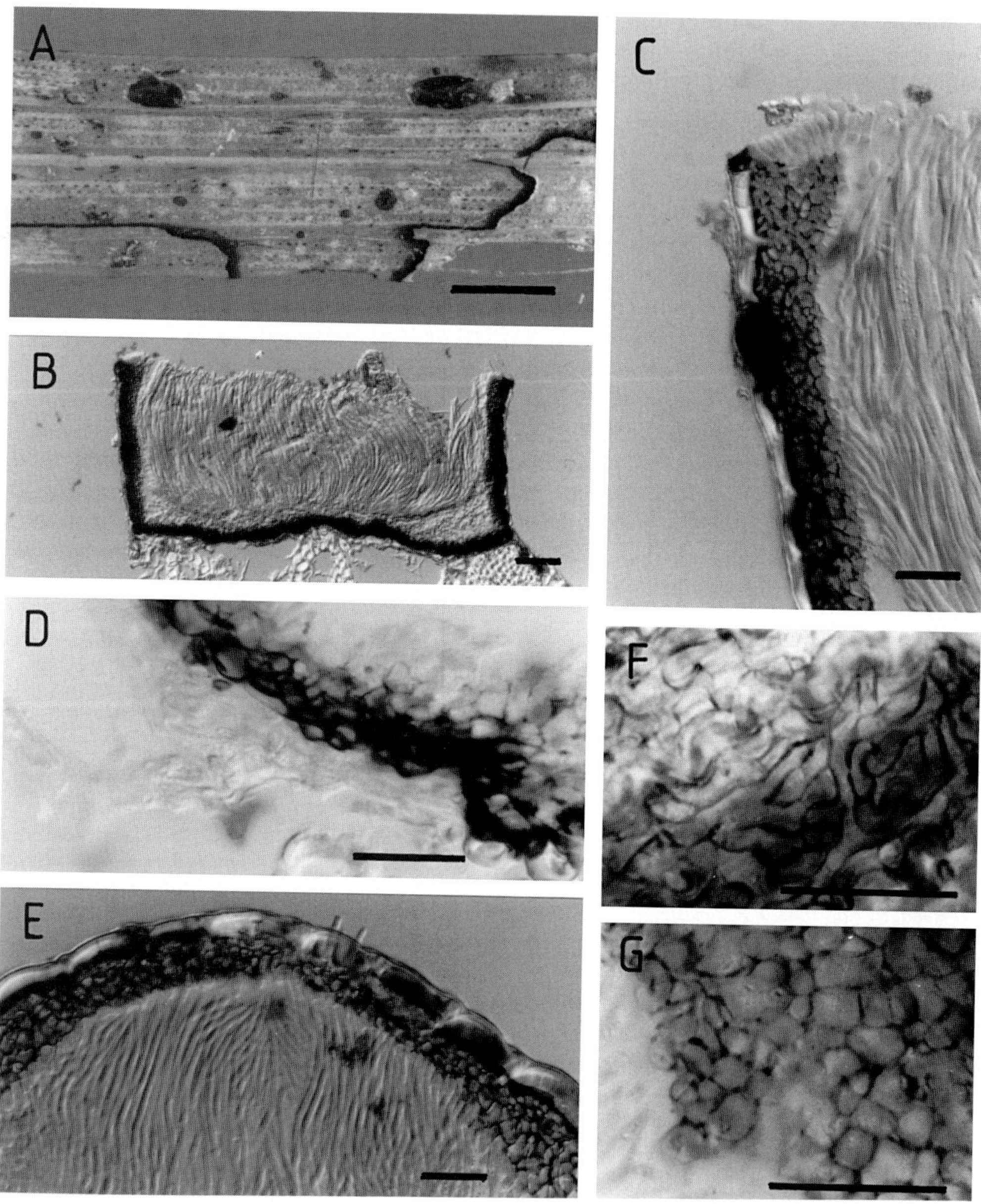

Fig. 117. *Lophodermium unciniae* (**PDD** 54534). **A**, ascomata (bar = 1 mm); **B**, ascoma in vertical section (bar = 50 µm); **C**, upper wall of ascoma in vertical section; **D**, lower wall of ascoma in vertical section; **E**, upper wall of unopened ascoma in vertical section, with slightly paler cells along future line of opening; **F**, clypeus adjacent to split epidermal cells, horizontal section; **G**, cells within upper wall of ascoma, horizontal section (bars = 20 µm).

comprising very dark, very thick-walled, angular cells, 5–7 μm diam., with few pale, thin-walled cells sometimes present along edge of opening slit, but lacking well-differentiated lip cells. *Lower wall* 10–12 μm thick in vertical section, comprising two or three rows of brown to dark brown, thick-walled, angular cells, 4–8 μm diam. *Paraphyses* 2 μm diam., undifferentiated or slightly swollen to 2·5–4 μm diam. at ± clavate apex, forming regular palisade-like epithecium 10 μm beyond asci. *Asci* 95–120 × 14–16·5 μm, subfusoid to subsaccate, tapering gradually to subtruncate apex with slightly thinner wall, 8-spored, spores extending to base of ascus, with no basal foot. *Ascospores* not seen released, 70–85 × 3–4 μm, tapering to narrow base, non-septate, gelatinous sheath not seen. *Conidiomata*-like structures associated with ascomata, round in outline, 0·1–0·2 mm diam., wall very pale brown with dark brown margin, immersed in host leaf, lenticular. *Conidia* and *conidiogenous cells* not seen.

Brazil: RIO DE JANEIRO: Serra de Itatiaia, on *Tillandsia* (as *Vriesea*), Jan. 1896, *E. Ule* 2139 (**HBG**!, holotype).

Host: *Tillandsia* (*Bromeliaceae*).

Distribution: Brazil.

Illustrations: See JOHNSTON 1993, figs 1, 2.

Notes: The relationship of *L. vrieseae* to other monocotyledon-inhabiting species of *Lophodermium* remains uncertain. The subepidermal ascomata, the very simple upper wall with no differentiated cells associated with the opening slit and the poorly-developed excipulum resemble the condition in two other South American species, *Lophodermium subtropicale* Speg. and *L. platyplacum* (Berk. & Curtis) Sacc. (JOHNSTON, 1989*a*). These can be distinguished from *L. vrieseae* by ascus shape and size, ascospore size and shape of the apex of the paraphyses.

Two species of *Rhytismataceae*, distinguishable both macroscopically and microscopically, are present on the piece of leaf which constitutes the type specimen of *L. vrieseae*. The second species is a *Hypoderma* with bifusiform ascospores, similar in many ways to *H. carinatum* P.R. Johnst. and *H. cookianum* P.R. Johnst. (JOHNSTON, 1990*d*).

REFERENCES

Arx, J.A. von & Müller, E. (1975). A Re-evaluation of the Bitunicate Ascomycetes With Keys to Families and Genera. *Studies in Mycology* **9**: 159 pp.

Baral, H.O. (1992). Vital versus herbarium taxonomy: morphological differences between living and dead cells of ascomycetes, and their taxonomic implications. *Mycotaxon* **44**: 333–390.

Barklund, P. (1987). Occurrence and pathogenicity of *Lophodermium piceae* appearing as an endophyte in needles of *Picea abies*. *Transactions of the British Mycological Society* **89**: 307–313.

Bryant, H.N. (1991). The polarization of character transformations in phylogenetic systematics: role of axiomatic and auxiliary assumptions. *Systematic Zoology* **40**: 433–445.

Campbell, R. & Syrop, M. (1975). Light and electron microscope study of hysterothecium development in *Lophodermella sulcigena*. *Transactions of the British Mycological Society* **64**: 209–214.

Cannon, P.F. & Minter, D.W. (1983). The nomenclatural history and typification of *Hypoderma* and *Lophodermium*. *Taxon* **32**: 572–583.

Cannon, P.F. & Minter, D.W. (1986). The *Rhytismataceae* of the Indian Subcontinent. *Mycological Papers* **155**: 123 pp.

Cannon, P.F.; Hawksworth, D.L. & Sherwood-Pike, M.A. (1985). *The British Ascomycotina* An Annotated Checklist. Farnham Royal, UK: Commonwealth Agricultural Bureaux.

Carroll, G.C. & Carroll, F.E. (1978). Studies on the incidence of coniferous needle endophytes in the Pacific Northwest. *Canadian Journal of Botany* **56**: 3034–3043.

Chevallier, F.F. (1822). Essai sur les Hypoxylons lichenoides. *Journal de Physique, de Chimie, d'Histoire Naturelle et des Arts* **94**: 28–61.

Chevallier, F.F. (1826). *Flore Générale des Environs de Paris* **1**. Paris: Ferra Jeune.

Choi, D. & Simpson, J.A. (1995). Ascospore germination and appressorium formation by *Cyclaneusma minus*. *Mycotaxon* **54**: 455–459.

Christoffersen, M.L. (1995). Cladistic taxonomy, phylogenetic systematics, and evolutionary ranking. *Systematic Biology* **44**: 440–454.

Clark, C. & Curran, D.J. (1986). Outgroup analysis, homoplasy, and global parsimony: a response to Maddison, Donoghue, and Maddison. *Systematic Zoology* **35**: 422–426.

Connor, H.E. & Edgar, E. (1987). Name changes in the indigenous New Zealand flora, 1960–1986 and Nomina Nova IV, 1983–1986. *New Zealand Journal of Botany* **25**: 115–170.

Dallwitz, M.J.; Paine, T.A. & Zurcher, E.J. (1995). *User's Guide to INTKEY* A Program for Interactive Identification and Information Retrieval. Canberra: CSIRO Division of Entomology

Darker, G.D. (1932). The *Hypodermataceae* of Conifers. *Contributions from the Arnold Arboretum* **1**: 131 pp.

Darker, G.D. (1967). A revision of the genera of the *Hypodermataceae*. *Canadian Journal of Botany* **45**: 1399–1444.

Dennis, R.W.G. (1980). New or critical fungi from the Highlands and Islands. *Kew Bulletin* **35**: 343–361.

De Notaris, G. (1847). Prime linee di una nuova disposizione de Pirenomiceti Isterini. *Giornale Botanico Italiano* **2**: 5–52.

Desmazières, J.B.H.J. (1843). Nouvelle notice sur quelques plantes cryptogames. *Mémoires de la Société Royale des Sciences de l'Agriculture et des Arts de Lille* 1842: 108–150.

Desmazières, J.B.H.J. (1847). Quatorzième notice sur les plantes cryptogames récement découvertes en France. *Annales des Sciences Naturelles Botanique* Serié 3, **8**: 172–192.

Di Cosmo, F.; Nag Raj, T.R. & Kendrick, W.B. (1984). A revision of the *Phacidiaceae* and related anamorphs. *Mycotaxon* **21**: 1–234.

Duby, J.E. (1862). Mémoire sur la tribu des Hystérinées de la famille des Hypoxylées (Pyrénomycètes). *Mémoires de la Société de Physique et d'Histoire Naturelle de Genève* **16**: 15–70.

Edgar, E. & Connor, H.E. (1983). Nomina Nova III, 1977–1982. *New Zealand Journal of Botany* **21**: 421–441.

Eldredge, N. & Cracraft, J. (1980). *Phylogenetic Patterns and the Evolutionary Process.* New York: Columbia University Press.

Ellis, J.B. & Everhart, B.M. (1892). *The North American Pyrenomycetes* A Contribution to Mycological Botany. Newfield, NJ: Ellis & Everhart.

Eriksson, B. (1970). On Ascomycetes on *Diapensiales* and *Ericales* in Fennoscandia. *Symbolae Botanicae Upsalienses* **19**(4): 1–71.

Eriksson, O. (1981). The Families of Bitunicate Ascomycetes. *Opera Botanica* **60**: 220 pp.

Eriksson, O.E. & Hawksworth, D.L. (1993). Outline of the Ascomycetes – 1993. *Systema Ascomycetum* **12**: 51–57.

Farr, D.F.; Bills, G.F.; Chamuris, G.P. & Rossman, A.Y. (1989). *Fungi on Plants and Plant Products in the United States.* St Paul, MN: APS Press.

Ferraris, T. (1902). Materiali per una flora micologica del Piemonte. Miceti della valle d'Aosta. *Malpighia* **16**: 441–481.

Fries, E. (1822-1823). *Systema Mycologicum* **2**. Lund: Ex Officina Berlingiana. 620 pp.

Fries, E. (1828). *Elenchus Fungorum* Commentarium in Systema Mycologicum. **2**. Griefswald: Ernesti Mauritii. 154 pp.

Fuckel, L. (1870). Symbolae Mycologicae. Beitrage zur Kenntnis der Rheinischen Pilze. *Jahrbücher des Nassauischen Vereins Naturkunde* **23–24**: 459 pp.

Fuckel, L. (1873). Symbolae Mycologicae. Beitrage zur Kenntnis der Rheinischen Pilze. Zweiter Nachtrag. *Jahrbücher des Nassauischen Vereins Naturkunde* **27–28**: 99 pp.

Gordon, C.C. (1966). Ascocarp centrum ontogeny of species of *Hypodermataceae. American Journal of Botany* **53**: 319–327.

Graniti, A. (1952). Contributo alla conoscenza delle specie graminicole Italiane del genere *Lophodermium. Nuovo Giornale Botanico Italiano* N.S. **59**: 27–42.

Greuter, W.; Barrie, F.R.; Burdet, H.M.; Chaloner, W.G.; Demoulin, V.; Hawksworth, D.L.; Jørgensen, P.M.; Nicolson, D.H.; Silva, P.C.; Trehane, P. & NcNeill, J. (eds) (1994). International Code of Botanical Nomenclature (Tokyo Code). *Regnum Vegetabile* **131**: 389 pp.

Greuter, W. & Nicolson, D.H. (1993). On the threshold to a new nomenclature? *Taxon* **42**: 925–927.

GROVE, W.B. (1937). *British Stem- and Leaf-Fungi (Coelomycetes)* **2**. *Sphaeropsidales*. Cambridge: Cambridge University Press.

HAWKSWORTH, D.L.; KIRK, P.M.; SUTTON, B.C. & PEGLER, D.N. (1995). *Ainsworth and Bisby's Dictionary of the Fungi*. Edn 8. Wallingford, UK: CAB INTERNATIONAL.

HAZSLINSKY, F. (1886). Magyarhon es tarsorszagainak szabalyos discomycetjei (discomycetes). *Mathematikai es Termeszettudomanyi Kozlemenyek Vonatkozoglag a Hazai Viszonyokra* **21**: 183–191.

HEDBERG, O. (1995). Cladistics in taxonomic botany – master or servant? *Taxon* **44**: 3–11.

HENNINGS, P. (1895). Mykologische notizen I. *Abhandlungen des Botanischen Vereins der Provinz Brandenburg* **37**: 1–14.

HENNINGS, P. (1900). Fungi. *Monsunia* **1**: 137–207.

HILITZER, A. (1929). Monographiká Studie o Ceskych Druzíck rádu *Hysteriales* a o Sypavkách Jimi Pusobenych. *Vědecké Spisy Vydávané Československou Akademií Zemědělskou* **3**: 162 pp.

HÖHNEL, F. VON (1906). Revision von 292 der von J. Feltgen aufgestellen Ascomycetenformen auf Grund der Originalexemplare. *Sitzungsberichten der Kaiserliche Akademie der Wissenschaften in Wien* **115**: 139 pp.

HÖHNEL, F. VON (1917*a*). Mycologische Fragmente CXXXIII. Über die Gattung *Lophodermium* Chevallier. *Annales Mycologici* **15**: 311–313.

HÖHNEL, F. VON (1917*b*). System der *Phacidiales* v. H. *Berichte der Deutsche Botanische Gesellschaft* **35**: 416–422.

HUMPHRIES, C.J. & FUNK, V.A. (1984). Cladistic methodology. *In* HEYWOOD, V.H. & MOORE, D.M. (eds), Current Concepts in Plant Taxonomy. London: Academic Press. *Systematics Association Special Volume* **25**: 323–362.

JACOBS, S.W.L. (1982). Comments on Proposal 520: Conservation of *Notodanthonia* Zotov (*Gramineae*). *Taxon* **31**: 737–743.

JOHNSTON, P.R. (1986). *Rhytismataceae* in New Zealand 1. Some foliicolous species of *Coccomyces* de Notaris and *Propolis* (Fries) Corda. *New Zealand Journal of Botany* **24**: 89–124.

JOHNSTON, P.R. (1988*a*). An undescribed pattern of ascocarp development in some non-coniferous *Lophodermium* species. *Mycotaxon* **31**: 383–394.

JOHNSTON, P.R. (1988*b*). A new species of *Meloderma* (*Rhytismataceae*), with notes on *Meloderma* and related genera. *Mycotaxon* **33**: 423–436.

JOHNSTON, P.R. (1989*a*). *Lophodermium* (*Rhytismataceae*) on *Clusia*. *Sydowia* **41**: 170–179.

JOHNSTON, P.R. (1989*b*). *Rhytismataceae* in New Zealand 2. The genus *Lophodermium* on indigenous plants. *New Zealand Journal of Botany* **27**: 243–274.

JOHNSTON, P.R. (1990*a*). *Lophodermium* (*Rhytismataceae*) reassessed. *In* REISINGER, A. & BRESINSKY, A. (eds), *Abstracts, Fourth International Mycological Congress, IMC4*, p. 26. Regensburg: University of Regensburg.

JOHNSTON, P.R. (1990*b*). Biogeography of New Zealand *Rhytismataceae*. *In* REISINGER, A. & BRESINSKY, A. (eds), *Abstracts, Fourth International Mycological Congress, IMC4*, p. 129. Regensburg: University of Regensburg.

JOHNSTON, P.R. (1990*c*). *Hypohelion* gen. nov. (*Rhytismataceae*). *Mycotaxon* **39**: 219–227.

JOHNSTON, P.R. (1990*d*). *Rhytismataceae* in New Zealand 3. The genus *Hypoderma*. *New Zealand Journal of Botany* **28**: 159–183.

Johnston, P.R. (1991). *Rhytismataceae* in New Zealand 4. *Pureke zelandicum* gen. and sp. nov. plus additional species in *Hypoderma*, *Lophodermium*, and *Propolis*. *New Zealand Journal of Botany* **29**: 395–404.

Johnston, P.R. (1992). *Rhytismataceae* in New Zealand 6. Checklist of species and hosts, with keys to species on each host genus. *New Zealand Journal of Botany* **30**: 329–351.

Johnston, P.R. (1993). Three species of *Rhytismataceae* from bromeliads. *Sydowia* **45**: 21–33.

Johnston, P.R. (1994). Ascospore sheaths of some *Coccomyces*, *Hypoderma* and *Lophodermium* species (*Rhytismataceae*). *Mycotaxon* **52**: 221–239.

Johnston, P.R. (1997) Tropical *Rhytismatales*. *In* Hyde, K.D. (ed.), *Biodiversity of Tropical Microfungi*, pp. 241–254. Hong Kong: University of Hong Kong Press.

Karsten, P.A. (1872). Fungi in insulus Spetsbergen et Beeren Eilend collecti Examinat, enumerat. *Öfversigt af Königlich Vetenskaps-Akadamiens Förhandlingar* **29**(2): 91–108.

Kitching, I.J. (1992). The determination of character polarity. *In* Forey, P.L.; Humphries, C.J.; Kitching, I.J.; Scotland, R.W.; Siebert, D.J. & Williams, D.M. (eds), *Cladistics* A Practical Course in Systematics, pp. 22–43. Oxford: Clarendon Press.

Korf, R.P. (1958). Japanese discomycete notes I–VIII. *Science Reports of the Yokohama National University* Section 2, **7**: 7–35.

Korf, R.P. (1988). Report (N.S. 1) of the Committee for Fungi and Lichens on proposals to conserve and/or reject names. *Taxon* **37**: 450–463.

Kuntze, O. (1898). *Revisio Generum Plantarum* **3**(3): 576 pp.

Lamarck, J. & Candolle, A.P. de (1805). *Flore Française* Edn 3, **2**: 600 pp. Paris: Desray.

Lamarck, J. & Candolle, A.P. de (1815). *Flore Française* Edn 3, **6**: 662 pp. Paris: Desray.

Langer, E. (1994). Die Gattung *Hyphodontia* John Eriksson. *Bibliotheca Mycologica* **154**: 298 pp.

Letrouit-Galinou, M.A.; Parguey-Leduc, A. & Janex-Favre, M.C. (1994). Ascoma structure and ontogenesis in ascomycete systematics. *In* Hawksworth, D.L. (ed.), *Ascomycete Systematics* Problems and Perspectives in the Nineties, pp. 23–36. New York: Plenum Press.

Livsey, S. & Minter, D.W. (1994). The taxonomy and biology of *Tryblidiopsis pinastri*. *Canadian Journal of Botany* **72**: 549–557.

Losa, M. (1947). Micromicetos del pirines Español. *Anales del Jardín Botánico de Madrid* **8**: 297–338.

Maddison, W.P.; Donoghue, M.J. & Maddison, D.R. (1984). Outgroup analysis and parsimony. *Systematic Zoology* **33**: 83–103.

Maire, R. (1906). Étude des champignons récoltés en asie mineure. *Bulletin des Séances de la Société des Sciences de Nancy* Série 3, **7**: 165–188.

Meacham, C.A. (1984). The role of hypothesised direction of characters in the estimation of evolutionary history. *Taxon* **33**: 26–38.

Meacham, C.A. & Duncan, T. (1987). The necessity of convex groups in biological classifications. *Systematic Botany* **12**: 78–90.

Minter, D.W. (1981). *Lophodermium* on Pines. *Mycological Papers* **147**: 54 pp.

MINTER, D.W. (1984). *Hypoderma rubi. CMI Descriptions of Pathogenic Fungi and Bacteria* **79** (no. 781): [2] pp.

MINTER, D.W. (1985). Some members of the *Rhytismataceae* (ascomycetes) on conifer needles from Central and North America. *In* PETERSON, G.W. (ed.), *Recent Research on Conifer Needle Diseases* Proceedings of the International Union of Forestry Research Organisations, Working Party on Needle Diseases Conference, 14–18 October, 1984, Gulfort, Massachusetts. *USDA Forest Service General Technical Report* **WO-50**: 71–106.

MINTER, D.W. & CANNON, P.F. (1984). Ascospore discharge in some members of the *Rhytismataceae. Transactions of the British Mycological Society* **83**: 65–92.

MISHLER, B.D. & DONOGHUE, M.J. (1982). Species concepts: a case for pluralism. *Systematic Zoology* **31**: 491–503.

MORGAN-JONES, J.F. & HULTON, R.L. (1977). Ascocarp development in *Lophodermium nitens. Canadian Journal of Botany* **55**: 2605–2612.

MORGAN-JONES, J.F. & HULTON, R.L. (1979). Ascocarp development in *Lophodermium pinastri. Mycologia* **71**: 1043–1052.

MORTON, J.B. (1990). Evolutionary relationships among arbuscular mycorrhizal fungi in the *Endogonaceae. Mycologia* **82**: 192–207.

MUELLER, G.M. (1992). Systematics of *Laccaria* (*Agaricales*) in the Continental United States and Canada, with Discussions on Extralimital Taxa and Descriptions of Extant Types. *Fieldiana* **30**: 158 pp.

MÜLLER, E. (1977). Zur Pilzflora des Aletschwaldreservats (Kt. Wallis, Schweiz). *Beiträge zur Kryptogamenflora der Schweiz* **15**(1): 126 pp.

NANNFELDT, J.A. (1932). Studien über die Morphologie und Systematik der nicht-lichenisierten inoperculaten Discomyceten. *Nova Acta Regiae Societatis Scientiarum Upsaliensis* Series 4, **8**(2): 368 pp.

NIXON, K.C. & CARPENTER, J.M. (1993). On outgroups. *Cladistics* **9**: 413–426.

NOGRASEK, A. & MATZER, M. (1994). Nicht-pyrenokarpe Ascomyceten auf Gefässpflanzen der Polsterseggenrasen II. Arten auf *Cyperaceae* und *Poaceae. Nova Hedwigia* **58**: 1–48.

OSORIO, M. & STEPHAN, B.R. (1989). Ascospore germination and appressorium formation *in vitro* of some species of the *Rhytismataceae. Mycological Research* **93**: 439–451.

OSORIO, M. & STEPHAN, B.R. (1991). Morphological studies of *Lophodermium piceae* (Fuckel) v. Hohnel on Norway spruce needles. *European Journal of Forest Pathology* **21**: 389–403.

OUELLETTE, G.B. & MAGASI, L.P. (1966). *Lophomerum*, a new genus of *Hypodermataceae. Mycologia* **58**: 275–280.

PETRAK, F. (1922). Beiträge zur Pilzflora von Albanien und Bosnien. *Annales Mycologici* **20**: 1–28.

PETRAK, F. (1947). Kleine Beiträge zur Pilzflora von Australien und Polynesien. *Sydowia* **1**: 264–276.

PETRAK, F. (1963). Mykologische Beiträge zur österreichischen Flora. *Sydowia* **16**: 155–198.

PETRINI, L. (1980). Untersuchungen an Grasbewohnenden Arten von *Lophodermium* (Ascomyceten). Thesis (M.Sc.), Institut für Spezielle Botanik, ETH, Zürich.

PIMENTEL, R.A. & RIGGINS, R. (1987). The nature of cladistic data. *Cladistics* **3**: 201–209.

POWELL, P.E. (1973). The genera *Duplicaria* (*Rhytismataceae*) and *Crandallia* (*Leptostromataceae*). *Mycologia* **65**: 1356–1370.

Powell, P.E. (1974). Taxonomic Studies in the Genus *Hypoderma*. Thesis (Ph.D.), Cornell University, Ithaca.

Ranojević, N. (1910). Zweiter Beitrag zur Pilzflora Serbiens. *Annales Mycologici* **8**: 347–402.

Rehm, H. (1881). Ascomyceten. *Berichte der Naturhistorischen Vereins in Augsburg* **26**: 132 pp.

Rehm, H. (1887). Ascomyceten: Hysteriaceen und Discomyceten. *Rabenhorst's Kryptogamen-Flora von Deutschland, Oesterreich und der Schweiz* Edn 2. **1**(3): 1–64.

Rehm, H. (1896). Ascomyceten: Hysteriaceen und Discomyceten. *Rabenhorst's Kryptogamen-Flora von Deutschland, Oesterreich und der Schweiz* Edn 2. **1**(3): 1105–1275.

Rehm, H. (1912). Zur Kenntnis der Discomyceten Deutschlands, Deutsch-Österreichs und der Schweiz. *Bericht der Bayernischen Botanischen Gessellschaft* **13**: 102–206.

Reynolds, D.R. (1986). Foliicolous ascomycetes 7. Phylogenetic systematics of the *Capnodiaceae*. *Mycotaxon* **27**: 377–403.

Roumegère, C. (1892). Fungi exsiccati precipué Gallici, LXII centurie. *Revue Mycologique* **14**: 168–178.

Saccardo, P.A. (1882). Fungi Gallici. *Michelia* **2**: 583–648.

Saccardo, P.A. (1899). *Sylloge Fungorum* **14**: 1273 pp.

Saho, H. & Zinno, Y. (1972). *Soleella cunninghamiae* sp. nov., causing needle blight of *Cunninghamia lanceolata*. *Nippon Ringaku Kaishi* [*Journal of the Japanese Forestry Society*] **54**: 346–349.

Samuels, G.J. (1985). *An Annotated Index to the Writings of Franz Petrak* **4**, H–L. *DSIR Research Bulletin* **230**: 417 pp.

Schweinitz, L.D. von (1832). Synopsis fungorum in America Boreali media degentium. *Transactions of the American Philosophical Society* Series 2, **4**(2): 141–316.

Sherwood, M.A. (1977). Taxonomic studies in the *Phacidiales*: *Propolis* and *Propolomyces*. *Mycotaxon* **5**: 320–330.

Sherwood, M.A. (1980). Taxonomic Studies in the *Phacidiales*: The Genus *Coccomyces* (*Rhytismataceae*). *Occasional Papers of the Farlow Herbarium* **15**: 120 pp.

Slowinski, J.B. (1993). 'Unordered' versus 'ordered' characters. *Systematic Zoology* **42**: 155–165.

Spegazzini, C. (1887). Fungi Fuegiani. *Boletín de la Academia Nacional de Ciencias en Córdoba* **11**: 135–308.

Spegazzini, C. (1924). Relacíon de un paseo hasta al Cabo de Hoorn. *Boletín de la Academia Nacional de Ciencias en Córdoba* **27**: 321–404.

Spooner, B.M. (1981). New records and species of British microfungi. *Transactions of the British Mycological Society* **76**: 265–301.

Spooner, B.M. (1991). *Lophodermium* and *Hypoderma* (*Rhytismatales*) from Mt Kinabalu, Sabah. *Kew Bulletin* **46**: 73–100.

Stevens, F.L. (1925). Hawaiian Fungi. *Bernice P. Bishop Museum Bulletin* **19**: 189 pp.

Stevens, P.F. (1991). Character states, continuous variation and phylogenetic analysis: a review. *Systematic Botany* **16**: 553–583.

Suto, Y. (1983). A new species of *Hypoderma* causing dieback in *Chamaecyparis obtusa* and *Thuja orientalis*. *Transactions of the Mycological Society of Japan* **24**: 419–424.

Swofford, D.L. (1993). *PAUP: Phylogenetic Analysis Using Parsimony* Version 3.1. Champaign, IL: Illinois Natural History Survey.

Swofford, D.L. & Begle, D.P. (1993). *PAUP: Phylogenetic Analysis Using Parsimony* Version 3.1. Users Manual. Washington, DC: Laboratory of Molecular Systematics, Smithsonian Institute.

Tehler, A. (1994). Cladistic analysis in ascomycete systematics: theory and practice. *In* Hawskworth, D.L. (ed), *Ascomycete Systematics* Problems and Perspectives in the Nineties, pp. 185–197. New York: Plenum Press.

Tehon, L. R. (1918). Systematic relationships of *Clithris*. *Botanical Gazette* **65**: 552–555.

Tehon, L.R. (1935). A Monographic Rearrangement of *Lophodermium*. *Illinois Biological Monographs* **13**(4): 151 pp.

Tehon, L. R. (1939). New species and taxonomic changes in the *Hypodermataceae*. *Mycologia* **31**: 674–692.

Terrier, C.A. (1942). Essai sur la systématique des *Phacidiaceae* (Fr.) *sensu* Nannfeldt (1932). *Beiträge zur Kryptogamen-Flora der Schweiz* **9**(2): 1–99.

Uecker, F.A. & Staley, J.M. (1973). Development of the ascocarp and cytology of *Lophodermella morbida*. *Mycologia* **65**: 1015–1027.

Walker, J. (1980). *Gaeumannomyces*, *Linocarpon*, *Ophiobolus* and several other genera of scolecospored ascomycetes and *Phialophora* conidial states, with a note on hyphopodia. *Mycotaxon* **11**: 1–129.

Weese, J. (1933). Euromycetes selecti exsiccati 26. *Mitteilungen aus dem Botanischen Institut der Technischen Hochschule in Wien* **10**: 76–80.

Wiley, E.O. (1981). *Phylogenetics*. New York: J. Wiley.

Ziller, W.G. (1968). Studies of hypodermataceous needle diseases. I. *Isthmiella quadrispora* sp. nov., causing needle blight of alpine fir. *Canadian Journal of Botany* **46**: 1377–1381.

APPENDIX

List of extra-limital species and collections examined to gather data for information presented in the descriptions of ascomatal structure and ontogeny and for the cladistic analyses.

Bifusella linearis (Peck) Höhn.: **USA:** NEW HAMPSHIRE: Weirs, on *Pinus strobus*, 4 Jul. 1934, *G.D. Darker* 5024 (Sherwood, *Phacidiales exsiccati* no. 11, **TNS** F194477).

Coccomyces globosus P.R. Johnst.: **Australia:** TASMANIA: Hartz National Park, on *Eucalyptus coccifera*, 19 May 1988, *P.R. Johnston & A. Mills* (**PDD** 48988). **New Zealand:** COROMANDEL: Little Barrier I, East Cape, track 16, on *Nestegis lanceolata*, 16 Jun. 1984, *P.R. Johnston et al.*(**PDD** 55250); SOUTHLAND: Catlins, Chloris Pass, Cairns Rd, on *Weinmannia racemosa*, 14 Oct. 1991, *P.R. Johnston* (**PDD** 59484).

Coccomyces lauraceus P.R. Johnst.: **New Zealand:** AUCKLAND: Waitakere Ranges, Home Track, on *Litsea calicaris*, 5 Oct. 1982, *P.R. Johnston* (**PDD** 44658).

Coccomyces limitatus (Berk. & M.A. Curtis) Sacc.: **New Zealand:** AUCKLAND: Waitakere Ranges, Sharps Bush, on *Knightia excelsa*, 28 Nov. 1983, *P.R. Johnston* (**PDD** 44673); Waitakere Ranges, Opanuku Rd, Taumata Track, on *Ripogonum scandens*, 2 Aug. 1987, *P.R. Johnston & E.M. Gibellini* (**PDD** 53644); COROMANDEL: Moehau, Te Hope stream track, on *Knightia excelsa*, 28 Aug. 1984, *P.R. Johnston* (**PDD** 46226); TARANAKI: Mt Egmont, Waiweranui Track, on *Weinmannia racemosa*, 25 Jun. 1983, *P.R. Johnston et al.*(**PDD** 44671); WESTLAND: Fox Glacier, Mt Fox Track, on *Pseudopanax colensoi* var. *ternatus*, 8 Apr. 1983, *P.R. Johnston et al.* (**PDD** 44666).

Coccomyces radiatus Sherwood: **New Zealand**: COROMANDEL: nr Port Charles, track between Stoney Bay and Fletcher Bay, on *Dracophyllum traversii*, 22 Apr. 1989, *P.R. Johnston & M. Rajchenburg* (**PDD** 55369); TARANAKI: Mt Egmont, Puniho Track, on *Knightia excelsa*, 25 Apr. 1983, *P.R. Johnston et al.* (**PDD** 44684); WAIKATO: nr Ngaruawahia, Hakarimata Walkway, on *Rubus cissoides*, 18 May 1989, *P.R. Johnston* (**PDD** 55591).

Colpoma* cf. *quercina (Fr.) Wallr.: **France:** La Preste, on *Populus tremula*, 12 May 1994, *G. Gilles* (**PDD** 64750).

***Colpoma* sp.**: **New Zealand:** NELSON: Nelson Lakes National Park, Lake Rotoiti, on *Nothofagus* SP., 17 May 1994, *P.R. Johnston* (**PDD** 63401).

Duplicaria acuminata Ellis & Everh.: **USA:** WASHINGTON: Marysville, on *Juncus burformis*, Jul. 1928, *J.M. Grant* (**FH**).

Hypoderma bihospitum P.R. Johnst.: **New Zealand:** COROMANDEL: Moehau, Te Hope Stream Track, on *Uncinia* sp., 28 Aug. 1984, *P.R. Johnston* (**PDD** 49301); MACKENZIE: Mt Cook National Park, nr Mt Cook Village, Red Tarns Track, on *Anisotome*, 13 Feb. 1989, *P.R. Johnston* (**PDD** 57142).

Hypoderma campanulatum P.R. Johnst.: **New Zealand:** GISBORNE: Huiarau Range, Maungapohatu Track, on *Dracophyllum traversii*, 28 May 1983, *P.R. Johnston et al.* (**PDD** 49283).

Hypoderma liliensis P.R. Johnst.: **New Zealand:** AUCKLAND: Waitakere Ranges, Kauri Knoll Track, on *Collospermum hastatum*, 29 Apr. 1987, *P.R. Johnston* (**PDD** 45561).

Hypoderma obtectum P.R. Johnst.: **New Zealand:** TAUPO: Tongariro National Park, Lake Rotopounamu Track, on *Weinmannia racemosa*, 22 Feb. 1984, *P.R. Johnston* (**PDD** 49264).

Hypoderma rubi (Pers.) DC. ex Chevall.: **New Zealand:** AUCKLAND: Waitakere Ranges, Swanson, University Hut, on *Olearia furfuracea*, 31 Mar. 1983, *P.R. Johnston* (**PDD** 48976); BULLER: nr Murchison, Maruia Saddle, Warbeck Scenic Reserve, on *Rubus fruticosus*, 16 Apr. 1983, *P.R. Johnston et al.*(**PDD** 48974); *idem loc.*, on *Rubus cissoides*, 16 Apr. 1983, *P.R. Johnston et al.* (**PDD** 48973); COROMANDEL: Little Barrier I, Summit Track, on *Olearia furfuracea*, 14 Jun. 1984, *P.R. Johnston* (**PDD** 48975); NELSON: Karamea, Granite Creek Rd, on *Griselinia*, 14 Apr. 1983, *P.R. Johnston et al.* (**PDD** 48968); NORTH CANTERBURY: Arthurs Pass National Park, Old Coach Rd, on *Nothofagus solandri*, 5 May 1989, *P.R. Johnston et al.* (**PDD** 55525); WESTLAND: Fox Glacier, Lake Matheson, on *Griselinia littoralis*, 7 Apr. 1983, *P.R. Johnston et al.* (**PDD** 48959); *idem loc.*, on *Coprosma*, 9 Apr. 1983, *P.R. Johnston et al.* (**PDD** 48949).

Hypoderma sigmoideum P.R. Johnst.: **New Zealand:** MID CANTERBURY, Mt Somers, Sharplin Falls Track, on *Nothofagus solandri*, 6 May 1995, *P.R. Johnston* (**PDD** 64961).

Hypodermella laricis Tubeuf: **USA:** IDAHO: Priest River, on *Larix occidentalis*, 3 Oct. 1915, *J.R. Wier* (**PDD** 132).

Lophodermium aucupariae (Schleich.) Darker: **Sweden:** UPPLAND: Dalby Parish, forest edge of the farm Jeriko, on *Sorbus aucuparia*, 7 May 1985, *K. & L. Holm* (**PDD** 59628).

Lophodermium brunneolum P.R. Johnst.: **New Zealand:** AUCKLAND: Waitakere Ranges, Home Track, on *Knightia excelsa*, 30 Nov. 1990, *P.R. Johnston* (**PDD** 58144); DUNEDIN: Mt Cargill, on *Dracophyllum longifolium*, 14 May 1984, *P.R. Johnston* (**PDD** 49324); SOUTHLAND: nr Invercargill, Awaroa Bog, on *Dracophyllum*, 9 May 1984, *P.R. Johnston* (**PDD** 49317).

Lophodermium foliicola (Fr.) P.F. Cannon & Minter: **Germany:** Thüringen, Arnstadt-Plaue, on *Pyrus communis*, 14 May 1906, *H. Sydow* (Sydow, *Mycotheca Germanica* no. 493, **PDD** 55994, as *L. hysterioides*).

Lophodermium maculare (Fr.) De Not.: **Canada:** QUÉBEC: Gaspe, Mt Albert, on *Vaccinium*, 22 Jul. 1963, *R.F. Cain* (**DAOM** 40781).

Lophodermium medium P.R. Johnst.: **New Zealand:** WESTLAND: Haast Pass, on *Nothofagus menziesii*, 12 Apr. 1983, *P.R. Johnston et al.* (**PDD** 44763).

Lophodermium melaleucum (Fr.) De Not.: **Austria:** TIROL: Ötztal, on *Vaccinium vitis-idaea*, Jul. 1940, *F. Petrak* (Petrak, *Reliquiae Petrakianae* no. 1224, **PDD** 61638). **Canada:** BRITISH COLUMBIA: Vancouver I, Ucluelet, on *Vaccinium ovatum*, 6 Oct. 1957, *W.G. Ziller* (**PDD** 18830 ex **DAOM** 57452).

Lophodermium minus (Tehon) P.R. Johnst.: **New Zealand:** NORTHLAND: Mangamuka Summit Track, on ?*Geniostoma ligustrifolium*, 15 Jul. 1982, *P.R. Johnston & G.J. Samuels* (**PDD** 42788, as *L. multimatricum*); WESTLAND: Westland National Park, Fox Glacier, Lake Matheson, on *Dracophyllum longifolium*, 7 May 1989, *P.R. Johnston et al.* (**PDD** 56091).

Lophodermium nigrofactum P.R. Johnst.: **New Zealand:** COROMANDEL: Little Barrier I, Summit Ridge, on *Dracophyllum pyrimidale*, 13 Jun. 1984, *P.R. Johnston* (**PDD** 45651).

Lophodermium pinastri (Schrad.) Chevall.: **New Zealand:** SOUTHLAND: Fiordland, nr Te Anau, Blackmount Forest, on *Pinus radiata*, 2 Mar. 1992, *P.R. Johnston* (**PDD** 59952). **Scotland:** nr Pitlochry, southern shore of Lake Rannoch, on *Pinus contorta*, 7 Jun. 1986, *P.R. Johnston & E.M. Gibellini* (**PDD** 45576).

Lophodermium pyrolae Parmelee: **USA:** ALASKA: Valdez, on *Pyrola*, 1 Sep. 1940, *D.V. Baxter* (**BPI**).

Lophodermium tindalii P.R. Johnst.: **New Zealand:** AUCKLAND: Waitakere Ranges, Destruction Gully Track, on *Dracophyllum sinclairii*, 30 Dec. 1983, *P.R. Johnston* (**PDD**

45655); COROMANDEL: Little Barrier I, Summit Ridge, on *Dracophyllum pyrimidale*, 13 Jun. 1984, *P.R. Johnston* (**PDD** 45652); SOUTHLAND: Longwood Forest, Bald Hill, nr summit, on *Dracophyllum*, 7 May 1984, *P.R. Johnston et al.* (**PDD** 49313).

Propolis desmoschoeni P.R. Johnst.: **New Zealand:** COROMANDEL: Waikawau Bay farm park, on *Desmoschoenus spiralis*, 21 Aug. 1988, *P.R. Johnston & E.M. Gibellini* (**PDD** 54306).

Propolis emarginata (Cooke & Massee) Sherwood: **New Zealand:** AUCKLAND: Waitakere Ranges, Huia Dam, on *Metrosideros excelsa*, 27 Apr. 1982, *P.R. Johnston* R52 (**PDD** 43952); Waitakere Ranges, Kaitarakihi Beach, on *Metrosideros excelsa*, 20 Jul. 1983, *P.R. Johnston* (**PDD** 43958).

Rhytisma acerinum (Pers.) Fr.: **England:** CUMBRIA: Windy Ridge Walk, on *Acer*, 21 Sep. 1985, *P.R. Johnston & E.M. Gibellini* (**PDD** 45575); YORKSHIRE: Thornton le Dale, on *Acer pseudoplanatus*, Apr. 1952, *J.M. Dingley* (**PDD** 11692).

Rhytisma punctatum (Pers.) Fr.: **Canada:** ONTARIO: Renfrew Co.: Burns Lake, on *Acer pennsylvanicum*, 24 Sep. 1988, *J.A. Parmelee & S. Kaneko* (**PDD** 58924 ex **DAOM** 198649). **England:** LINCOLNSHIRE: *s. loc.*, on *Acer pseudoplanatus*, 10 Aug. 1988, *H.J. Houghton* (**IMI** 327191). **USA:** OREGON: Portland Heights, on *Acer macrophyllum*, 27 Sep. 1920, *J.S. Boyce* (**IMI** 23191).

Rhytisma salicinum (Pers.) Fr.: **Canada:** YUKON: Dempster Highway, on *Salix*, 8 Aug. 1980, *P.M. de Carteret* (**PDD** 58927 ex **DAOM** 189552).

LIST OF NEW TAXA AND NEW COMBINATIONS

Duplicaria antarctica (Speg.) P.R. Johnst., **comb. nov.**

Lophodermium alienum P.R. Johnst., **sp. nov.**
L. eucalypti (Rodway) P.R. Johnst., **comb. nov.**
L. fusiforme P.R. Johnst., **sp. nov.**
L. grandialpinum P.R. Johnst., **sp. nov.**
L. nitidum P.R. Johnst., **sp. nov.**
L. nonramosum P.R. Johnst., **sp. nov.**
L. petriniae P.R. Johnst., **sp. nov.**

Terriera arundinacea (Penz. & Sacc.) P.R. Johnst., **comb. nov.**
T. asteliae (P.R. Johnst.) P.R. Johnst., **comb. nov.**
T. breve (Berk.) P.R. Johnst., **comb. nov.**
T. clithris (Starbäck) P.R. Johnst., **comb. nov.**
T. dracaenae (W. Phillips & Harkn.) P.R. Johnst., **comb. nov.**
T. fourcroyae (Berk. & Broome) P.R. Johnst., **comb. nov.**
T. fuegiana (Speg.) P.R. Johnst., **comb. nov.**
T. javanica (Penz. & Sacc.) P.R. Johnst., **comb. nov.**
T. latiascus P.R. Johnst., **sp. nov.**
T. longissima P.R. Johnst., **sp. nov.**
T. nematoidea (P.R. Johnst.) P.R. Johnst., **comb. nov.**
T. pandani (Tehon) P.R. Johnst., **comb. nov.**
T. javanica var. **pandani** (Penz. & Sacc.) P.R. Johnst., **comb. nov.**
T. sacchari (Lyon) P.R. Johnst., **comb. nov.**
T. samuelsii P.R. Johnst., **sp. nov.**
T. stevensii P.R. Johnst., **sp. nov.**

INDEX OF FUNGAL NAMES

TAXA in **bold** are accepted names; those in *italic* are synonyms. NUMERALS in **bold** refer to main entries, whilst those marked with an asterisk* refer to a figure.

HOST LIST

Species are listed only when accepted as valid following examination of herbarium material; thus, species such as *L. ambiguum*, reported in the literature from *Poa* but with no material available to examine, have been excluded from the list. Also, species are listed only when the host genus is known; thus, species known only to occur on '*Poaceae*' or '*Cyperaceae*' are not listed.

Monocotyledonous hosts only are listed for the few species found also on dicotyledons. The reported host range is based largely on data provided on herbarium packets by the original collectors, thus accuracy cannot be guaranteed.

Lophodermium is abbreviated to L.; *Terriera* to T.; all other generic names are given in full.

Agavaceae
Dracaena
T. dracaenae
Furcraea
T. fourcroyae
Alliaceae
Allium
L. alliaceum
Arecaceae
Euterpe
T. latiascus
Asteliaceae
Astelia
T. asteliae
Bambusaceae
Bambusa
T. arundinacea
Bromeliaceae
Tillandsia
L. vrieseae
Cyperaceae
Carex
L. alpinum, T. breve, L. caricinum, L. fusiforme, L. gramineum, L. cf. luzulae, L. nitidum, L. nonramosum, L. unciniae
Gahnia
T. breve, L. hauturuanum, L. inclusum, T. nematoidea
Scirpus
L. raapianum, L. tumidulum
Uncinia
T. breve, L. unciniae
Vincentia
T. stevensii
Heliconiaceae
Heliconia
T. latiascus
Iridaceae
Iris
L. iridicolum
Juncaceae
Juncus
L. culmigenum, Rhytisma juncicola, L. cf. juncinum, L. tumidulum, L. unciniae
Luzula
L. cf. luzulae, L. tumidulum
Rostkovia
Duplicaria antarctica, T. fuegiana
Liliaceae
Convallaria
L. herbarum (Fr.) Fuckel
Xerophyllum
L. eucalypti
Pandanaceae
Pandanus
T. pandani, T. javanicum var. pandani
Phormiaceae
Xeronema
L. agathidis
Poaceae
Agropyron
L. culmigenum, L. robergei
Agrostis
L. alpinum, L. culmigenum, L. fusiforme, L. gramineum, L. petriniae
Aira